Reactor Physics

Reactor Physics

P. F. Zweifel
University Professor
Virginia Polytechnic Institute
and State University

McGraw-Hill Book Company
New York St. Louis San Francisco Düsseldorf Johannesburg
Kuala Lumpur London Mexico Montreal New Delhi
Panama Rio de Janeiro Singapore Sydney Toronto

Reactor Physics

Library of Congress Catalog Card Number 72–174628

07–073597–2

1 2 3 4 5 6 7 8 9 0 DODO 7 9 8 7 6 5 4 3 2

This book was set in Times New Roman, and printed and bound by R. R. Donnelley & Sons Company. The drawings were done by F. W. Taylor Company. The editors were B. J. Clark and M. E. Margolies. John A. Sabella supervised production.

To Kathy

Contents

Preface

The responsibility of the reactor physicist is to calculate and measure the properties of the nuclear reactor. These properties, which must be thoroughly understood before a reactor is constructed, include not only the critical size and shape but also the power distribution (and associated necessary heat-removal capacity), the kinetic behavior and safety analysis, the long-term reactivity changes, and the necessary shielding. Some of these quantities involve only physics, others involve engineering aspects of the nuclear system, while others are determined primarily by economic factors. But in general, there is considerable feedback among physics, engineering, and economic factors, so that the problems of reactor design become highly nonlinear (in this sense).

In this book, we shall attempt only to describe how reactor physics calculations are carried out; in fact, we shall stress the fundamentals, leaving the detailed techniques for books which are more concerned with design methods. Here we must confess that our own definition of physics is being used. One usually finds, for example in the physics divisions of some laboratories or in various reactor physics conferences which have been held from time to time, a great deal of work carried out and reported on which we would classify as nuclear engineering rather than physics. For this reason, it is perhaps appropriate that we should define our own concept of reactor physics.

Our working definition will be the following: Reactor physics is that area of (applied) physics which describes the neutron distribution in a nuclear reactor, or in an experimental assembly which is designed to provide information concerning such reactors. This definition is rather limited. For example, the calculation or measurement of microscopic cross-section information—certainly important for nuclear reactor design—is excluded. On the other hand, the conversion of basic cross-section information to a form which is useful in reactor design, i.e., multigroup constants, is appropriately a part of reactor physics. The calculation of the power density in a reactor core, which is determined by the neutron distribution, is included, then, within our definition, but any discussion of how this heat is to be removed, or utilized, is not.

Kinetics problems are also included within our definition, because they involve the time dependence of the neutron distribution in a reactor. However, we consider the application of kinetics calculations to safety analysis another area of nuclear engineering, and so it is not included. Primarily then, the book should appeal to the person who wishes to learn the physical bases upon which reactor designs are made. He will have to go elsewhere to learn the direct applications to nuclear power plants. Furthermore, this book is intended to be elementary. For this reason, diffusion theory is used almost exclusively throughout. Neutron transport has been discussed in a more sophisticated manner elsewhere,[1] but because there are few real applications of this approach in elementary reactor considerations, little discussion will be included here. In still another book,[2] the analysis of slowing-down effects in nuclear reactors has been described. Some of that material will also be described here, primarily for the sake of completeness. But by and large, that reference is of a more advanced and detailed scope than is intended for this book, so the overlap will not be very great.

Since all practical reactor design calculations are carried out today with the aid of high-speed digital computers, the person who intends to design reactors should be as familiar with the computer as his predecessors, say in the 1940s, were with the slide rule. In this way also the context of this book is determined. Complicated analytical methods will not be discussed; rather, those fundamentals and techniques which lay an appropriate ground work for the future use of the computer will be stressed.

In order to provide what we think to be the best basis for physical understanding, the first seven chapters of this book are devoted entirely to one-group theory, or, as it is referred to sometimes, the *one-speed approximation*. Thus, problems of neutron distribution, criticality of bare and reflected reactors, kinetics, long-term reactivity effects, etc., are completely treated in this approximation. Then, in Chapter 8, the extension of these results to the multigroup (i.e., the practical) treatment is described. In Chapter 9, there is a discussion of how the parameters which must be used in multigroup calculations may be obtained. There is an attempt to make the coverage sufficiently broad that subsequent application may be made by the reader to various types of reactor systems, such as fast or thermal, low or

[1] K. M. Case and P. F. Zweifel: "Linear Transport Theory," Addison-Wesley Publishing Company, Inc., Reading, Mass, 1967.

[2] Joel Ferziger and P. F. Zweifel: "The Theory of Neutron Slowing Down in Nuclear Reactors," Pergamon Press, New York, 1966.

high enrichment, and H_2O or D_2O. As the occasion demands, we include in the text descriptions of theories applicable to specific types of reactors. Chapter 10 is in fact devoted to a discussion of some of the problems peculiar to reactors of different types.

It is assumed that the reader has a sound knowledge of mathematics, including such topics as partial differential equations and boundary-value problems, with a smattering of complex analysis and linear algebra. Since such topics are generally covered today in most good undergraduate engineering and science curricula, the present book might appropriately be used at the advanced undergraduate or first-year graduate level.

In addition, a rudimentary knowledge of nuclear and modern physics is assumed—for example, at a level of Kaplan's excellent treatise.[3] For this reason, very little introductory nuclear physics material is included.

The material in this book is based largely on a one-semester course in reactor physics given by the author over a period of years at the University of Michigan, Department of Nuclear Engineering.

The actual writing was carried out in a number of stages. It began when the author was delivering a series of lectures on reactor physics at the Pakistan Institute of Nuclear Science and Technology under the auspices of the International Atomic Energy Agency. It continued after the author left Michigan in 1968 to join the Physics Department, Virginia Polytechnic Institute. Most of the writing was completed during the summer of 1969 when the author was a visiting scholar in the Physics Department of the University of Michigan. The contributions of all these organizations are gratefully acknowledged.

This book would never have become a reality except for the contributions of Professor George Summerfield, formerly of the VPI Physics Department and now Professor of Nuclear Engineering, University of Michigan. In addition to yeomen service in proofreading, preparing tables, graphs, etc., Professor Summerfield prepared essentially all the material in Sections 5.4, 6.6, 7.4, 7.5, 8.3, 8.8, 9.5, and 9.6. To him and to Professors Thomas Pigford and Milton Edlund, who reviewed the manuscript, go my sincere gratitude.

In addition, I want to acknowledge the important contributions of Mrs. Lisa Joyner, who typed the final draft of the manuscript, and to Mrs. Sharon Vaught, who corrected, proofread, edited, and in general made herself invaluable.

P. F. Zweifel

[3] Irving Kaplan: "Nuclear Physics," Addison-Wesley Publishing Company, Inc., Reading, Mass., 1955.

1
Introductory Remarks

1.1 BASIC CONCEPTS OF REACTOR PHYSICS

A nuclear reactor is an assembly containing a sufficient supply of *fissionable isotopes*,[1] for example, U^{235} or Pu^{239}, that a self-sustaining chain reaction is possible. The basic physics is very simply understood. When a fissionable nucleus is struck by a neutron, in a certain percentage of cases the nucleus will undergo fission. For example,

$$n + U^{235} \rightarrow A + B + \nu n'. \tag{1.1}$$

Here a U^{235} nucleus is struck by a neutron yielding two lighter nuclei called fission products, and denoted in Eq. (1.1) by the symbols A and B, plus additional neutrons whose average number is denoted by ν. This number ν is a (slight) function of the energy of the neutron which causes the fission and a stronger function of the fissionable isotope. For essentially zero-energy (*thermal*) neutrons striking uranium, $\nu \approx 2.42$.

[1] Also referred to as *fuel*.

Table 1.1 Value of α as a function of incident neutron energy[a]

Isotope	*Energy*	α
U^{235}	Thermal	0.18
	30 eV	0.65
	100 eV	0.52
	1,200 eV	0.47
	15 keV	0.41
Pu^{239}	Thermal	0.42
	100 eV	0.81
	1,200 eV	0.60
	15 keV	0.45

[a] Reactor Physics Constants, *AEC Rept. ANL* 5800, 2d ed; Alvin M. Weinberg and Eugene P. Wigner: "The Physical Theory of Neutron Chain Reactions," The University of Chicago Press, Chicago, 1958.

In addition to reaction (1.1), radiative capture is possible.

$$n + U^{235} \rightarrow U^{236} + \gamma. \tag{1.2}$$

The ratio of radiative capture events to fission events is generally denoted by the symbol α. It is, in effect, the ratio of the microscopic cross sections[1] for capture, σ_c, and for fission, σ_f,

$$\alpha = \frac{\sigma_c}{\sigma_f}. \tag{1.3}$$

In Table 1.1 we show α as a function of bombarding neutron energy for the two isotopes U^{235} and Pu^{239}. Another quantity often used in the reactor physics literature is the average number of neutrons emitted per neutron absorbed by the fissionable isotope, denoted by the symbol η. We see from our previous definitions that

$$\eta = \frac{\nu}{1 + \alpha}, \tag{1.4a}$$

$$\eta = \frac{\nu\sigma_f}{\sigma_c + \sigma_f}, \tag{1.4b}$$

$$\eta = \frac{\nu\sigma_f}{\sigma_a}, \tag{1.4c}$$

[1] *Cross section* is defined in some detail in Sec. 2.1.

where we have introduced the total-absorption cross section σ_a as

$$\sigma_a = \sigma_c + \sigma_f . \tag{1.5}$$

Curves of η versus incident neutron energy are shown in Fig. 1.1.

A chain reaction occurs when some of the ν neutrons that are emitted in the fission process strike additional fissionable nuclei, which produces more fissions, etc. For the reactor to operate at a steady fission rate (i.e., be *critical*), it is necessary that an exact balance be established between neutrons produced in fission and those lost. The losses include not only the neutron which produces the next fission and those which disappear through reaction (1.2)—radiative capture in the fuel—but also those which are captured by the other nuclei present in the system (structural material, coolant, moderator) and those which "leak" from the boundaries of the system, never to return. One of the purposes of reactor physics is to calculate the balance exactly by adjusting the size and composition until a steady-state situation is attained.

The cross section for fission is a strong function of the energy of the bombarding neutron. From Fig. 1.2 we see that the neutrons with energy of 1 eV or less are much more effective in maintaining the chain reaction than *fast* neutrons are. On the other hand, from Fig. 1.3 we note that the neutrons are emitted in the fission process with energies of the order of 2 MeV.[1] To reduce the energy of these neutrons, a light material called a *moderator* may be added to the reactor. Such a material reduces the neutron energy through elastic collisions. Clearly, light elements will absorb more energy when they are struck by neutrons than heavy elements will. Thus, usually the moderators contain H, D, Be, or C. Hydrogen and deuterium are generally used in compounds H_2O and D_2O, although sometimes organic compounds are used as well. In addition to moderating neutrons these materials also capture neutrons, a fact which must be taken into account in reactor design.

Reactors which are built without moderators are what are called *fast reactors*. Because of the dependence of the cross section on energy, such reactors require a considerably higher inventory of fuel than do reactors with

[1] An excellent fit to the fission spectrum of various fissionable isotopes is given by the Watt formula, $N(E) = aE^{1/2}\, e^{(-E/k_B T_f)}$ (E in MeV). The parameters a and T are:

Isotope	a	$k_B T_f$
U^{233}	1.872	1.290
U^{235}	1.888	1.306
Pu^{239}	2.121	1.333

The average energy of the neutrons is 1.93 MeV for U^{235} and U^{233}, and 2.00 MeV for Pu^{239}.

UNCHARTED DATA (MTR)

E, eV	$\eta(E)$	E, eV	$\eta(E)$
4.60	0.47	6.45	0.84
4.80	0.52	7.45	1.04
5.00	1.08	9.25	1.22
6.00	0.70		

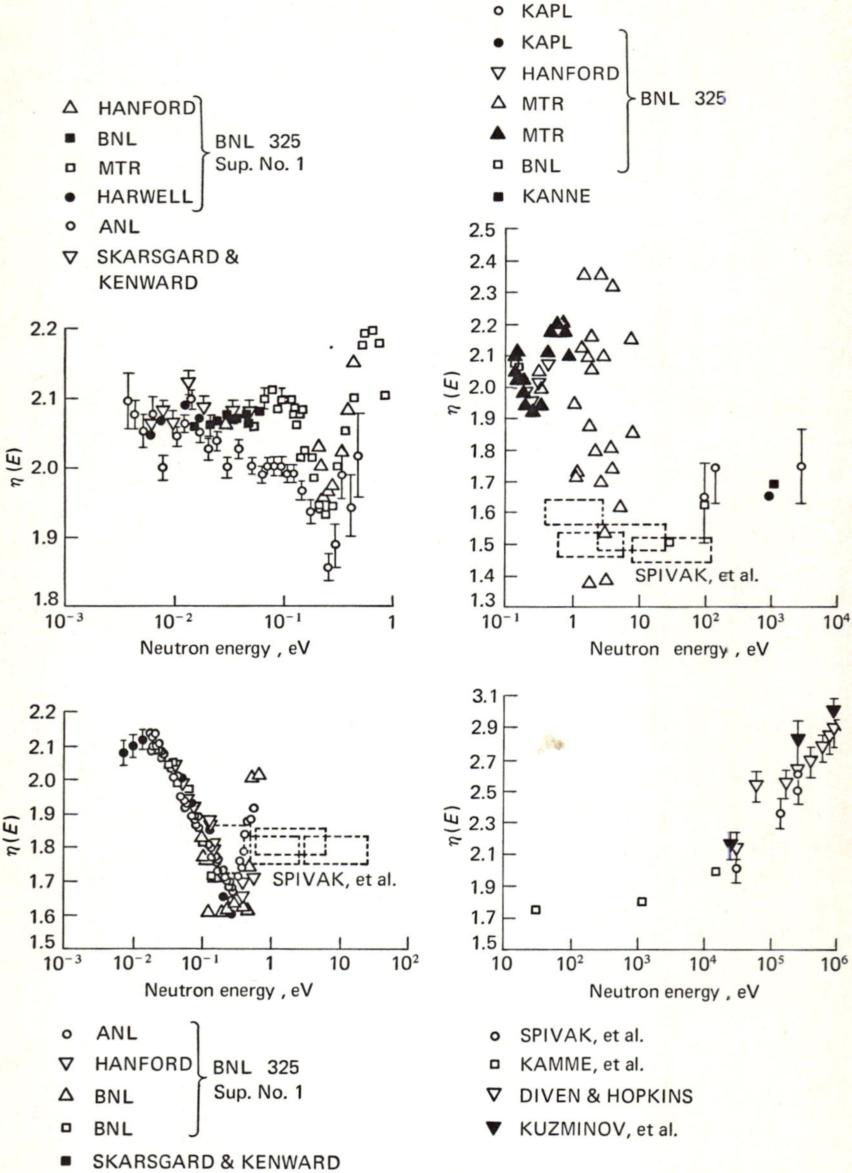

Fig. 1.1 Neutron yield per neutron absorbed (η) for U^{235} (upper curves) and Pu^{239} (lower curves) as a function of energy. (*Source: ANL–5800, figs. 2.3, 2.4, 2.6, and 2.8.*)

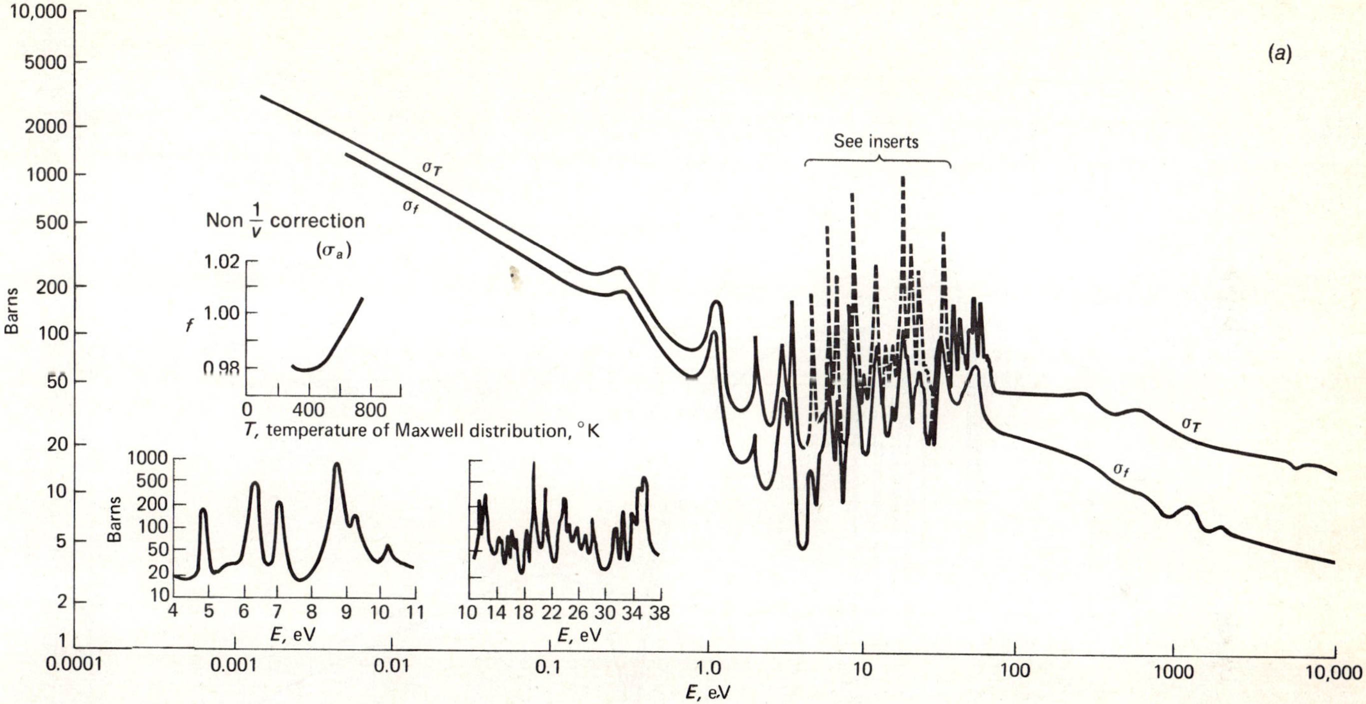

Fig. 1.2 Fission cross section as a function of incident neutron energy for (*a*) U^{235} and (*b*) Pu^{239}. (*Source: Alvin M. Weinberg and Eugene P. Wigner, The Physical Theory of Neutron Chain Reactors, figs. 5.4 and 5.5, University of Chicago Press, Chicago, 1958.*)

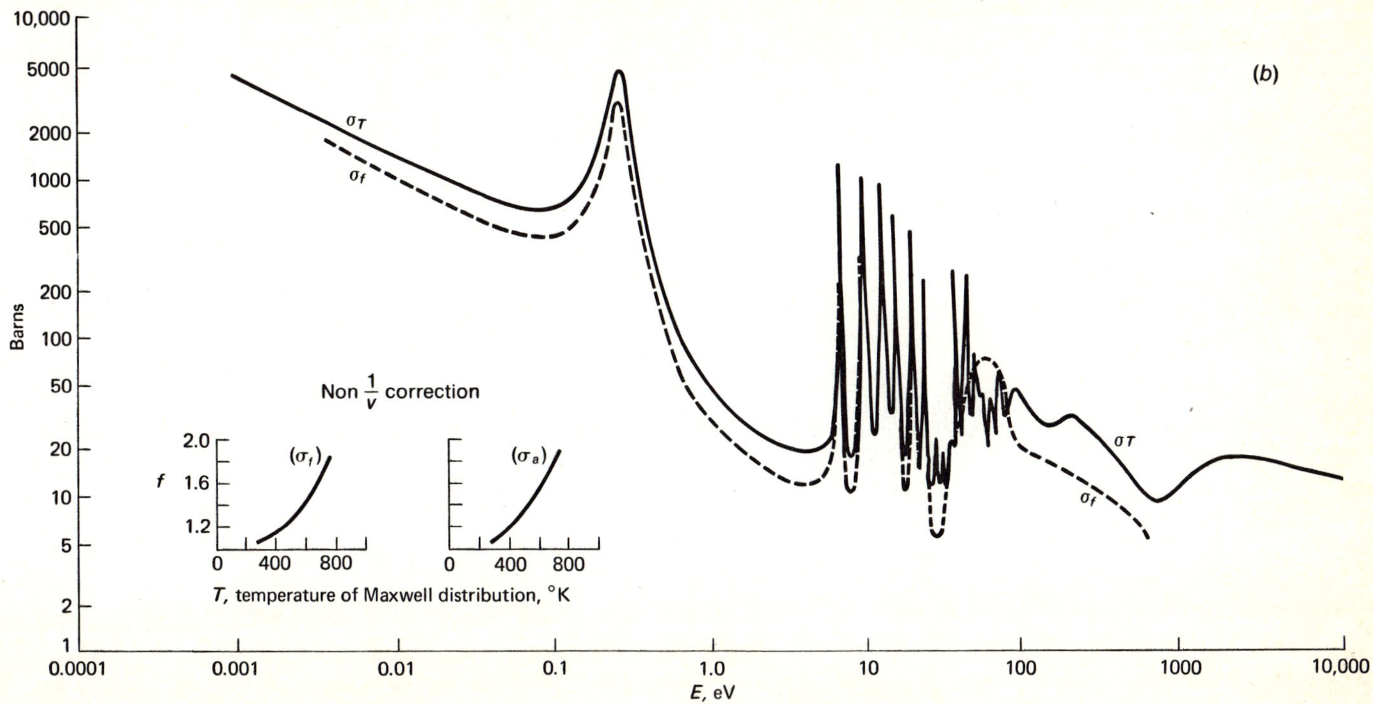

Fig. 1.2 (*Continued*)

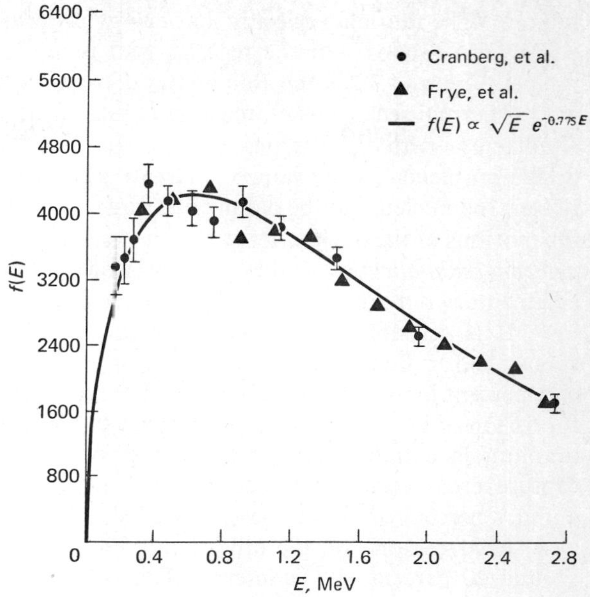

Fig. 1.3 Spectrum of fission neutrons emitted by U^{235}. (*Source: Weinberg and Wigner, op. cit., fig. 5.2. The dotted points are from Lawrence Cranberg and Norris G. Nereson, Fission Neutron Spectrum of* U^{235} *from 0.2 to 3 MeV, Los Alamos Scientific Laboratory, LA-1916, May 1955. The triangles are from Glenn M. Frye, Jr., Juanita H. Gammel, and Louis Rosen, Energy Spectrum of Neutrons from Thermal Neutron Fission of* U^{235} *and from An Untamped Multiplying Assembly of* U^{235}, *Los Alamos Scientific Laboratory, LA-1670, May 1954.*)

moderators. However, even in fast reactors the neutron spectrum is moderated to some extent by inelastic collisions with the fuel and the structural materials, as well as by elastic and inelastic collisions with the coolant (usually liquid sodium). In fact, the average energy of the neutrons in a completely unmoderated fast reactor is of the order of 200 keV or less, whereas the mean energy of these neutrons, when emitted, is of the order of 2 MeV, as we have seen.

Since uranium metal is not so metallurgically desirable a fuel as are oxides and carbides, fast reactors are being designed today containing a considerable amount of light elements. As a result, the energy spectrum tends to be degraded to the order of 1 to 10 keV, a fact which complicates the design procedures. If more moderator is added to the reactor, the neutron spectrum degrades further, through the *intermediate* region to the *thermal* region. The thermal region is that energy range comparable to the thermal energy of the atoms in the reactor, that is, $k_B T$, where T is the (absolute) temperature and k_B is the Boltzmann constant. (At 20°C, $k_B T \approx 0.025$ eV.)

Thermal reactors, i.e., reactors in which the neutrons come into quasi-equilibrium with the thermal motions of the atoms, present some unique design problems. For example, in treating high-energy neutron collisions with nuclei, the nucleus can be assumed at rest. For thermal neutrons, however, the motions of the nucleus must also be considered. This problem, known as *neutron thermalization*, is beyond the scope of this book except for a few rudimentary remarks made in Chaps. 9 and 10.

It is certainly well known that the isotope of uranium which is responsible for fissions is U^{235}, which occurs with an abundance of only 0.71 percent in natural uranium.[1] However, because of the large cross section for fission of U^{235} by thermal neutrons, a chain reaction is possible in natural uranium in certain cases. (In the case of H_2O moderator, the hydrogen capture cross section is so large that uranium slightly enriched in U^{235}, about 1 percent or slightly less, is needed.)

Fast reactors, on the other hand, require uranium enriched to at least around 20 percent U^{235} content. This is a great disadvantage of fast reactors. However, fast reactors do have certain advantages which we shall discuss presently.

First, we should say something about the energy production in reactors. The fission reaction (1.1) yields approximately 200 MeV of useful energy.[2] Converted to more practical units, the complete fission of 1 g of uranium yields about 24,000 kWh of useful energy. Most of the useful energy is kinetic energy of the fission fragments. These fragments are very massive and hence have a short range; thus they are stopped in the fuel elements in which they were produced. As a result, the fuel element heats up, and the heat must be removed by a coolant of some sort. The details of how this heat is converted into electrical energy, say, are identical in most respects[3] with the procedures used in coal-fired boilers, and so will not be discussed further.

Returning now to the question of what virtues, if any, fast reactors may have, we must consider the production of Pu^{239}, which proceeds according to the chain

$$n + U^{238} \rightarrow Np^{239} + \gamma, \tag{1.6a}$$

$$Np^{239} \rightarrow Pu^{239} + \beta^- + \bar{\nu}, \tag{1.6b}$$

[1] Except for fast fission in U^{238}, which gives a relatively minor contribution to the chain reaction.

[2] The reason that not all the energy is useful is that some is carried away by high-energy particles such as neutrinos.

[3] Except for the possibility that the coolant becomes radioactive in passing through the reactor core.

(where $\bar{\nu}$ represents an antineutrino). This reaction is the only way in which Pu^{239} can be obtained; it does not occur in nature. The reason that it is desirable to produce Pu^{239} is that there is only a limited amount of U^{235} in the world. Estimates of the exact amount differ, but it is quite clear that there is 140 times as much U^{238} as U^{235}. Thus, if all the U^{238} could be converted into Pu^{239}, there would be much more nuclear fuel available. Clearly, in any reactor containing substantial amounts of U^{238} there will be some production of Pu^{239}. The production of this isotope (and higher isotopes such as Pu^{240}) is an important factor, both in design and economics, of many reactor types. However, eventually even this Pu^{239} will be used up[1] unless more Pu^{239} atoms are created than fuel atoms (U^{235} or Pu^{239}) are destroyed. The ratio of Pu^{239} atoms created to fuel atoms destroyed is called the *breeding ratio*; so one desires a breeding ratio greater than unity.

Now, recall the quantity η, introduced previously [Eq. (1.4)]. Of the η neutrons created when a neutron is absorbed in fuel, one must be used to maintain the chain reaction, and one to form a Pu^{239} nucleus. This leaves $\eta - 2$ neutrons to be absorbed in the moderator, coolant, etc., and to be lost by leakage. Only if $\eta - 2 > 0$ can one conceivably achieve a breeding ratio greater than unity. Refer back to Fig. 1.1, in which η was given as a function of energy. Note the much higher value of η for Pu^{239} than for U^{235}, and also how η increases in the *fast* region for both isotopes. This is the reason for the recent attention being given to fast reactors, particularly Pu-fueled.

To this point, one fissionable isotope has not been discussed, namely, U^{233}. This also does not occur in nature, but is produced by a radioactive chain following neutron capture in thorium. Some attention, particularly at the Oak Ridge National Laboratory, has been given to building a thorium converter, and many experts feel that such an approach is definitely more promising than the fast breeder reactors, principally because of safety problems associated with the latter.

This has been a necessarily brief discussion of some of the principal aspects of reactor physics and how these effect the choice of a reactor type. There will be considerably more detailed discussion in the subsequent chapters where such aspects as control rods, reflectors, and lifetime analyses, which were ignored here, will be developed. The discussion of the present section has made it clear that the proper study of the neutron distribution in a reactor requires the solution of an energy-dependent problem. This is generally treated, in most practical calculations, in what is called the *multigroup* scheme, in which it is assumed that the neutrons, instead of having a continuous energy spectrum, occupy only a set of discrete energy levels. This approximation, which greatly simplifies reactor calculations, has been found

[1] Admittedly, at the time of this writing, there is no hard design for a Pu^{239}-fueled power reactor.

quite reliable. In the first part of the book, however, we assume that all neutrons have a single energy. This is called the *one-group* (or *one-speed*) approximation. Clearly, this approach must be very bad for practical calculations. But it turns out that a large part of the physics and mathematics of reactor calculations can be understood thoroughly in the one-speed case. The transition to the multigroup scheme then becomes almost trivial. Such an approach is followed largely for pedagogical reasons.

Throughout the book, the absence of many discussions which have become traditional in reactor physics texts will be noted. This is in line with our policy, described earlier, of orienting the discussion to the computer. Even such standard calculations as solutions of the Fermi age equation are introduced only as a means of generating multigroup constants which are then used in computer solutions. While the traditionalists might object to such an approach, we feel it to be in line with current practice.

1.2 PHYSICAL DESCRIPTION OF THE CHAIN REACTION

In the previous section, some rather general remarks were made concerning the neutron chain reaction and its influence on various reactor types. On the basis of these remarks it is possible to express, in symbolic form, the condition of reactor criticality.

We begin by defining a quantity denoted by k and called the *criticality factor*. For a critical reactor, $k = 1$. If $k > 1$, the reactor is said to be *supercritical*, and physically this means that the neutron population increases as a function of time. If, on the other hand, $k < 1$, the reactor is said to be *subcritical*, and the neutron population decreases. Physically, k can be thought of as the ratio of neutrons present in a given *generation* to those in the preceding generation.

It is convenient to consider a reactor which is so large that there is no neutron loss by leakage. For such an *infinite* reactor the criticality factor is denoted by the symbol k_∞. Clearly,

$$k_\infty > k. \tag{1.7}$$

Consider a single neutron which is absorbed by a fuel nucleus. If the reactor is critical, then there should be exactly one neutron in the next generation available for absorption in a fuel nucleus. The original neutron produces η fission neutrons. Restricting our attention at the moment to the one-group, infinite-reactor case, a fraction $\Sigma_a^{\text{fuel}}/\Sigma_a$ of these neutrons will be reabsorbed in the fuel. (Here, Σ_a^{fuel} is the macroscopic absorption cross section of the fuel, while Σ_a represents the total absorption cross section of all materials in the reactor, including the fuel.) Then in this simplified model we have

$$k_\infty = \eta f, \quad \text{one group,} \tag{1.8}$$

where we have introduced the standard definition of the *thermal utilization*

$$f = \frac{\Sigma_a^{\text{fuel}}}{\Sigma_a}. \tag{1.9}$$

Incidentally, η is sometimes referred to as the *fuel efficiency*.

If we now admit the possibility of leakage to our problem, we have

$$k = \eta f P_T, \quad \text{one group,} \tag{1.10}$$

where P_T is simply the probability that a neutron does *not* leak from the reactor (the *nonescape probability*).

Consider now the energy-dependent case. We assume that fast neutrons are being produced which are then slowed to thermal energy before they produce fission (such a simplified model is useful only for thermal reactors). Some of the fast neutrons will produce fission in U^{238}. The ratio of neutrons which slow down past the U^{238}-fission threshold (about 1.2 MeV) to those originally produced is denoted by ε, called the *fast fission factor*. As these neutrons slow down through the resonance region, a fraction p will escape resonance capture. This is called the *resonance escape probability*. Thus,

$$k_\infty = \eta \varepsilon p f, \quad \text{thermal reactor,} \tag{1.11}$$

which is the *four-factor formula*. If we now include the possibility of leakage,

$$k = \eta \varepsilon p f P_f P_T, \tag{1.12}$$

where P_f is the nonleakage probability for fast neutrons.

Such a description of the chain reaction is grossly oversimplified. In the first place, the quantities p, f, P_f, and P_T are strongly dependent upon the construction of the reactor and are not at all trivial to calculate except for very idealized models.[1] In the second place, this description assumes that all fissions are caused by thermal neutrons or by fast neutrons reacting in U^{238}. For this reason, formulas of this type are useful only for qualitative considerations. They do, however, give some insight into the physical processes going on in the reactor. In addition, as we shall see later in this book, calculations of some of the quantities entering these formulas are useful in determining the parameters (group constants) which are to be used in more practical calculations. Thus, the physical meaning of these various parameters should be clearly understood.

[1] They represent spatial and energy averages of basic reaction rates.

It is possible to modify the various formulas given above to include the effect of, say, epithermal fission. Then one would have

$$k = k_e + k_T, \tag{1.13}$$

where k_e is the contribution to k from fissions produced by epithermal neutrons, and k_T the contribution from thermal neutrons. This could even be extended to a multigroup scheme. However, such formulas are not terribly useful and so will not be discussed further.

2
The Neutron Diffusion Equation

2.1 INTRODUCTION AND DEFINITIONS

The basic physical processes governing the behavior of a nuclear reactor include changes in the distribution of neutrons in the reactor. The mathematical description of the neutron distribution is based on an equation of neutron balance called the *neutron transport equation*, or simply the *transport equation*. It is actually a linearized form of the Boltzmann equation familiar in the description of gases. (For this reason, the neutron transport equation is sometimes referred to as the Boltzmann equation.) Actually, the transport equation is so complicated that solutions have been found only for rather simplified physical models, certainly not for practical reactor systems. For this reason an approximation to the transport equation, known as the *diffusion equation*, is generally used in reactor analysis, and the approximation will be adopted throughout this book. However, we shall first derive the complete transport equation and then show how the diffusion equation may be obtained as an approximation to it.

Before deriving the transport equation, a number of definitions are needed.

DEFINITION 2.1 THE NEUTRON ANGULAR DENSITY $\psi(\mathbf{r},\mathbf{u},t)$

Consider a medium in which neutrons are diffusing. In any small-volume element d^3r, located at position **r** *at the time t, there will be a certain number of neutrons of all velocities. Consider only those neutrons whose velocity lies within the element in velocity space d^3u located about velocity* **u**.[1] *The number of such neutrons is equal to $\psi(\mathbf{r},\mathbf{u},t)\, d^3r\, d^3u$. Putting it another way $\psi(\mathbf{r},\mathbf{u},t)$ represents the density of neutrons per unit volume in velocity space at the point* **r** *with velocity* **u** *at time t.*

DEFINITION 2.2 THE AVERAGE NEUTRON VELOCITY $\mathbf{v}(\mathbf{r},t)$

This average is to be understood in the kinetic theory sense, i.e., we expect the results to be valid statistically. Then

$$\mathbf{v}(\mathbf{r},t) = \frac{\int \mathbf{u}\psi(\mathbf{r},\mathbf{u},t)\, d^3u}{\int \psi(\mathbf{r},\mathbf{u},t)\, d^3u}. \tag{2.1a}$$

Note that for an isotropic neutron distribution (ψ dependent only on u, the magnitude of **u**), $\mathbf{v} \equiv 0$. It is important not to confuse the average velocity **v** with the velocity variable **u**.

In considering energy-dependent problems we shall have occasion to use the average velocity for neutrons of a given speed $\mathbf{v}(\mathbf{r},u,t)$. This is defined in a manner analogous to (2.1*a*), where $d^3u = u^2\, du\, d\Omega$.

$$\mathbf{v}(\mathbf{r},u,t) = \frac{\int \mathbf{u}\psi(\mathbf{r},\mathbf{u},t)\, d\Omega}{\int \psi(\mathbf{r},\mathbf{u},t)\, d\Omega}, \tag{2.1b}$$

DEFINITION 2.3 THE AVERAGE SPEED $\bar{v}(\mathbf{r},t)$

$$\bar{v}(\mathbf{r},t) = \frac{\int d^3u\, u\psi(\mathbf{r},\mathbf{u},t)}{\int d^3u\, \psi(\mathbf{r},\mathbf{u},t)}. \tag{2.2}$$

The average speed is the average of the magnitude of **u**. *Clearly*, $\bar{v}$ is not just $v(\mathbf{r},t)$, the magnitude of $\mathbf{v}(\mathbf{r},t)$. For example, for an isotropic distribution $\mathbf{v}(\mathbf{r},t)$ is zero, whereas $\bar{v}(\mathbf{r},t)$ is not zero. Note also that a quantity $\bar{v}(\mathbf{r},u,t)$ could be defined in analogy with (2.1*b*).

[1] Later (in Chap. 9) we shall use the symbol u to represent lethargy. No confusion should arise between lethargy u and speed $u = |\mathbf{u}|$.

DEFINITION 2.4 THE NEUTRON FLUX $\phi(\mathbf{r},t)$

$$\phi(\mathbf{r},t) \equiv \int d^3u\, u\psi(\mathbf{r},\mathbf{u},t) \tag{2.3a}$$

$$= \bar{v}(\mathbf{r},t) \int d^3u\, \psi(\mathbf{r},\mathbf{u},t) \tag{2.3b}$$

$$= \bar{v}(\mathbf{r},t)\rho(\mathbf{r},t). \tag{2.3c}$$

Here we have introduced the symbol $\rho(\mathbf{r},t)$ to represent the neutron density, i.e., the total density of neutrons at $\mathbf{r}$ independent of velocity.

The energy-dependent flux $\phi(\mathbf{r},u,t)$ has a similar definition,

$$\phi(\mathbf{r},u,t) = \int d\mathbf{\Omega}\, u\psi(\mathbf{r},\mathbf{u},t). \tag{2.3d}$$

DEFINITION 2.5 THE NEUTRON CURRENT $\mathbf{J}(\mathbf{r},t)$

$$\mathbf{J}(\mathbf{r},t) = \mathbf{v}(\mathbf{r},t)\rho(\mathbf{r},t), \tag{2.4a}$$

or

$$\mathbf{J}(\mathbf{r},t) = \int d^3u\, \mathbf{u}\psi(\mathbf{r},\mathbf{u},t). \tag{2.4b}$$

We see that the definition of $\mathbf{J}$ is quite similar to that of ϕ except that in the former the average velocity $\mathbf{v}$ enters rather than the speed[1] [c.f. Eqs. (2.3c) and (2.4a)]. Thus, the units of flux and current are the same (neutrons per square centimeter per second), but the current is a vector quantity, while the flux is a scalar.

Both flux and current have specific physical meanings. Consider first the current.

Theorem 2.1 *The net number of neutrons crossing an area* $d\mathbf{S}$ *per unit time is given by* $\mathbf{J} \cdot d\mathbf{S}$.

Before proving this rather simple theorem, we should define what is meant by *net* number. It is, in a sense, an algebraic meaning. Suppose one neutron crosses $d\mathbf{S}$ in a direction denoted by the unit vector $\mathbf{\Omega}$, while in the same time interval, another neutron crosses in the opposite direction $-\mathbf{\Omega}$. These two neutrons algebraically add to zero, so that if no other neutrons crossed $d\mathbf{S}$ in the time interval under consideration, the *net* number would be equal to zero, although the total number crossing $d\mathbf{S}$ in this time interval would be two neutrons. Thus, the quantity $\mathbf{J} \cdot d\mathbf{S}$ represents a *leakage* of neutrons across $d\mathbf{S}$, and as such will enter the diffusion equation.

[1] Hence, in an isotropic neutron distribution $\mathbf{J}(\mathbf{r},t) = 0$, although $\phi(\mathbf{r},t) \neq 0$.

Proof We first consider the number of neutrons crossing $d\mathbf{S}$ with velocity $\mathbf{u}$ (see Fig. 2.1). Construct a cylinder with axis $\mathbf{u}$ and base $d\mathbf{S}$. The volume of this cylinder is $\mathbf{u} \cdot d\mathbf{S}$ cm^3, and the number of neutrons of velocity $\mathbf{u}$ in this cylinder is the product of the volume $\mathbf{u} \cdot d\mathbf{S}$ and the angular density $\psi(\mathbf{r},\mathbf{u},t)$,

$$n(\mathbf{u}) = \mathbf{u} \cdot d\mathbf{S}\ \psi(\mathbf{r},\mathbf{u},t). \tag{2.5}$$

Since the cylinder is u cm long, this is also the number of neutrons with velocity $\mathbf{u}$ which cross $d\mathbf{S}$ in one second. Thus the net number of neutrons which cross $d\mathbf{S}$ in 1 s is obtained by integration over all $\mathbf{u}$,

$$n = \int \mathbf{u} \cdot d\mathbf{S}\ \psi(\mathbf{r},\mathbf{u},t)\, d^3u, \tag{2.6a}$$

$$n = \mathbf{v}(\mathbf{r},t) \cdot d\mathbf{S}\ \rho(\mathbf{v},t), \tag{2.6b}$$

where Eqs. (2.1) and (2.3) have been used. Equation (2.6*b*) is the desired result since $\mathbf{J} = \mathbf{v}\rho$.

Before attempting to assign a physical meaning to the flux ϕ, it is necessary to introduce the concept of the *neutron mean free path.* We shall be dealing, in reactor physics, with the distribution of neutrons within a material medium of some sort. This distribution is of course modified by collisions between the neutrons and the material nuclei.[1] These collisions are of various types, e.g., scattering, absorption, and fission.

[1] In principle, neutrons also collide with each other. However, because of the low neutron densities ($\approx 10^{10}$ cm^{-3}) compared with material densities ($\approx 10^{22}$ cm^{-3}) in a typical reactor system, neutron-neutron collisions can be ignored. Of course, we must also assume the neutron-neutron cross section to be much less than 10^{12} times the neutron-nucleus cross section. In fact, the neutron-neutron cross section is known to be comparable to the neutron-nucleus cross section. This is fortunate, since it means that equations determining the neutron distribution will be linear.

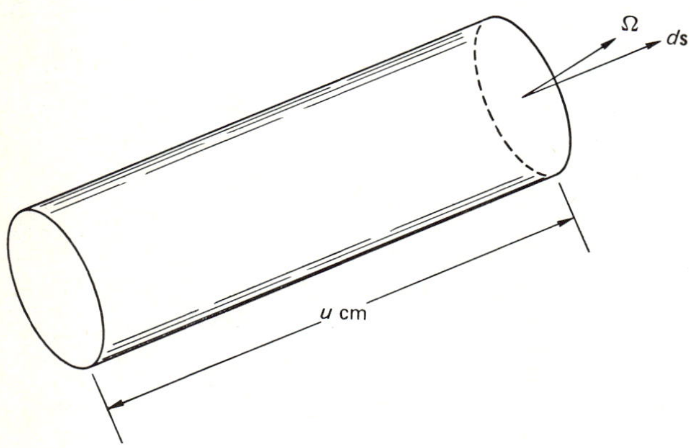

Fig. 2.1 Calculation of leakage.

DEFINITION 2.6 MEAN FREE PATH λ_i

With each process i, we associate a quantity λ_i such that the probability ΔP_i that a neutron will undergo such a process in traversing a small length dx is given by dx/λ_i. Symbolically,

$$\Delta P_i = \frac{dx}{\lambda_i}. \tag{2.7}$$

Thus λ_i is called the mean free path and $1/\lambda_i$ is a probability per unit length, for small path lengths, that a neutron will undergo a reaction of type *i*. Clearly, λ_i has the units of length. We note again that the subscript *i* refers to the process being considered, such as absorption λ_a, scattering λ_s, or fission λ_f.

We might next ask the question, what is the probability $P_i(l)$ that a neutron will travel distance l without suffering an interaction of type i?

To answer this, suppose one begins with n_0 neutrons, all traveling in a given direction, and ask how this neutron beam is attenuated by collisions of the *i*th type. Denoting by $n(l)$ the number of neutrons remaining at a distance from the starting point, we have, using Eq. (2.7),

$$\frac{dn(l)}{dl} = \frac{-n(l)}{\lambda_i}, \tag{2.8a}$$

so that

$$n(l) = n_0 e^{-l/\lambda_i}. \tag{2.8b}$$

The neutron beam is exponentially attenuated.

DEFINITION 2.7 THE CROSS SECTION

The reciprocal of λ_i is called the macroscopic cross section, and is denoted by

$$\Sigma_i = \frac{1}{\lambda_i}. \tag{2.9}$$

The macroscopic cross section is a weighted sum, over all the nuclei present, of the microscopic cross sections associated with the νth type of nucleus present.

$$\Sigma_i = \sum_\nu N_\nu \sigma_i^\nu, \tag{2.10}$$

where N_ν is the number density of the νth nucleus (note that Σ_i has units of inverse length, while σ_i^ν has units of area; hence, the term cross section).

Specifically,

$$N_\nu = \frac{n_\nu L_0 Z_\nu}{M_\nu}, \tag{2.11}$$

where n_ν is the density of the νth material (in grams per cubic centimeter), L_0 is Avogadro's number (0.602×10^{24} molecules/mole), M_ν is the molecular weight (in grams per mole), and Z_ν is the number of atoms per molecule. Thus, for example, to calculate the density N_{H} of hydrogen nuclei in water, we use the values $n_{\mathrm{H}} = 1$, $M_{\mathrm{H}} = 18$, and $Z_{\mathrm{H}} = 2$, so that $N_{\mathrm{H}} = 0.067 \times 10^{24}$ cm^3. The exponent 10^{24} is customarily retained because the normal unit of σ is a barn, equal to 10^{-24} cm^2.

The sum of the cross sections for all possible processes i is called the *total* cross section Σ_T (the total microscopic cross section is of course denoted by σ_T). Microscopic cross sections are measured, as well as calculated. The best tabulation[1] is found in Atomic Energy Commission (AEC) Report BNL 325, although various other tabulations are available.

It is important that the concept of cross section, both partial and total, macroscopic and microscopic, be understood fully at this time, because of the extensive use which will be made of all these quantities.

In general the cross sections Σ_i are strong functions of the magnitude of the neutron speed u. In highly anisotropic materials such as single crystals, Σ_i can also be a function of the direction of $\mathbf{v}$. However, such anisotropic materials are rarely used in reactors.

DEFINITION 2.8 THE COLLISION RATE

We consider neutrons diffusing in a material medium, and ask for the rate R_i at which reactions of type i occur. The complete definition of $R_i(\mathbf{r},\mathbf{u},t)$ is that $R_i(\mathbf{r},\mathbf{u},t)\, d^3u\, d^3r\, dt$ is the number of neutrons in d^3r at position $\mathbf{r}$ with velocities in d^3u at $\mathbf{u}$, undergoing reactions of type i in the time interval dt at time t. The probability that a neutron with velocity $\mathbf{u}$ will undergo a reaction of type i in a unit time is just the distance it travels in a unit time, u cm, times the probability per unit length, Σ_i cm^{-1}, that a collision of type i will occur (see Definition 2.6). Then R_i is just the number of neutrons per unit volume in phase space times this probability for a single neutron,

$$R_i(\mathbf{r},\mathbf{u},t) = \Sigma_i(\mathbf{r},\mathbf{u},t)u\psi(\mathbf{r},\mathbf{u},t) \tag{2.12a}$$

$$= \frac{u\psi(\mathbf{r},\mathbf{u},t)}{\lambda_i(\mathbf{r},\mathbf{u},t)}. \tag{2.12b}$$

We note, in passing, that there is an absorption rate $\Sigma_a u\psi$, a scattering rate $\Sigma_s u\psi$, a fission rate $\Sigma_f u\psi$, and in fact a total collision rate $\Sigma_T u\psi$.

[1] In the author's opinion. Actually, cross-section compilations of reactor materials today are available on magnetic tapes for direct insertion in the computer. Regular revision service is available. This system, known as EDNF-B, is available through Brookhaven National Laboratory.

Sometimes, one asks for the rate at which reactions occur in a specific type of nucleus, e.g., absorption in uranium. Denoting this quantity by $R_a{}^{\mathrm{U}}$, we have

$$R_a{}^{\mathrm{U}} = N_{\mathrm{U}} \sigma_a{}^{\mathrm{U}} u\psi(\mathbf{r},\mathbf{u},t). \tag{2.13}$$

We may, if we wish, denote the product $N_{\mathrm{U}}\sigma_a{}^{\mathrm{U}}$ by the symbol $\Sigma_a{}^{\mathrm{U}}$ and call it the macroscopic cross section for absorption in uranium. It is merely the reciprocal of the mean free path for the specific process, absorption of neutrons in uranium. As we can see from (1.5), the neutron cross sections for independent processes are additive. Also, from (2.10) we see that the macroscopic cross sections for different types of nuclei are additive. Clearly, these additive properties of the cross sections imply that the mean free paths are combined as the reciprocal of the sum of reciprocals. This is most easily seen by referring to (2.9).

Suppose we wish to know the total reaction rate for all neutron velocities. To obtain this result we simply integrate (2.12*a*) over velocities,

$$R_i(\mathbf{r},t) = \Sigma_i(\mathbf{r},t)\phi(\mathbf{r},t) \tag{2.14a}$$

$$= \frac{\bar{v}\rho(\mathbf{r},t)}{\lambda_i(\mathbf{r},t)}, \tag{2.14b}$$

where we have defined

$$R_i(\mathbf{r},t) = \int d^3u\, R_i(\mathbf{r},\mathbf{u},t), \tag{2.15a}$$

$$\Sigma_i(\mathbf{r},t) = \int d^3u\, u \frac{\Sigma_i(\mathbf{r},\mathbf{u},t)\psi(\mathbf{r},\mathbf{u},t)}{\phi(\mathbf{r},t)}, \tag{2.15b}$$

$$\frac{1}{\lambda_i(\mathbf{r},t)} = \int d^3u \frac{u\psi(\mathbf{r},\mathbf{u},t)}{\lambda_i(\mathbf{r},\mathbf{u},t)\phi(\mathbf{r},t)}. \tag{2.15c}$$

If we are interested in the reaction rate for all velocities at a given speed u (or energy $E = \frac{1}{2}mu^2$), the quantities $\phi(\mathbf{r},t)$, $\Sigma(\mathbf{r},t)$, $\bar{v}(\mathbf{r},t)$, and so forth, are simply replaced by $\phi(\mathbf{r},u,t)$, $\Sigma(\mathbf{r},u,t)$, and so forth.

The connection between the reaction rate and the flux given in (2.14*a*) is, in effect, the physical meaning of the flux to which we alluded earlier. Of course, this interpretation will give a precise physical meaning to the flux only if we choose a cross section that is independent of $\mathbf{u}$, in which case the definition of $\Sigma_i(\mathbf{r},t)$ in (2.15*b*) is independent of the flux.

DEFINITION 2.9 THE DIFFERENTIAL SCATTERING CROSS SECTION $\Sigma_s(\mathbf{u}' \to \mathbf{u},\mathbf{r},t)$

One of the collision processes referred to in Definition 2.8 *was the scattering process. Suppose a neutron is initially traveling with a velocity* $\mathbf{u}'$, *whereupon it scatters from a nucleus, ending up with a velocity* $\mathbf{u}$ (*see Fig.* 2.2). *If* $u' = u$,

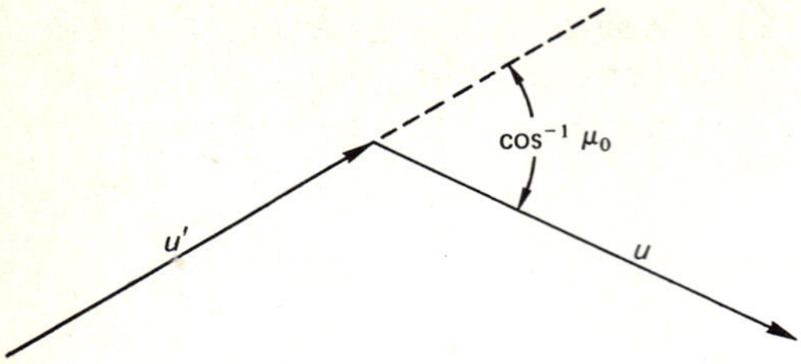

Fig. 2.2 The scattering process.

no energy is exchanged between the neutron and the nucleus. In general, we do not expect u′ and u to be equal since it is possible for the neutron to lose energy while the nucleus gains energy.[1] *If only the kinetic energy of the nucleus is increased, the process is called elastic. If the discrete, internal levels of the nucleus are excited in the collision, the collision is called inelastic. The nuclear levels are a few kiloelectronvolts or more in energy; consequently the inelastic and elastic processes are quite different. Clearly, inelastic processes will occur only for neutrons with energies above the lowest nuclear level. (We could include the fission process in our definition of* $\Sigma_s(\mathbf{u}' \to \mathbf{u})$ *as well,*[2] *but usually it is treated as a separate source contribution.)*

Now we can define the differential scattering cross section such that $[\Sigma_s(\mathbf{u}' \to \mathbf{u})\, d^3u]^{-1}$ *is the mean free path for the process in which a neutron of velocity* $\mathbf{u}'$ *is scattered into the volume element* d^3u *at* $\mathbf{u}$. *Thus, according to* (2.12) *the collision rate for the scattering process* $(\mathbf{u}' \to \mathbf{u})$ *is*

$$dR_s(\mathbf{u}' \to \mathbf{u},\mathbf{r},t) = u'\psi(\mathbf{r},\mathbf{u}',t)\Sigma_s(\mathbf{u}' \to \mathbf{u})\, d^3u. \tag{2.16}$$

Let us recall the notation $\mathbf{u} = u\mathbf{\Omega}$ (so that $\mathbf{\Omega}$ is a unit vector in the direction of neutron velocity). General considerations of rotational symmetry can be used to show that in essentially all cases of interest in reactor physics, $\Sigma_s(\mathbf{u}' \to \mathbf{u})$ is a function only of u', u, and the cosine of the scattering angle $\mu_0 = \mathbf{\Omega}' \cdot \mathbf{\Omega}$ (see Fig. 2.2).

The *total scattering* cross section is obtained by integrating over final velocities since, as we have seen in Definition 2.8, macroscopic cross sections are additive,

$$\Sigma_s(\mathbf{u}') = \int d^3u\, \Sigma_s(\mathbf{u}' \to \mathbf{u}). \tag{2.17}$$

Isotropic scattering means that all values of μ_0 are equally likely, so that $\Sigma_s(\mathbf{u}' \to \mathbf{u})$ is *not* a function of μ_0. In general, however, the scattering in reactor systems is not isotropic.

[1] Since the nuclei have thermal motions, it is also possible for the neutron to gain energy. However, such processes are only important for neutrons with energies below 1 eV.
[2] We shall not always indicate explicitly the dependence of the cross section on **r** and *t*.

In the one-speed approximation which is used throughout the first seven chapters of this book, it is assumed that there is no exchange of energy between the neutron and the nuclei, that is, $u' = u$. Then a scattering event can change only the direction of a neutron's velocity. This requires that $\Sigma_s(\mathbf{u}' \to \mathbf{u})$ can be written in the form

$$\Sigma_s(\mathbf{u}' \to \mathbf{u}) = \frac{\delta(u' - u)}{u^2} \Sigma_s(\mathbf{\Omega}' \cdot \mathbf{\Omega}). \tag{2.18}$$

2.2 THE NEUTRON TRANSPORT EQUATION

A derivation of the neutron transport equation has been made by Osborn and Yip.[1] Here, we shall present a much simpler phenomenological derivation. We consider the balance of the number of neutrons in a small volume element $d^3r\, d^3u$ of phase space.

Consider, then, the change dN in time dt of the number of neutrons with velocities in d^3u about $\mathbf{u}$ and positions of d^3r about $\mathbf{r}$. We denote by S the surface bounding the volume V.

Referring to the definition of the angular density (Definition 2.1), we have

$$dN = d^3u\, dt \int \frac{\partial}{\partial t} \psi(\mathbf{r},\mathbf{u},t)\, d^3r. \tag{2.19}$$

Clearly, (2.19) follows simply from the definitions of dN and $\psi(\mathbf{r},\mathbf{u},t)$. However, we also can list all the different possible ways by which dN can change. This provides us with the balance relation.

$dN = -$ (*a*) net number of neutrons leaking across S in dt

$-$ (*b*) number of neutrons suffering collisions in dt

$+$ (*c*) number of secondary, scattered neutrons produced with velocities in d^3u in dt

$+$ (*d*) number of neutrons produced by sources in dt.

Term (*a*) accounts for those neutrons which enter and leave d^3r with no change in velocity. Term (*b*) represents neutrons leaving d^3u because of a velocity change in scattering and those neutrons lost completely by absorption. Term (*c*) represents neutrons which are scattered into d^3u from all other velocities. Finally, term (*d*), which has not been considered previously, represents the number of neutrons appearing in $d^3u\, d^3r$ from extraneous sources and fission.

[1] R. K. Osborn and S. Yip: "The Foundations of Neutron Transport Theory," Gordon and Breach, Science Publishers, Inc., New York, 1966.

Equation (2.6*a*) gives the net number of neutrons crossing any surface per unit time; so we can write term (*a*) as

$$(a) = d^3u\,dt \int \psi(\mathbf{r},\mathbf{u},t)\mathbf{u}\cdot d\mathbf{S}. \tag{2.20a}$$

The integral in (2.20*a*) can be converted to a volume integral by an application of Gauss' theorem,

$$(a) = d^3u\,dt \int d^3r\,\mathbf{u}\cdot\nabla\psi(\mathbf{r},\mathbf{u},t). \tag{2.20b}$$

The collision rate is discussed in Definition 2.8, and from Eq. (2.12*b*) we see immediately that term (*b*), the loss due to collisions, is

$$(b) = d^3u\,dt \int d^3r\,\frac{u\psi(\mathbf{r},\mathbf{u},t)}{\lambda_T(\mathbf{r},\mathbf{u},t)} \tag{2.21a}$$

$$= d^3u\,dt \int d^3r\,u\Sigma_T(\mathbf{r},\mathbf{u},t)\psi(\mathbf{r},\mathbf{u},t). \tag{2.21b}$$

Our analytical description of the scattering process contributing to term (*c*) is given in Definition 2.9. In particular, (*c*) is given by the integral of (2.16) over all initial velocities $\mathbf{u}'$,

$$(c) = d^3u\,dt \int d^3u' \int d^3r\,u'\psi(\mathbf{r},\mathbf{u}',t)\Sigma_s(\mathbf{u}'\to\mathbf{u}). \tag{2.22}$$

Finally, if we denote the source density by $S(\mathbf{r},\mathbf{u},t)$, then

$$(d) = d^3u\,dt \int d^3r\,S(\mathbf{r},\mathbf{u},t). \tag{2.23}$$

Substituting (2.19), (2.20*b*), (2.21*b*), (2.22), and (2.23) in our basic balance relation, we have (after canceling the differential factors $d^3u\,dt$)

$$\int d^3r \left\{\frac{\partial}{\partial t}\psi(\mathbf{r},\mathbf{u},t) + \mathbf{u}\cdot\nabla\psi(\mathbf{r},\mathbf{u},t) + u\Sigma_T(\mathbf{r},\mathbf{u},t)\psi(\mathbf{r},\mathbf{u},t) - S(\mathbf{r},\mathbf{u},t) - \int d^3u'\,u'\psi(\mathbf{r},\mathbf{u}',t)\Sigma_s(\mathbf{u}'\to\mathbf{u},\mathbf{r},t)\right\} = 0. \tag{2.24}$$

Since the volume of integration is arbitrary, the integrand in (2.24) must vanish at every point $\mathbf{r}$. This gives the neutron transport equation

$$\frac{\partial}{\partial t}\psi(\mathbf{r},\mathbf{u},t) + \mathbf{u}\cdot\nabla\psi(\mathbf{r},\mathbf{u},t) + u\Sigma_T(\mathbf{r},\mathbf{u},t)\psi(\mathbf{r},\mathbf{u},t) = \int d^3u'\,u'\psi(\mathbf{r},\mathbf{u}',t)\Sigma_s(\mathbf{u}'\to\mathbf{u}) + S(\mathbf{r},\mathbf{u},t). \tag{2.25}$$

This equation can also be written in terms of the angular flux $\Phi(\mathbf{r},\mathbf{u},t)$,

$$\Phi(\mathbf{r},\mathbf{u},t) = u\psi(\mathbf{r},\mathbf{u},t). \tag{2.26}$$

In terms of Φ, the transport equation is

$$\frac{1}{u}\frac{\partial}{\partial t}\Phi(\mathbf{r},\mathbf{u},t) + \mathbf{\Omega}\cdot\nabla\Phi(\mathbf{r},\mathbf{u},t) + \Sigma_T(\mathbf{r},\mathbf{u},t)\Phi(\mathbf{r},\mathbf{u},t)$$

$$= \int d^3u'\,\Phi(\mathbf{r},\mathbf{u}',t)\Sigma_s(\mathbf{u}'\to\mathbf{u}) + S(\mathbf{r},\mathbf{u},t), \tag{2.27a}$$

where the unit vector $\mathbf{\Omega}$ has been defined in the previous section (under Definition 2.9). In the same section, we noted that for all the systems which we shall consider, $\Sigma_s(\mathbf{u}'\to\mathbf{u},\mathbf{r},t)$ will be a function of $\mu_0 = \mathbf{\Omega}\cdot\mathbf{\Omega}'$, rather than the directions $\mathbf{\Omega}$ and $\mathbf{\Omega}'$ individually This also implies that all cross sections (such as Σ_a) are functions of speed rather than velocity $\mathbf{u}$. Physically these conditions imply that the probability of reaction is independent of the direction of the neutron which causes the reaction. Furthermore, we shall assume all cross sections to be independent of time. Strictly speaking, this assumption is not true (cf. xenon poisoning or long-term reactivity changes). However, these changes occur over rather long times relative to interesting changes in the neutron distribution. Consequently, the time dependence of the cross section is treated in a different fashion.

With these remarks taken into account, Eq. (2.27*a*) can be rewritten in the form

$$\frac{1}{u}\frac{\partial}{\partial t}\Phi(\mathbf{r},\mathbf{u},t) + \mathbf{\Omega}\cdot\nabla\Phi(\mathbf{r},\mathbf{u},t) + \Sigma_T(\mathbf{r},u)\Phi(\mathbf{r},\mathbf{u},t)$$

$$= S(\mathbf{r},\mathbf{u},t) + \int d^3u'\,\Phi(\mathbf{r},\mathbf{u}',t)\Sigma_s(u'\to u,\,\mathbf{\Omega}'\cdot\mathbf{\Omega}). \tag{2.27b}$$

Finally, we note that in many applications we shall be concerned with time-independent neutron distributions. For this, steady-state case (2.27*b*) reduces to

$$\mathbf{\Omega}\cdot\nabla\Phi(\mathbf{r},\mathbf{u}) + \Sigma_T(\mathbf{r},u)\Phi(\mathbf{r},\mathbf{u}) = S(\mathbf{r},\mathbf{u}) + \int d^3u'\,\Phi(\mathbf{r},\mathbf{u}')\Sigma_s(u'\to u,\,\mathbf{\Omega}'\cdot\mathbf{\Omega}). \tag{2.27c}$$

2.3 THE ONE–SPEED TRANSPORT EQUATION

The transport equation as derived in the previous section is too general for our needs in the first portion of this book. We shall be using one-speed theory exclusively. However, the conversion of Eq. (2.27) is trivial. We simply assume that there is no energy exchanged when neutrons collide with nuclei, so that we can use (2.18) in the scattering integral in (2.27*b*).

$$\frac{1}{u}\frac{\partial}{\partial t}\Phi(\mathbf{r},\boldsymbol{\Omega},t) + \boldsymbol{\Omega}\cdot\nabla\Phi(\mathbf{r},\boldsymbol{\Omega},t) + \Sigma_T(\mathbf{r})\Phi(\mathbf{r},\boldsymbol{\Omega},t)$$
$$= S(\mathbf{r},\boldsymbol{\Omega},t) + \int d\boldsymbol{\Omega}'\,\Phi(\mathbf{r},\boldsymbol{\Omega}',t)\Sigma_s(\mathbf{r},\,\boldsymbol{\Omega}'\cdot\boldsymbol{\Omega}). \tag{2.28}$$

Here we have replaced $\mathbf{u}$ with $\boldsymbol{\Omega}$ in the arguments of Φ and S. We note that u enters the one-speed equations only as a parameter and plays no essential part in the theory.

In the one-speed approximation it is convenient to use a flux and current defined as integrals only over $\boldsymbol{\Omega}$ rather than $\mathbf{u}$ as was done in (2.3) and (2.4),

$$\phi(\mathbf{r},t) = \int d\boldsymbol{\Omega}\,\Phi(\mathbf{r},\boldsymbol{\Omega},t), \tag{2.29a}$$

$$\mathbf{J}(\mathbf{r},t) = \int d\boldsymbol{\Omega}\,\boldsymbol{\Omega}\Phi(\mathbf{r},\boldsymbol{\Omega},t). \tag{2.29b}$$

At this point, we might discuss briefly the validity of the one-speed theory to be used so extensively in the first portion of the book. As we pointed out in the introduction, the main reason for the use of the one-speed theory is pedagogical. However, there is some technical merit in one-speed theory if the cross sections are taken to be suitably averaged functions of the neutron energy spectrum as in (2.14*b*). (This is true particularly in fast systems, where the neutron energy spectrum covers a relatively narrow energy range.) Thus, some qualitative features of practical systems can be understood from one-speed theory, although admittedly, one should not push this interpretation too far.

2.4 ANOTHER FORM OF THE BALANCE RELATION

The transport equation (2.28) can be written in another form in terms of the dependent variables ϕ and $\mathbf{J}$. We can obtain this new relation by integrating (2.28) over the direction of the velocity.

$$\frac{1}{u}\frac{\partial}{\partial t}\int d\boldsymbol{\Omega}\,\Phi(\mathbf{r},\boldsymbol{\Omega},t) + \nabla\cdot\int d\boldsymbol{\Omega}\,\boldsymbol{\Omega}\Phi(\mathbf{r},\boldsymbol{\Omega},t) + \Sigma_T(\mathbf{r})\int d\boldsymbol{\Omega}\,\Phi(\mathbf{r},\boldsymbol{\Omega},t)$$
$$= \int d\boldsymbol{\Omega}\,S(\mathbf{r},\boldsymbol{\Omega},t) + \int d\boldsymbol{\Omega}'\,\Phi(\mathbf{r},\boldsymbol{\Omega}',t)\int d\boldsymbol{\Omega}\,\Sigma_s(\mathbf{r},\,\boldsymbol{\Omega}'\cdot\boldsymbol{\Omega}). \tag{2.30}$$

Clearly, (2.30) is just a balance relation for neutrons of all velocities in a small volume of space d^3r. Since, we are using the one-speed approximation, we can determine the integral $\int d\boldsymbol{\Omega}\,\Sigma_s(\mathbf{r},\,\boldsymbol{\Omega}'\cdot\boldsymbol{\Omega})$ by using (2.18) in (2.17)

$$\Sigma_s(\mathbf{r}) = \int d\boldsymbol{\Omega}\,\Sigma_s(\mathbf{r},\,\boldsymbol{\Omega}'\cdot\boldsymbol{\Omega}). \tag{2.31a}$$

With definitions (2.29) and a definition of a total source as

$$S(\mathbf{r}) = \int d\boldsymbol{\Omega}\; S(\mathbf{r},\boldsymbol{\Omega}), \tag{2.31b}$$

we can write (2.30) in terms of ϕ and $\mathbf{J}$ as follows:

$$\frac{1}{u}\frac{\partial}{\partial t}\,\phi(\mathbf{r},t) + \nabla \cdot \mathbf{J}(\mathbf{r},t) + \Sigma_a(\mathbf{r})\phi(\mathbf{r},t) = S(\mathbf{r},t). \tag{2.32}$$

We have, of course, used the definition of the total cross section given in Definition 2.7 to write

$$\Sigma_a = \Sigma_T - \Sigma_s\,. \tag{2.33}$$

This new balance relation appears to be much simpler in form than (2.28). However, (2.28) with suitable boundary conditions is sufficient to determine $\psi(\mathbf{r},\mathbf{u},t)$ uniquely.[1] Equation (2.32), by itself, is not sufficient to determine both $\phi(\mathbf{r},t)$ and $\mathbf{J}(\mathbf{r},t)$ uniquely. We can obtain unique solutions of (2.32) only with the help of an (approximate) relation between ϕ and $\mathbf{J}$.

2.5 FICK'S RULE

Equation (2.32) is one equation in the four unknowns, ϕ and the three components of $\mathbf{J}$. Our next step will be to derive three more equations (or if you prefer, one vector equation) relating ϕ and $\mathbf{J}$. These equations, known as *Fick's rule*, are only approximate. When they are substituted in the (exact) balance relation (2.32), the diffusion equation is obtained. For this reason, Fick's rule is considered the basic relation of diffusion theory.

The basic idea is to expand $\Phi(\mathbf{r},\boldsymbol{\Omega},t)$ in a series of the form[2]

$$\Phi(\mathbf{r},\boldsymbol{\Omega},t) = \frac{1}{4\pi}\left[A(\mathbf{r},t) + 3\mathbf{B}(\mathbf{r},t)\cdot\boldsymbol{\Omega} + \cdots\right]. \tag{2.34}$$

The truncation of this series, after only two terms, is the reason that Fick's rule is only approximate. If the angular density were isotropic (i.e., independent of $\boldsymbol{\Omega}$), only the first term in this expansion would be present. The inclusion of the second term means that we are assuming the neutron distribution to be "almost isotropic."

If Eq. (2.34) is integrated over Ω, we have, recalling (2.29*a*),

$$\phi(\mathbf{r},t) = A(\mathbf{r},t), \tag{2.35}$$

[1] K. M. Case and P. F. Zweifel: "Linear Transport Theory," Addison-Wesley Publishing Company, Inc., Reading, Mass., 1967.

[2] The numerical factors $1/4\pi$ and 3 are inserted for convenience.

since

$$\int d\mathbf{\Omega} = 4\pi \tag{2.36}$$

and

$$\int \mathbf{\Omega} \cdot \mathbf{B}\, d\mathbf{\Omega} = 0. \tag{2.37}$$

Thus, we see that $A(\mathbf{r},t)$ is actually the neutron flux.

Next, we multiply Eq. (2.34) by Ω_z and integrate. Using (2.29*b*), we have

$$J_z(\mathbf{r},t) = B_z(\mathbf{r},t), \tag{2.38}$$

where we have used

$$\int \Omega_z\, d\mathbf{\Omega} = \int \Omega_z \Omega_x\, d\mathbf{\Omega} = \int \Omega_z \Omega_y\, d\mathbf{\Omega} = 0 \tag{2.39a}$$

and

$$\int \Omega_z^{\,2}\, d\mathbf{\Omega} = \frac{4\pi}{3}. \tag{2.39b}$$

[Equations (2.37) and (2.39) can be proved easily by expressing the components of $\mathbf{\Omega}$ in spherical polar coordinates: $\Omega_z = \cos\theta$, $\Omega_x = \sin\theta\cos\phi$, and $\Omega_y = \sin\theta\sin\phi$.]

In a similar fashion, we can prove

$$J_x(\mathbf{r},t) = B_x(\mathbf{r},t), \tag{2.40a}$$

$$J_y(\mathbf{r},t) = B_y(\mathbf{r},t), \tag{2.40b}$$

so that, in fact,

$$\mathbf{J}(\mathbf{r},t) = \mathbf{B}(\mathbf{r},t). \tag{2.40c}$$

Then, Eq. (2.34), the angular expansion of $\Phi(\mathbf{r},\mathbf{\Omega},t)$ becomes

$$\Phi(\mathbf{r},\mathbf{\Omega},t) = \frac{1}{4\pi}\left[\phi(\mathbf{r},t) + 3\mathbf{\Omega}\cdot\mathbf{J}(\mathbf{r},t)\right] + \cdots. \tag{2.41}$$

If we insert this expansion into the transport equation (2.28), we obtain

$$\begin{aligned}\frac{1}{4\pi u}\frac{\partial}{\partial t}\left[\phi(\mathbf{r},t) + 3\mathbf{\Omega}\cdot\mathbf{J}(\mathbf{r},t)\right] &+ \frac{1}{4\pi}\Sigma_T(\mathbf{r})\left[\phi(\mathbf{r},t) + 3\mathbf{\Omega}\cdot\mathbf{J}(\mathbf{r},t)\right] \\ &+ \frac{1}{4\pi}\mathbf{\Omega}\cdot\nabla\left[\phi(\mathbf{r},t) + 3\mathbf{\Omega}\cdot\mathbf{J}(\mathbf{r},t)\right] = \frac{1}{4\pi}S(\mathbf{r},t) + \frac{1}{4\pi}\Sigma_s(\mathbf{r})\phi(\mathbf{r},t) \\ &+ \frac{3}{4\pi}\int d\mathbf{\Omega}'\,\mathbf{\Omega}'\cdot\mathbf{J}(\mathbf{r},t)\Sigma_s(\mathbf{r},\mathbf{\Omega}'\cdot\mathbf{\Omega}),\end{aligned} \tag{2.42}$$

where we have assumed that the source is isotropic, and recalling (2.31*b*), we have written $S(\mathbf{r},\mathbf{\Omega}) = S(\mathbf{r})/4\pi$. Now, if we multiply (2.42) by Ω_z, say, and integrate over $\mathbf{\Omega}$, we obtain

$$\frac{3}{4\pi u}\frac{\partial}{\partial t}\int d\mathbf{\Omega}\,\Omega_z\,\mathbf{\Omega}\cdot\mathbf{J}(\mathbf{r},t) + \frac{3}{4\pi}\Sigma_T(\mathbf{r})\int d\mathbf{\Omega}\,\Omega_z\,\mathbf{\Omega}\cdot\mathbf{J}(\mathbf{r},t)$$

$$+\frac{1}{4\pi}\int d\mathbf{\Omega}\,\Omega_z\,\mathbf{\Omega}\cdot\nabla\phi(\mathbf{r},t) = \frac{3}{4\pi}\int d\mathbf{\Omega}'\,\mathbf{\Omega}'\cdot\mathbf{J}(\mathbf{r},t)\int d\mathbf{\Omega}\,\Omega_z\,\Sigma_s(\mathbf{r},\mathbf{\Omega}'\cdot\mathbf{\Omega}), \tag{2.43}$$

where we have used (2.37) and

$$\int d\mathbf{\Omega}\,\Omega_\mu\,\Omega_\nu\,\Omega_\gamma = 0, \tag{2.44}$$

where μ, ν, and γ refer to the x, y, and z components. Using the definition of μ_0 (Fig. 2.2), we can write

$$\int d\mathbf{\Omega}\,\Omega_z\,\Sigma_s(\mathbf{r},\mathbf{\Omega}'\cdot\mathbf{\Omega}) = 2\pi\int_{-1}^{1} d\mu_0\,\mu_0\,\Sigma_s(\mathbf{r},\mu_0)\Omega_z' \tag{2.45a}$$

$$= \Sigma_s(\mathbf{r})\overline{\mu(\mathbf{r})}\Omega_z'. \tag{2.45b}$$

Clearly, $\overline{\mu(\mathbf{r})}$ is just the average of the scattering angle μ_0 over the cross section

$$\bar{\mu} = 2\pi\int_{-1}^{1}\frac{d\mu_0\,\mu_0\,\Sigma_s(\mathbf{r},\mu_0)}{\Sigma_s(\mathbf{r})}. \tag{2.45c}$$

If we insert (2.45*b*) into (2.43) and use Eq. (2.39) to perform the integrals over $\mathbf{\Omega}'$, we obtain

$$\frac{1}{u}\frac{\partial}{\partial t}J_z(\mathbf{r},t) + [\Sigma_T(\mathbf{r}) - \Sigma_s(\mathbf{r})\overline{\mu(\mathbf{r})}]\,J_z(\mathbf{r},t) = -\frac{1}{3}\frac{\partial\phi(\mathbf{r},t)}{\partial z}. \tag{2.46}$$

We can, of course, derive similar equations for the x and y components of $\mathbf{J}$. The resulting set of three equations can be written in vector notation as

$$\frac{1}{u}\frac{\partial}{\partial t}\mathbf{J}(\mathbf{r},t) + \Sigma_{tr}(\mathbf{r})\mathbf{J}(\mathbf{r},t) = -\frac{1}{3}\nabla\phi(\mathbf{r},t), \tag{2.47}$$

where we have introduced the definition

$$\Sigma_{tr} = \Sigma_T - \Sigma_s\bar{\mu}; \tag{2.48}$$

Σ_{tr} is referred to as the *transport cross section.*

Equation (2.47) does provide us with a connection between $\mathbf{J}$ and ϕ. In fact, (2.47) and (2.32) are combined to give the telegrapher's equation. The diffusion equation is obtained by neglecting the term $(1/u)(\partial/\partial t)\mathbf{J}$ in

Eq. (2.47). The justification for neglecting this term stems from the magnitude of u ($\approx 10^5$ cm/s for thermal neutrons). Typically, Σ_{tr} is approximately 1 cm^{-1}, and $\phi(\mathbf{r},t)$ changes significantly over distances of about 1 cm. Thus, in order for $(1/u)(\partial/\partial t)\mathbf{J}$ to be significant (for thermal neutrons), $\mathbf{J}$ would have to change significantly during times of about 10^{-5} s or less. Such rapid time changes would be unusual. For fast neutrons, u is larger and the change $\mathbf{J}$ would have to be correspondingly faster. We discuss the telegrapher's equation and the approximations inherent in the diffusion equation in somewhat greater detail in Appendix F.

Neglecting the first term, then, in (2.47) we have Fick's law connecting ϕ and $\mathbf{J}$,

$$\mathbf{J}(\mathbf{r},t) = -D(\mathbf{r})\nabla\phi(\mathbf{r},t), \tag{2.49a}$$

where we have introduced the diffusion coefficient $D(\mathbf{r})$,

$$D(\mathbf{r}) = \frac{1}{3\Sigma_{tr}(\mathbf{r})}. \tag{2.49b}$$

Inserting (2.49) into (2.32), we obtain, finally, the diffusion equation

$$\frac{1}{u}\frac{\partial}{\partial t}\phi(\mathbf{r},t) - \nabla \cdot D(\mathbf{r})\nabla\phi(\mathbf{r},t) + \Sigma_a(\mathbf{r})\phi(\mathbf{r},t) = S(\mathbf{r},t). \tag{2.50}$$

2.6 BOUNDARY CONDITIONS

The time-dependent diffusion equation (2.50) requires the specification of an initial condition plus boundary values before a complete solution can be determined. There is no difficulty in specifying the initial condition, but the appropriate boundary conditions must be determined from physical considerations. We consider three types of boundary, and discuss below the conditions appropriate to each type. They are:

1. Boundaries at infinity
2. Boundaries between two dissimilar material media
3. Boundaries between a material medium and vacuum

The boundary condition at infinity is simplest; we specify that the neutron density must remain finite.[1] As a matter of fact, the neutron density must remain finite everywhere, except in the immediate vicinity of a source. Needless to say, physical reality also requires that a flux never be negative.

At the interface between two material media, the rigorous transport theoretical boundary condition is that the angular density $\psi(\mathbf{r},\mathbf{u},t)$ should be a

[1] Certain exceptional cases will be discussed in context.

continuous function of $\mathbf{r}$ for every $\mathbf{u}$. This condition cannot be met exactly in the diffusion approximation. However, we can require that the flux, $\phi(\mathbf{r},t)$ and the current $\mathbf{J}(\mathbf{r},t)$ be continuous.

It is interesting to see how this continuity condition is modified if the boundary contains a source. Consider a medium which is infinite in the y, z directions with an infinite plane source of strength Q_s/cm^2 s^{-1} located at the position $x = 0$. Then from symmetry,[1]

$$\frac{\partial \phi}{\partial y} = \frac{\partial \phi}{\partial z} = 0; \tag{2.51a}$$

also,

$$J_y = J_z = 0, \tag{2.51b}$$

so that the conservation equation (2.50) becomes[2]

$$\frac{1}{u}\frac{\partial \phi}{\partial t}(x,t) = Q_s\,\delta(x) - \Sigma_a \phi(x,t) - \frac{\partial J_x}{\partial x}(x,t). \tag{2.52}$$

If this equation is now integrated from $x - \varepsilon$ to $x + \varepsilon$, we obtain (ε is infinitesimal and positive)

$$J_x(0+) - J_x(0-) = Q_s. \tag{2.53}$$

The current is discontinuous by an amount equal to the source strength. Equation (2.53) is generally called the *jump condition*. This result is general; its diffusion theory version of course replaces J_x with

$$J_x = -D\frac{\partial \phi}{\partial x}. \tag{2.54}$$

Finally, we consider the appropriate condition for boundary type 3 above, i.e., the interface between a material medium and vacuum. Again, for simplicity, we should consider the plane symmetric type, so that the flux and current depend only upon a single coordinate x. The correct boundary condition is clearly that there be no neutrons returned from the vacuum to the material medium. Again, this condition cannot be fulfilled exactly in the diffusion approximation, so we choose the weaker condition[3] (x_s

[1] Or from group-theoretical arguments. See Case and Zweifel: *op. cit.*

[2] See Appendix B for a discussion of the Dirac delta distribution $\delta(x)$.

[3] This implies that the net number of neutrons returned from the vacuum into the medium is zero. Clearly this is true, but it is a weaker condition than the exact condition that no neutrons return at all. The latter condition, of course, must be placed on the angular density $\psi(\mathbf{r},\mathbf{u},t)$. In diffusion theory, we deal only with the flux and current rather than the complete angular density, and, as a consequence, we cannot satisfy the exact boundary condition.

represents any point on the interface),

$$\int_{\Omega_x>0} d\Omega\, \Omega_x \Phi(x_s, \Omega_x) = 0.$$

By using Eqs. (2.41) and (2.49*a*) this gives[1]

$$\frac{1}{\phi}\frac{\partial \phi}{\partial x} = -\frac{1}{2D}. \tag{2.55}$$

If in the vicinity of the boundary we replace $\phi(x)$ with a linear function

$$\phi(x) = \phi(0)(1 + Kx), \tag{2.56}$$

we see, by comparing Eqs. (2.37) and (2.38), that $K = -3/2\Sigma_{tr}$ so that

$$\phi(x) = \phi(0)\left(1 - \frac{3x\Sigma_{tr}}{2}\right). \tag{2.57}$$

Note that the flux does not vanish at $x = 0$. Rather, it extrapolates linearly to zero at a point

$$z_0 = \frac{2\lambda_{tr}}{3} = \frac{2}{3\Sigma_{tr}} \tag{2.58}$$

outside the boundary. The quantity z_0 is called the *linear extrapolated end point*. In general, the flux near the boundary is not linear; if the actual flux shape were extrapolated to zero, a different extrapolated end point would be obtained. (Cf. Fig. 2.3).

[1] Problem 2.9.

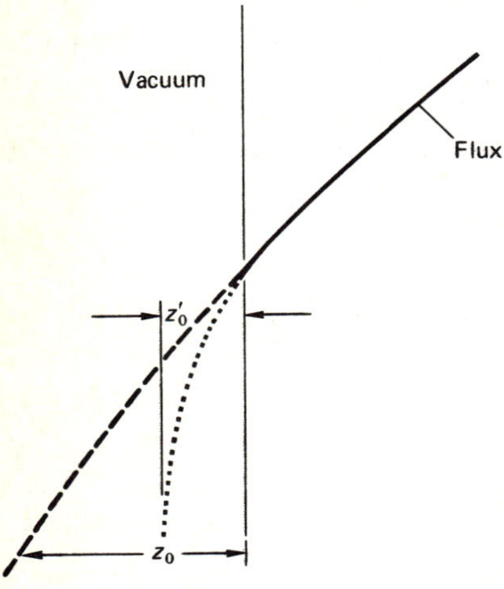

Fig. 2.3 Linear extrapolated end point z_0 and actual z_0'.

The calculation of the extrapolated end point is discussed in detail elsewhere,[1] where it is found to be a function of the relative amounts of fission, absorption, and scattering in the material medium. For pure scattering (i.e., no fission or absorption) it is found that the flux near a boundary far from any other boundary or source is indeed linear; z_0 is then given by an expression which depends upon the anisotropy of the scattering; for isotropic scattering

$$z_0 = 0.7104\lambda.\ddagger \tag{2.59}$$

Note that $\lambda_{tr} = \lambda_s = \lambda_T = \lambda$ for this special boundary condition.

For curved boundaries the calculation of the extrapolated end point is much more complicated. However, if the radius of curvature is large, even curved boundaries may be taken to be plane, to a good approximation. Furthermore, the extrapolated boundary correction is usually a small fraction of the dimension of the reactor. If this is not the case, generally the assembly is not sufficiently large for the diffusion approximation to be valid. For this reason, we shall adopt the simple condition

$$z_0 = 0.7\lambda_{tr} \tag{2.60}$$

for all ratios of scattering, absorption, and fission and for all boundary shapes, plane or curved. The boundary condition is that the extrapolated flux vanish at a distance z_0 outside the actual physical boundary of the surface.

Let us recapitulate the boundary conditions which we shall apply to the solutions of the neutron diffusion equation.

1. The flux shall remain finite everywhere except in the immediate vicinity of a neutron source.
2. The flux and current shall be continuous everywhere in the interior of a diffusing medium, except at a source, where the current is discontinuous [cf. Eq. (2.53), which applies for a plane].
3. At an interface between a diffusing medium and vacuum, the flux shall be calculated as though it vanished at a distance $0.7\lambda_{tr}$ beyond the boundary.

A final word of caution should be noted in connection with condition 3. Needless to say, the flux does not actually vanish at a distance $0.7\lambda_{tr}$ beyond the boundary. If it did, there would be no need for reactor shields! Rather,

[1] Case and Zweifel, *op. cit.*

‡ E. Richard Cohen, in *Nucl. Sci. Eng.*, **42**, 23 (1970), has calculated z_0 to 25 decimal places! Dr. Cohen has been quoted (perhaps apocryphally) as referring to this calculation as a typical example of "computer overkill." In case anyone is interested, the result is $z_0 = 0.7104460895987630727325241$.

the extrapolated end-point condition is a mathematical fiction which gives the correct neutron distribution *inside the reactor* only. If the distribution outside the reactor is needed, as for calculations of biological hazards, additional techniques must be introduced. Also, if the condition $0.7\lambda_{tr}$ is felt to be not sufficiently accurate because of absorption or curvature effects, methods of accounting for these effects on z_0 can be found in various references.[1]

2.7 VALIDITY OF DIFFUSION THEORY

On the basis of the derivation of the diffusion equation, as described earlier, some general remarks can be made on its applicability. First, we recall that the flux Φ was expanded in a Taylor's series in $\mathbf{\Omega}$ by using only zeroth- and first-order terms [Eq. (2.41)]. Thus, if the flux is strongly anisotropic, we should expect the diffusion approximation to break down. This occurs in a medium with a high-absorption cross section, or in any medium sufficiently close (a few mean free paths) to a boundary. If the cross sections are rapidly varying functions of position, we should expect diffusion theory to break down. Actually, in all of these instances the flux would change appreciably over short distances. If the flux varies rapidly in space, it is highly anisotropic.

These general conclusions have been verified by actual numerical comparison of solutions of the diffusion equation with essentially exact transport—theoretical solutions,[2] by using a one-speed approximation. In one-speed transport theory it is convenient to introduce a parameter c called the *mean number of secondaries per collision*, or to put it another way, the number of neutrons given off, on the average, for each collision which takes place. For a medium containing no fissionable isotope,

$$c = \frac{\Sigma_s}{\Sigma_T}, \tag{2.61}$$

i.e., the ratio of scattering to total cross sections. If a fissionable material is present, then

$$c = \frac{\Sigma_s + \nu\Sigma_f}{\Sigma_T}. \tag{2.62}$$

It is found that diffusion theory is adequate for $c \to 1$. Actually, the exact neutron distribution is composed of two components, an asymptotic portion which holds far from boundaries and sources, plus a transient portion introduced near boundaries and sources. Diffusion theory describes only the

[1] For example, Case and Zweifel, *op. cit.*, chap. 6.
[2] Case and Zweifel, *op. cit.*

asymptotic portion of the neutron distribution. For $c \approx 1$ the asymptotic portion is well described by the diffusion equation, and in addition, the transient portion is not very important. For c not close to unity, diffusion theory breaks down in two respects. First, it does not give an adequate description of the asymptotic part of the distribution, and second, the transient part becomes more and more important. In fact, for $c = 0$ there *is* no asymptotic distribution—it is all transient![1]

2.8 THE DIFFUSION EQUATION IN PLANE GEOMETRY

Now that the diffusion equation has been derived, and appropriate boundary conditions established, we can discuss solutions of the equation. Consider it first in plane symmetry. This has already been discussed in Sec. 2.6. In particular, we shall consider the situation of neutrons diffusing in a medium which is infinite in the y, z directions, in which cross sections and sources are functions only of x. Then the flux and current depend only on x and, in addition, the current has only an x component. The stationary diffusion equation becomes

$$\left[-\frac{d}{dx} D(x) \frac{d}{dx} + \Sigma_a(x)\right] \phi(x) = S(x). \tag{2.63}$$

Suppose we make a change of variable such that

$$dy = \frac{dx}{D(x)}, \tag{2.64a}$$

$$\phi(y)\, dy = \phi(x)\, dx, \tag{2.64b}$$

$$S(y)\, dy = S(x)\, dx. \tag{2.64c}$$

Then Eq. (2.63) can be written

$$\left[-\frac{d^2}{dy^2} + D(y)\Sigma_a(y)\right] \phi(y) = D(y)S(y), \tag{2.65}$$

In this equation, we note that distances are measured in units of $\lambda_{tr}/3$. Now Eq. (2.65) will be simpler to solve than (2.63) if the product $D(y)\Sigma_a(y)$ is a constant, even though D and Σ_a individually depend upon position. We shall assume this product is indeed constant, and denote

$$D(y)\Sigma_a(y) = \rho^2 \tag{2.66}$$

and

$$S(y)D(y) = S_0(y), \tag{2.67}$$

[1] Clearly, c must be a positive quantity, and if $\Sigma_s = 0$, then $c = k_\infty$, Eq. (1.8), a relation valid in one-speed theory.

a modified source. Then Eq. (2.65) becomes

$$\left(-\frac{d^2}{dy^2}+\rho^2\right)\phi(y)=S_0(y). \tag{2.68a}$$

For no source, this equation becomes

$$\left(-\frac{d^2}{dy^2}+\rho^2\right)\phi(y)=0. \tag{2.68b}$$

Consider various solutions of this equation.

a INFINITE-MEDIUM, PLANE-SOURCE, GREEN'S FUNCTION

We consider first the infinite-medium, plane-source, Green's function (see Fig. 2.4). This is the solution to Eq. (2.68*a*) in the infinite medium with

$$S_0(y)=D(0)\,\delta(y) \tag{2.69a}$$

$$=D\,\delta(y), \tag{2.69b}$$

representing a plane source in the $y=0$ plane. Then the solution, call it $G_\infty(0\to y)$, obeys the homogeneous equation (2.68*b*) everywhere except at the origin, where condition (2.53) must be applied.

The solution of the homogeneous equation (2.68*b*) which vanishes at infinity is

$$G_\infty(0\to y)=A_1e^{-\rho y},\qquad y>0, \tag{2.70a}$$

and

$$G_\infty(0\to y)=A_2\,e^{+\rho y},\qquad y<0. \tag{2.70b}$$

So that in general

$$G_\infty(0\to y)=A_1e^{-\rho|y|}, \tag{2.70c}$$

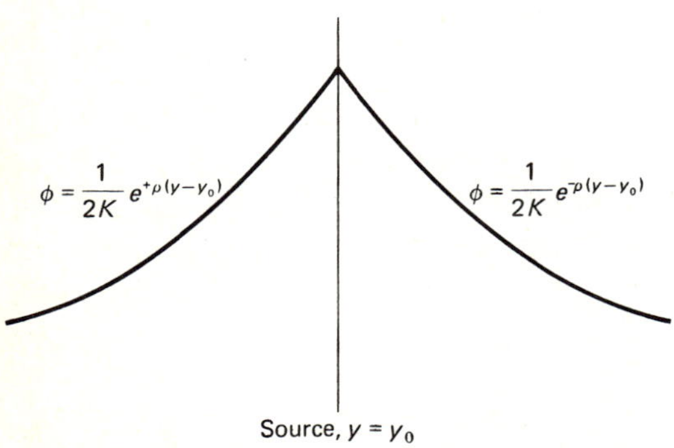

Fig. 2.4 Infinite-medium Green's function.

where we have used the fact that $A_1 = A_2$ since G must be continuous at $y = 0$. To evaluate A_1, we have, from Eq. (2.54)

$$J_x(0\pm) = \pm KA_1, \tag{2.71}$$

so that the jump condition (2.53) yields

$$A_1 = \frac{1}{2K}, \tag{2.72}$$

where we have introduced the standard notation

$$K = \sqrt{\frac{\Sigma_a}{D}} = \frac{\rho}{D}. \tag{2.73}$$

Hence,

$$G_\infty(0 \to y) = \frac{1}{2K} e^{-\rho|y|}. \tag{2.74a}$$

Now it is clear the the origin could be shifted to the plane y_0, say, by the substitution $y \to y - y_0$. Then the general form of the infinite-medium Green's function can be written

$$G_\infty(y_0 \to y) = \frac{1}{2K} e^{-\rho|y-y_0|}. \tag{2.74b}$$

It is interesting to transform back to the independent variable x when D is a constant. Using Eq. (2.64*b*), we have

$$G_\infty(y_0 \to y)\, dy = G_\infty(x_0 \to x)\, dx, \tag{2.75}$$

with dy/dx given by Eq. (2.64*a*), or

$$G_\infty(x_0 \to x) = \frac{1}{2DK} e^{-K|x_0-x|}. \tag{2.76}$$

The physical meaning of $G_\infty(x_0 \to x)$ is the neutron flux at x due to an infinite plane source located at x_0 *in an infinite medium*. In general, the Green's function has this meaning; the subscript ∞ indicates that an infinite medium is being considered. We have also explicitly noted that G_∞ is a function only of $|x_0 - x|$, a so-called displacement kernel.[1] We might calculate the mean square distance a neutron travels from the source before it is absorbed.

[1] Actually a displacement kernel may depend on $x_0 - x$ instead of $|x_0 - x|$. Thus G_∞ is a special kind of displacement kernel. All infinite medium kernels are of this form because of translational and rotational invariance.

This is

$$\langle x^2 \rangle = \frac{\int_{-\infty}^{\infty}(x-x_0)^2 G_\infty(x-x_0)\,dx}{\int_{-\infty}^{\infty} G_\infty(x)\,dx} \tag{2.77a}$$

$$= \frac{\int_0^\infty x^2 e^{-Kx}\,dx}{\int_0^\infty e^{-Kx}\,dx} \tag{2.77b}$$

$$= \frac{2}{K^2}. \tag{2.77c}$$

Because of its relation to $\langle x^2 \rangle$, generally $1/K^2$ is called the *diffusion area* and is denoted by L^2. The quantity L is called the *diffusion length.*

The Green's function can be used to compute the distribution due to a distributed source $S(x)$.[1]

$$\phi(x) = \int_{-\infty}^{\infty} S(x_0) G_\infty(x_0 - x)\,dx_0. \tag{2.78}$$

Suppose, for example, $S(x)$ is a constant source throughout all space, of magnitude Q neutrons/cm^3. Then

$$\phi(x) = Q \int_{-\infty}^{\infty} G_\infty(x_0 - x)\,dx_0 \tag{2.79}$$

$$= \frac{Q}{\Sigma_a}. \tag{2.80}$$

Equation (2.80) has a simple interpretation. It states that the absorption rate $\Sigma_a \phi$ is equal to the rate at which source neutrons are produced.

b FINITE-MEDIUM GREEN'S FUNCTION

We consider a problem similar to the infinite-medium problem discussed above, except that we seek the solution in a medium of finite width, say a (called a *slab*), Fig. 2.5. Take the origin to be at the center of the slab, and the source to be at an arbitrary point y_0. Then we have

$$G(y_0 \to y) = \begin{cases} A_1 e^{-\rho y} + A_2 e^{\rho y}, & y > y_0, \quad (2.81a) \\ A_3 e^{-\rho y} + A_4 e^{\rho y}, & y < y_0. \quad (2.81b) \end{cases}$$

This represents the general solution to Eq. (2.68*b*). We apply the following boundary conditions. First, the flux must vanish at the extrapolated end points,

$$G(y_0 \to a + z_0) = 0, \tag{2.82a}$$

$$G(y_0 \to -a - z_0) = 0. \tag{2.82b}$$

[1] This relation follows immediately from the physical meaning of the Green's function described earlier.

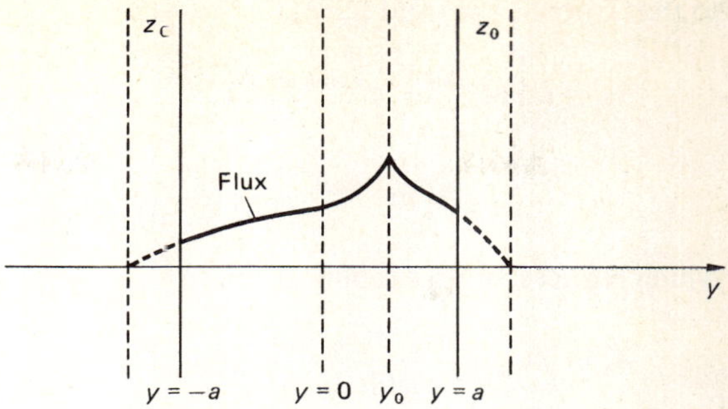

Fig. 2.5 Green's function for a slab.

Also, at $y = y_0$, the flux is continuous,

$$G(y_0 \to y_0+) = G(y_0 \to y_0-). \tag{2.83}$$

Finally, at $y = y_0$, the jump condition is obeyed,

$$\frac{\partial}{\partial y} G(y_0 \to y)\Big]_{y_0-}^{y_0+} = -D(y_0). \tag{2.84}$$

These four conditions are sufficient to determine the four coefficients A_1, A_2, A_3, and A_4. Calling $a + z_0 = a'$, we have

$$A_1 = -A_2 e^{2\rho a'} \tag{2.85a}$$

and

$$A_3 = -A_4 e^{-2\rho a'}. \tag{2.85b}$$

Thus,

$$G(y_0 \to y) = A_2(e^{\rho y} - e^{-\rho(y-2a')}), \quad y > y_0, \tag{2.86a}$$

and

$$G(y_0 \to y) = A_4(e^{\rho y} - e^{\rho(y+2a')}), \quad y < y_0. \tag{2.86b}$$

From (2.83),

$$A_2 = A_4 \frac{e^{\rho y_0} - e^{-\rho(y_0+2a')}}{e^{\rho y_0} - e^{-\rho(y_0-2a')}} \tag{2.87a}$$

$$= A_4 \frac{\sinh \rho(y_0 + a')}{\sinh \rho(y_0 - a')}. \tag{2.87b}$$

Finally, from (2.84),

$$D + A_2 \rho(e^{\rho y_0} + e^{-\rho(y_0-2a')}) = A_4 \rho(e^{\rho y_0} + e^{-\rho(y_0+2a')}), \tag{2.88}$$

so that from (2.87*b*) and (2.88),

$$A_4 = D\left[\rho(e^{\rho y_0} + e^{-\rho(y_0+2a')}) - \rho e^{-2\rho a'} \frac{\sinh \rho(y_0 + a')(e^{\rho y_0} + e^{-\rho(y_0-2a')})}{\sinh \rho(y_0 - a')}\right]^{-1}, \tag{2.89}$$

which reduces, upon simplification, to

$$A_4 = D\frac{e^{\rho a'}}{2\rho}\frac{\sinh \rho(y_0 - a')}{\sinh 2\rho a'}. \tag{2.90}$$

Thus,

$$G(y_0 \to y) = \begin{cases} \dfrac{e^{-\rho a'} \sinh \rho(y_0 + a')}{2K \sinh 2\rho a'}(e^{\rho y} - e^{-\rho(y-2a')}), & y > y_0, \quad (2.91a) \\ \dfrac{e^{\rho a'} \sinh \rho(y_0 - a')}{2K \sinh 2\rho a'}(e^{\rho y} - e^{-\rho(y+2a')}), & y < y_0. \quad (2.91b) \end{cases}$$

c MULTILAYERED SLABS

As a final example of the solution of the diffusion equation in slab geometry, we might consider a problem containing adjacent slabs (see Fig. 2.6). For simplicity we shall take a symmetric problem with the source at the center. We shall also assume that the cross sections are constant in each region. Then the solution to the diffusion equation will be

$$\phi_1 = A_1 e^{-K_1|x|} + A_2 e^{K_1|x|}, \quad x \text{ in region } 1, \tag{2.92a}$$

$$\phi_2 = A_3 e^{-K_2|x|} + A_4 e^{+K_2|x|}, \quad x \text{ in region } 2. \tag{2.92b}$$

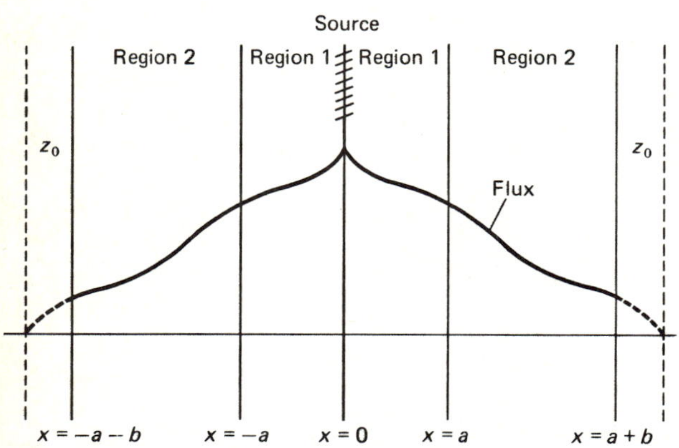

Fig. 2.6 Green's function for a multilayered slab.

These solutions apply for $x > 0$ and $x < 0$. Because of the symmetry, we have $\phi(-x) = \phi(x)$.

Now there are four conditions to determine the four unknown constants A_1, A_2, A_3, and A_4. They are:

1. $\phi_2(a + b + z_0) = 0$.
2. $\phi_1(a) = \phi_2(a)$; continuity of flux at material interface.
3. $-D_1 \left.\frac{\partial \phi_1}{\partial x}\right|_{x=a} = -D_2 \left.\frac{\partial \phi_2}{\partial x}\right|_{x=a}$; continuity of current at material interface.
4. The jump condition at $x = 0$.

The details will be left as an exercise to the reader. In exactly the same fashion, solutions could be found for any number of layers of slab, with or without symmetry conditions prevailing.

2.9 THE DIFFUSION EQUATION IN SPHERICAL GEOMETRY

We now turn to the study of the neutron diffusion equation in spherical symmetry. We shall assume the region under consideration to be composed of concentric spherical shells, in each of which D and Σ_a are constant. Then, the flux will depend only on r, the distance from the center of symmetry to the point $\mathbf{r}$, and current $\mathbf{J}$ will have a radial component only,

$$J_r = -D \frac{\partial \phi}{\partial r}. \tag{2.93a}$$

Also, if we assume that D does not depend on r, the term $\nabla \cdot D\nabla$ in the diffusion equation (2.50) becomes

$$\nabla \cdot D\,\nabla\phi(r) = D\,\nabla^2\phi(r) \tag{2.93b}$$

$$= D\frac{d^2\phi}{dr^2} - \frac{2}{r}\frac{d\phi}{dr}. \tag{2.93c}$$

Here we have introduced the expression for ∇^2 in spherical coordinates, dropping the angular derivatives since, by assumption, ϕ depends only upon r. Consider first the distribution in an infinite homogeneous medium for a shell source $S[\delta(r - r_0)/4\pi r^2]$ as shown in Fig. 2.7.

First we have to derive the analog of the jump condition (2.53). This is easily obtained by the same technique used in obtaining Eq. (2.53).

$$J_r(r_0+) - J_r(r_0-) = \frac{S}{4\pi r_0{}^2}. \tag{2.94}$$

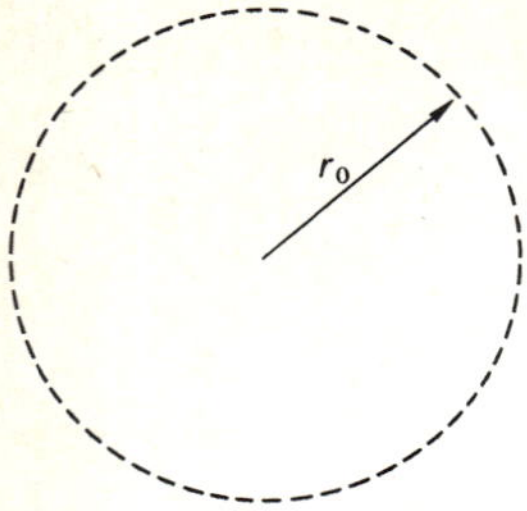

Fig. 2.7 Shell source.

The diffusion equation is

$$-\frac{d^2}{dr^2}\phi - \frac{2}{r}\frac{d}{dr}\phi + K^2\phi = \frac{S(r)}{D}. \tag{2.95}$$

The standard transformation

$$u(r) = r\phi(r) \tag{2.96}$$

reduces this equation to

$$-\frac{d^2}{dr^2}u(r) + K^2u(r) = \frac{rS(r)}{D}. \tag{2.97}$$

The solution to this equation for a shell source at $r = r_0$ is

$$u(r) = A_1e^{-Kr}, \quad r > r_0, \tag{2.98a}$$

and

$$u(r) = A_2 e^{-Kr} + A_3 e^{Kr}, \quad r < r_0. \tag{2.98b}$$

In (2.98*a*), the positive exponential is not included so that the flux will not diverge at infinity. In (2.98*b*), the flux $\phi = u/r$ will blow up at zero unless

$$A_2 = -A_3. \tag{2.99}$$

Thus

$$\phi(r) = \frac{A_1}{r} e^{-Kr}, \quad r > r_0, \tag{2.100a}$$

and

$$\phi(r) = \frac{A_2}{r}\sinh Kr, \quad r < r_0. \tag{2.100b}$$

Furthermore, the flux must be continuous at $r = r_0$,

$$A_1 = A_2 e^{Kr_0}\sinh Kr_0. \tag{2.101}$$

Finally, we have the jump condition

$$A_1 e^{-Kr_0}\left(\frac{1}{r_0{}^2}+\frac{K}{r_0}\right)-A_2\left(\frac{\sinh Kr_0}{r_0{}^2}-\frac{K\cosh Kr_0}{r_0}\right)=\frac{S}{4\pi r_0{}^2 D}. \tag{2.102}$$

From these we find

$$\phi=\frac{S}{4\pi D}\frac{\sinh Kr_0}{Kr_0 r}\frac{e^{-K(r-r_0)}}{\sinh Kr_0+\cosh Kr_0}, \quad r>r_0, \tag{2.103a}$$

and

$$\phi=\frac{S}{4\pi DKr_0 r}\frac{\sinh Kr}{\sinh Kr_0+\cosh Kr_0}, \quad r<r_0. \tag{2.103b}$$

In the limit $r_0 \to 0$, $S \to 1$, we obtain, from (2.103*a*), the infinite-medium, point-source Green's function

$$G_{\text{pt}}=\frac{1}{4\pi D}\frac{e^{-Kr}}{r}. \tag{2.104}$$

For a constant-source Q throughout space,

$$\phi = Q\int d^3r\, G_{\text{pt}}(r) \tag{2.105a}$$

$$=\frac{Q}{D}\int_0^\infty re^{-Kr}\,dr$$

$$=\frac{Q}{\Sigma_a}. \tag{2.105b}$$

This result, of course, is identical with Eq. (2.80), which was obtained by integration of the plane-source Green's function solution through all space.

The plane-source solution can be obtained by treating the point-source solution as a Green's function for the plane source, and integrating over the surface of an infinite plane (see Fig. 2.8). The flux at a distance x from the plane is composed of contributions from each area element on the plane,

$$G_\infty(0\to x)=\int_0^\infty 2\pi\rho\, d\rho\, G_{\text{pt}}(\sqrt{\rho^2+x^2}). \tag{2.106}$$

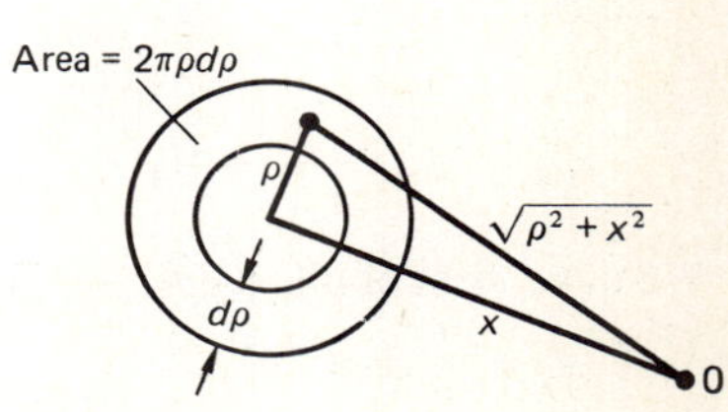

Fig. 2.8 Connection between plane- and point-source fluxes.

By using the change of variable

$$R^2 = \rho^2 + x^2 \tag{2.107}$$

so that

$$R\,dR = \rho\,d\rho, \tag{2.108}$$

Eq. (2.106) becomes

$$G_\infty(0 \to x) = 2\pi \int_x^\infty R\,dR\,G_{\text{pt}}(R). \tag{2.109}$$

Equation (2.109) may be inverted by differentation,

$$G_{\text{pt}}(R) = -\frac{1}{2\pi x}\frac{dG_\infty}{dx}(0 \to x)\bigg|_{x=R}. \tag{2.110}$$

It is left as an exercise to the reader to verify that Eqs. (2.109) and (2.110) are indeed satisfied by the plane-source Green's function, Eq. (2.76), and the point-source Green's function, Eq. (2.104).

Also of interest is a calculation of $\langle r^2 \rangle$, the mean square distance traveled by a neutron from a point source (before it is finally absorbed). In analogy with Eq. (2.77*a*), we can write

$$\langle r^2 \rangle = \frac{\int d^3r\,r^2 G_{\text{pt}}(r)}{\int d^3r\,G_{\text{pt}}(r)} \tag{2.111a}$$

$$= \frac{\int_0^\infty r^3 e^{-Kr}\,dr}{\int_0^\infty r e^{-Kr}\,dr}; \tag{2.111b}$$

$$\langle r^2 \rangle = \frac{6}{K^2} = 6L^2. \tag{2.111c}$$

Thus $\langle r^2 \rangle = 3\langle x^2 \rangle$. This is not surprising, since $r^2 = x^2 + y^2 + z^2$ and we expect $\langle x^2 \rangle = \langle y^2 \rangle = \langle z^2 \rangle$ because of the rotational invariance of the point-source infinite-medium system.

The calculation of the neutron density in a spherically symmetric system of finite size proceeds in the same manner as for a multilayered slab system. As an example, consider a sphere, region 1, surrounded by a spherical region 2 (see Fig. 2.9) with a point source at the center. We have

$$\phi_1 = A_1\frac{e^{-K_1 r}}{r} + A_2\frac{e^{K_1 r}}{r}, \quad \text{region 1}, \tag{2.112a}$$

$$\phi_2 = A_3\frac{e^{-K_2 r}}{r} + A_4\frac{e^{K_2 r}}{r}, \quad \text{region 2}, \tag{2.112b}$$

We have, first of all, $\phi_2(R_1 + z_0) = \phi_2(R_1') = 0$. Thus

$$A_3 = -A_4 e^{2K_2 R'_1}. \tag{2.113}$$

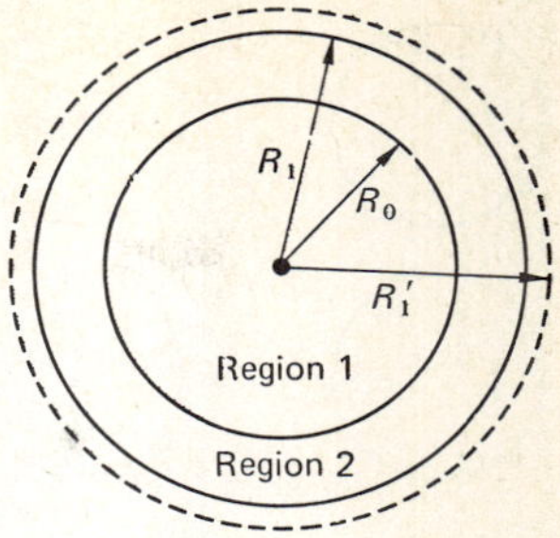

Fig. 2.9 Multishell system.

Furthermore, we have the continuity conditions at $r = R_0$,

$$\phi_1(R_0) = \phi_2(R_0), \tag{2.114a}$$

and

$$-D_1 \frac{d}{dr} \phi_1 \bigg|_{r=R_0} = -D_2 \frac{d}{dr} \phi_2 \bigg|_{r=R_0}. \tag{2.114b}$$

These three conditions serve to determine three of the four constants, say, A_1, A_2, and A_3. To determine the last constant, we have been in the habit of using a jump condition across a source. However, one cannot "jump" a point. We could, if we wished, solve this problem for a shell source of radius r_0 in region 1, and take the limit $r_0 \to 0$. A simpler procedure is to modify the jump condition for a point source. If we surround the source with a sphere of radius r_0, then the number of neutrons crossing this sphere in the limit as $r_0 \to 0$ is simply equal to the strength of the source,

$$\lim_{r_0 \to 0} - D_1 \frac{d}{dr} \phi_1 \bigg|_{r=r_0} = Q. \tag{2.115}$$

2.10 THE DIFFUSION EQUATION IN CYLINDRICAL GEOMETRY

Solutions of the diffusion equation in cylindrical geometry have more practical application, perhaps, than do their plane- and spherical-symmetry cousins because many reactors are built with cylindrical fuel elements surrounded by moderator. In analyzing such systems, the physical approximation is frequently made that the source of thermal neutrons is localized in the moderator, where most of the slowing-down occurs. The standard problem, in such a case, is to calculate the thermal utilization f in a reactor lattice (see Fig. 2.10).

To do this calculation, it is assumed that the lattice is infinite, and that the calculation can be carried out for a *unit cell.* From symmetry arguments,

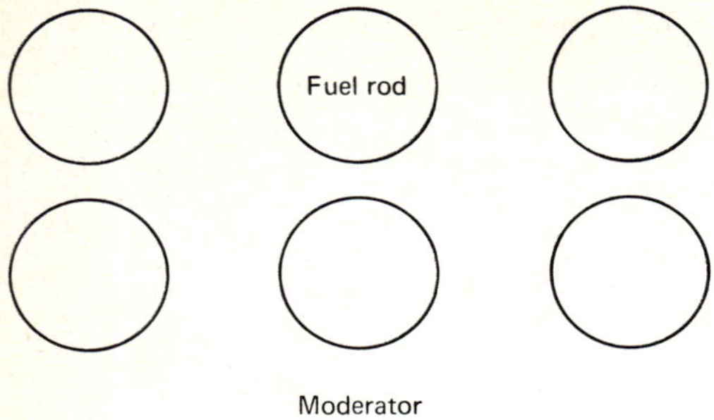

Fig. 2.10 Reactor lattice.

the current at the boundary of the cell must vanish, which serves as the boundary condition. To simplify the calculation, the rectangular (or hexagonal) boundary of the cell is replaced by a circle, so that the diffusion equation is solved in a region consisting of a cylindrical fuel rod surrounded by an annular region of moderator (see Fig. 2.11).

In cylindrical coordinates, assuming azimuthal symmetry, we have

$$\nabla^2 = \frac{\partial^2}{\partial z^2} + \frac{1}{\rho}\frac{\partial}{\partial \rho}\rho\frac{\partial}{\partial \rho}, \tag{2.116}$$

so that the diffusion equation becomes

$$\left[-D_M\left(\frac{\partial^2}{\partial z^2} + \frac{1}{\rho}\frac{\partial}{\partial \rho}\rho\frac{\partial}{\partial \rho}\right) + \Sigma_{aM}\right]\phi_M = q_M, \tag{2.117a}$$

$$\left[-D_F\left(\frac{\partial^2}{\partial z^2} + \frac{1}{\rho}\frac{\partial}{\partial \rho}\rho\frac{\partial}{\partial \rho}\right) + \Sigma_{aF}\right]\phi_F = 0. \tag{2.117b}$$

Here the subscripts M and F denote moderator and fuel regions, respectively. The flux is to be determined separately in each region by solving Eqs. (2.117)

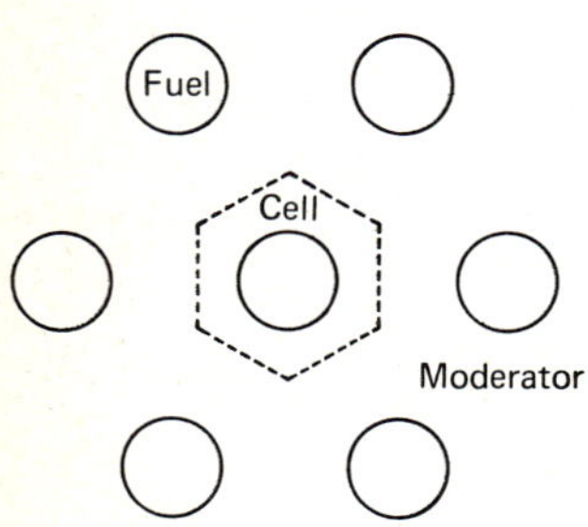

Fig. 2.11*a* Square lattice.

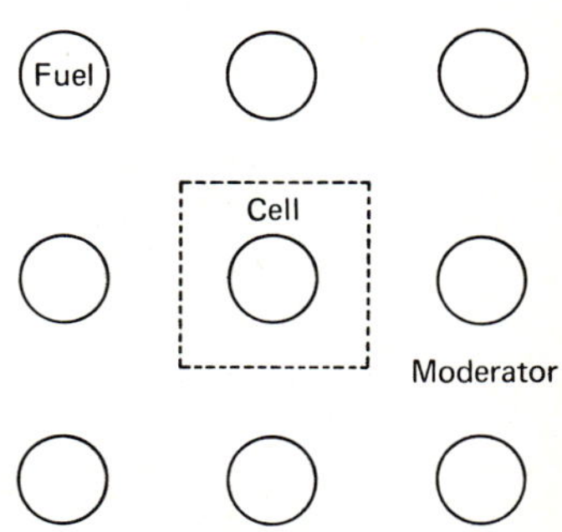

Fig. 2.11*b* Hexagonal lattice.

separately. Then the boundary conditions, continuity of the flux, and current at $\rho = R_0$ are to be applied.

We first solve Eq. (2.117*a*) subject to the restrictions (Fig. 2.12)

$$\phi_M(\rho, \pm H') = 0, \tag{2.118a}$$

$$-D_M \frac{\partial \phi_M}{\partial \rho}(\rho,z)\bigg|_{\rho=a} = 0. \tag{2.118b}$$

The first condition states that the flux vanishes on the extrapolated[1] top and bottom of the reactor, while the second is a statement, discussed earlier, of the fact that one expects the current to vanish on the boundary of the cell.

Before attempting to solve Eq. (2.117*a*), we should consider the form of $q_M(\rho,z)$. It is customary to assume it to be constant in ρ, but as even the slowing-down flux should vanish on the extrapolated top and bottom of the reactor, one generally takes

$$q_M(\rho,z) = q_M \cos \frac{\pi z}{2H'}, \tag{2.119}$$

which vanishes on the appropriate surfaces.

[1] We assume the extrapolation correction is the average of those appropriate to fuel and moderator.

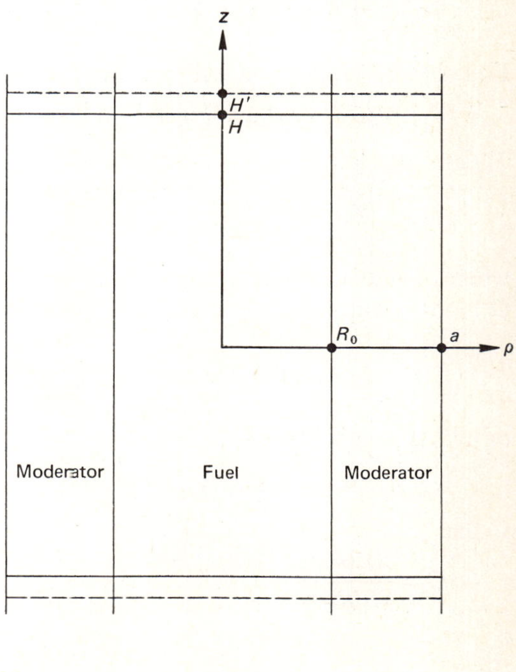

Fig. 2.12 Simplified unit cell (side view).

We attempt a solution to Eq. (2.117a) of the form

$$\phi_M(\rho,z) = \chi_M(\rho) \cos \frac{\pi z}{2H'}. \tag{2.120}$$

Substituting this *ansatz* in Eq. (2.117a) gives an equation for $\chi_M(\rho)$,

$$\left\{-D_M\left[-\left(\frac{\pi}{2H'}\right)^2 + \frac{1}{\rho}\frac{\partial}{\partial\rho}\,\rho\,\frac{\partial}{\partial\rho}\right] + \Sigma_{aM}\right\}\chi_M = q_M, \tag{2.121}$$

or

$$\left\{-\frac{1}{\rho}\frac{\partial}{\partial\rho}\,\rho\,\frac{\partial}{\partial\rho} + \left[\frac{\Sigma_{aM}}{D_M} + \left(\frac{\pi}{2H'}\right)^2\right]\right\}\chi_M = \frac{q_M}{D_M}. \tag{2.122}$$

The solution of this equation is the sum of a particular solution plus a general solution of the homogeneous equation. Evidently

$$\chi_p = \frac{q_M}{\Sigma_{aM} + D_M(\pi/2H')^2} \tag{2.123}$$

is a particular solution. The homogeneous equation is

$$-\left[\chi_H''(\rho) + \frac{1}{\rho}\chi_H'(\rho)\right] + \kappa_M{}^2\chi_H(\rho) = 0, \tag{2.124a}$$

where we have abbreviated

$$\kappa_M{}^2 = \frac{\Sigma_{aM}}{D_M} + \left(\frac{\pi}{2H'}\right)^2. \tag{2.124b}$$

Equation (2.124a) is Bessel's equation (see Appendix D). Its solution is

$$\chi_H(\rho) = AI_0(\kappa_M\rho) + CK_0(\kappa_M\rho). \tag{2.125a}$$

Then

$$\phi_M = [\chi_H(\rho) + \chi_p(\rho)] \cos \frac{\pi z}{2H'}, \tag{2.125b}$$

where A and C are arbitrary constants. We see that boundary condition (2.118a) is automatically satisfied. To satisfy (2.118b), we have

$$\left[A\frac{\partial}{\partial\rho}I_0(\kappa_M\rho) + C\frac{\partial}{\partial\rho}K_0(\kappa_M\rho)\right]_{\rho=a} = 0, \tag{2.126a}$$

or (cf. Appendix D)

$$\kappa_M[AI_1(\kappa_M\rho) - CK_1(\kappa_M\rho)]_{\rho=a} = 0, \tag{2.126b}$$

so that

$$C = \frac{AI_1(\kappa_M a)}{K_1(\kappa_M a)}. \tag{2.126c}$$

Then

$$\phi_M = \left\{\chi_p(\rho) + A\left[I_0(\kappa_M \rho) + \frac{I_1(\kappa_M a)}{K_1(\kappa_M a)} K_0(\kappa_M \rho)\right]\right\}\cos\frac{\pi z}{2H'}. \qquad (2.126d)$$

The solution of Eq. (2.117*b*) is carried out in much the same way except that:

1. There is no particular solution since Eq. (2.117*b*) is homogeneous.
2. The term proportional to K_0 is absent since $K_0 \to \infty$ as $\rho \to 0$. Thus,

$$\phi_F(\rho) = A' I_0(\kappa_F \rho) \cos\frac{\pi z}{2H'}, \qquad (2.127a)$$

where κ_F has a definition analogous to that of κ_M,

$$\kappa_F^2 = \frac{\Sigma_{aF}}{D_F} + \left(\frac{\pi}{2H'}\right)^2. \qquad (2.127b)$$

Here A' is an arbitrary constant. We now have two equations to determine the two unknown constants A and A'. Continuity of flux at $\rho = R_0$ gives

$$A' I_0(\kappa_F R_0) = A\left[I_0(\kappa_M R_0) + \frac{I_1(\kappa_M a)}{K_1(\kappa_M a)} K_0(\kappa_M R_0)\right] + \frac{q_M}{D_M \kappa_M^2}, \qquad (2.128a)$$

while continuity of current gives

$$A' D_F \kappa_F I_1(\kappa_F R_0) = A D_M \kappa_M\left[I_1(\kappa_M R_0) - \frac{I_1(\kappa_M a)}{K_1(\kappa_M a)} K_1(\kappa_M R_0)\right]. \qquad (2.128b)$$

From Eqs. (2.128*a*) and (2.128*b*), A and A' can be found. The fluxes in fuel and moderator are then given by (2.126*d*) and (2.127*a*). The details of obtaining the final expression will be left as an exercise for the reader.

A point demonstrated by Prob. 2.7 is quite important in reactor design. The reason that fewer neutrons are absorbed by the fuel when it is lumped than when it is spread homogeneously throughout the reactor is that the fuel in the interior of the rod is *shielded* from the neutron flux by the outer layers of the fuel. Putting it another way, the flux $\phi_F(\rho)$ is depressed within the the fuel, so that fewer neutrons are absorbed. (This is precisely the effect which is discussed in Sec. 7.3 on self-shielding.) The importance of this effect in reactor design is due to the fact that lumps of fuel, although they decrease f, also decrease resonance capture. In fact, the lattice can be designed so that the product pf and hence k_∞ is actually increased. Except for this, the first reactor CP1, a graphite-uranium pile, would have been subcritical.

2.11 MEASUREMENT OF DIFFUSION PARAMETERS

A number of experimental techniques have been developed to measure some of the parameters which enter the diffusion equation. One method, the pulsed-neutron technique, is described in Chap. 4. A straightforward method of measuring L^2 is to introduce a source of thermal neutrons into the material whose diffusion length it is desired to measure and to use some of the expressions derived in the previous section to deduce L^2.

A possible experimental arrangement is shown in Fig. 2.13.

Thermal neutrons are obtained by passing neutrons from a reactor through a *thermal column*, which is nothing more than a 6-ft, or longer, block of graphite which serves to moderate essentially all the fast neutrons that are incident upon it from the reactor. This serves as a plane source of neutrons which then pass into the moderator, a large tank of water, for example. Thermal-neutron detectors, perhaps Cd-covered U^{235} foils, are then spaced along the axis of the tank, the induced activity in a given foil being proportional to the neutron flux at the foil's position.

To analyze this experiment, we write the diffusion equation for the experiment shown in Fig. 2.13.

$$(-D\nabla^2 + \Sigma_a)\phi = S_0\,\delta(z)\cos\frac{\pi x}{2a'}\cos\frac{\pi y}{2b'}, \tag{2.129}$$

where we are assuming that the spatial dependence of the source in the x and y directions is a cosine which vanishes on the extrapolated boundaries $x = \pm a'$ and $y = \pm b'$. The assumption is that the thermal column is sufficiently long that *higher modes*, for example, $\cos(3\pi x/2a')$, have died away.

Setting $\phi = \chi(z)\cos \pi x/2a' \cos \pi y/2b'$ in Eq. (2.129) gives an equation for $\chi(z)$,

$$\left\{-D\left[-\left(\frac{\pi}{2a'}\right)^2 - \left(\frac{\pi}{2b'}\right)^2 + \frac{d^2}{dz^2}\right] + \Sigma_a\right\}\chi(z) = S_0\,\delta(z) \tag{2.130a}$$

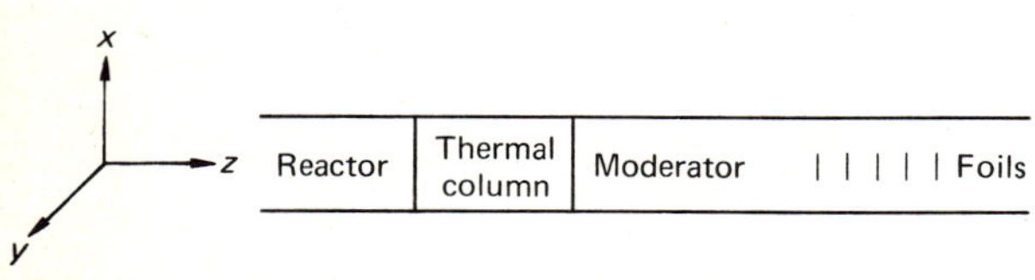

Fig. 2.13 An experimental apparatus for measurement of diffusion length.

or

$$-\chi''(z) + \frac{1}{D}\left[\Sigma_a + \left(\frac{\pi}{2a'}\right)^2 + \left(\frac{\pi}{2b'}\right)^2\right]\chi(z) = \frac{S_0}{D}\,\delta(z). \tag{2.130b}$$

The solution of Eq. (2.130) for $z > 0$ is, for the moderator infinitely long in the z direction

$$\chi = \chi_0\, e^{[-\Sigma_a/D-(\pi/2a')^2-(\pi/2b')^2]z}. \tag{2.131}$$

The arbitrary constant χ_0 could be determined from the source condition at $z = 0$, but it is not needed to determine L. We see that a plot of $\ln \chi(z)$ versus z has a slope

$$\text{Slope} = -\left[\frac{\Sigma_a}{D} + \left(\frac{\pi}{2a'}\right)^2 + \left(\frac{\pi}{2b'}\right)^2\right] \tag{2.132a}$$

$$= -\left[\frac{1}{L^2} + \left(\frac{\pi}{2a'}\right)^2 + \left(\frac{\pi}{2b'}\right)^2\right]. \tag{2.132b}$$

Since the activity induced in the detector foils is proportional to $\chi(z)$, it is only necessary to plot the activity on log paper and to determine the slope. L^2 is then determined from Eq. (2.132).

In performing this experiment, one should be careful to make measurements sufficiently far from the plane $z = 0$ for diffusion theory to be applicable. (Remember that diffusion theory is not valid close to a boundary where the flux is changing rapidly.) Also, we have assumed in writing Eq. (2.131) that the medium in which the experiments are carried out is infinite. Otherwise, a positive exponential term should be added to (2.131), with the coefficient adjusted to make $\chi(z)$ vanish at $z = c'$, the extrapolated end of the assembly. For this reason, measurements should not be made near the end of the experimental assembly where the positive exponential term becomes important.

PROBLEMS

2.1 Calculate the flux $\phi(x)$ from a constant source for the case that Σ_a and D are not constants. The Green's function is given by (2.74*b*).

2.2 Use $G(y_0 \to y)$ [Eq. (2.91)] to calculate the neutron flux in a slab containing a uniform source S. Compare this with the direct solution of the diffusion equation.

2.3 Calculate the solution of (2.92*a*) and (2.92*b*).

2.4 Calculate the disadvantage factor F defined as the ratio of the average flux in the fuel to the average flux in the entire cell. [The averages are spatial averages, i.e.,

$$\phi_{\text{fuel}} = \frac{2\pi}{V_{\text{fuel}}}\int \rho\, d\rho\, dz\, \phi_F(\rho,z),$$

and so forth.] Use the cylindrical geometry described in Sec. 2.10.

2.5 Calculate the thermal utilization f defined as the ratio of the number of neutrons absorbed in fuel to the total number absorbed. How is f related to the disadvantage factor? Again, use the geometry of Sec. 2.10.

2.6 Calculate the infinite-medium Green's function in cylindrical geometry, i.e., the flux from a cylindrical shell source of radius ρ_0. Obtain the line source flux as the limit $\rho_0 \to 0$.

2.7 In a homogeneous medium,

$$f_{\text{homo}} = \frac{\Sigma_{aF}}{\Sigma_{aF} + \Sigma_{aM}}.$$

Prove that the f calculated in Prob. 2.5 is always less than f_{homo}.

2.8 Rederive the expression for the measured diffusion length (2.132), taking into account the finite length of the experimental assembly.

2.9 Derive Eq. (2.55).

2.10 Complete the solution of the flux in a multilayered slab by calculating the constants A_1, A_2, A_3, and A_4 in Eqs. (2.92).

2 11 The same as Prob. 2.10 for the multishell system Eqs. (2.112).

3
One-Speed Reactor

3.1 INTRODUCTION: BASIC EQUATION

We now turn to a description of the nuclear reactor, in the one-group approximation. As was described in Sec. 1.2, a reactor consists of a mixture of a fissionable isotope, or fuel, with structural material, coolant, and, except in fast reactors, a moderator. As we saw, elastic and inelastic collisons always degrade the neutron energy below the fission energy ($\sim$2 MeV). For this reason, the one-group description of a reactor is not expected to provide highly reliable design information. However, it does provide a basic understanding which makes the transition to the practical (multigroup) scheme (discussed in Chap. 8) quite simple. Also, the information deduced from a one-group calculation is not *completely* meaningless, particularly for very fast or very thermal systems. Admittedly, however, if such calculations are expected to provide any useful information, the cross sections must be approximately averaged over the energy region in which the neutron population is highest. The proper method of obtaining such averages is discussed in Chap. 9.

In our one-speed approximation there is no need for a moderator to be present. Nevertheless, we shall assume, in order to make the calculations more meaningful, that mixed with the fissionable isotope is other material which may be *moderator*, coolant, structural material, etc. In this grossly simplified model, the neutrons are emitted in the process of fission—a number ν on the average per fission. Of these ν neutrons, some are absorbed in the nonfuel material in the reactor, some are absorbed in nonfission capture in the fuel itself, and some leak from the reactor system. All these processes are assumed to occur at a single energy. After these losses have been taken into account, there must be one neutron still present in the assembly to cause a new fission, or a self-sustaining chain reaction will not be possible. (These concepts have been discussed more fully in Sec. 1.2.)

Let us try to express these concepts mathematically. We wish to maintain a steady-state neutron distribution in a reactor in the absence of extraneous sources (such as those considered in the previous chapter) but in the presence of a fission source. This source is evidently given by the product of the fission rate $\Sigma_f \phi(\mathbf{r})$ with the mean number of neutrons emitted per fission,

$$\text{Reactor source} = \nu\Sigma_f(\mathbf{r})\phi(\mathbf{r}). \tag{3.1}$$

The time-dependent diffusion equation then may be written[1]

$$\frac{1}{\bar{v}}\frac{\partial}{\partial t}\phi + \Sigma_a\phi - \nabla \cdot D\nabla\phi = \nu\Sigma_f\phi. \tag{3.2}$$

If we assume a time dependence of the form

$$\phi(\mathbf{r},t) = \phi(\mathbf{r})e^{-\Lambda t}, \tag{3.3}$$

Eq. (3.2) becomes an eigenvalue equation[2]

$$(\Sigma_a - \nu\Sigma_f - \nabla \cdot D\nabla)\phi(\mathbf{r}) = \frac{\Lambda}{\bar{v}}\phi(\mathbf{r}). \tag{3.4}$$

We are interested in the conditions for a zero eigenvalue Λ, in which case, from Eq. (3.3), the neutron distribution is seen to be stationary in time. A reactor for which $\Lambda = 0$ is said to be *critical*. If $\Lambda < 0$, the neutron flux increases with time and the reactor is *supercritical*. If, on the other hand, we have $\Lambda > 0$, then the neutron density in the reactor decays exponentially with time, in which case we refer to the reactor as *subcritical*. Thus, we see that Λ is closely connected with the criticality factor k introduced in Sec. 1.3.

[1] A *one-group* problem really differs from a *one-speed* problem in that in the former, average cross sections, etc., must be used even though the formal mathematics looks the same. Thus, in Eq. (3.2) $\bar{v}$ appears, rather than u.

[2] The concept of *eigenvalue* is discussed in detail in Chap. 5.

The eigenvalue problem (3.4) is sometimes set up in the following different way. Consider a system which is, say, subcritical. If in some manner Nature could be persuaded to increase the value of ν by an appropriate amount, then the reactor could be made critical. At the moment, no one has discovered how to persuade Nature to make this change in ν, but we can imagine that it has been made, in which case $\phi(\mathbf{r})$ would obey the time-independent version of Eq. (3.1) with the different ν, call it ν_c:

$$(\Sigma_a - \nabla \cdot D\nabla)\phi(\mathbf{r}) = \nu_c \Sigma_f \phi(\mathbf{r}). \tag{3.5}$$

We then calculate the value of ν_c by solving this (eigenvalue) equation. If $\nu_c = \nu$, the reactor is actually critical. But if $\nu_c < \nu$, the reactor is supercritical (fewer neutrons are needed to sustain the reaction than are actually supplied by nature); whereas if $\nu_c > \nu$, the reactor is subcritical (more neutrons are needed). The quantity ν_c is related to the criticality factor by

$$k = \frac{\nu}{\nu_c}. \tag{3.6}$$

For $k = 1$, the neutron number $\nu = \nu_c$ and the reactor is critical; for $k < 1$, the reactor is subcritical; and for $k > 1$, the reactor is supercritical. Also,

$$\rho = \frac{k - 1}{k} \tag{3.7}$$

is a quantity frequently used, and is called the *reactivity*. For positive (negative) reactivity, a reactor is supercritical (subcritical).

By comparing solutions of Eqs. (3.4) and (3.5) we shall be able to relate the eigenvalues Λ and ν_c (or k and ρ). This will be done in the next section.

3.2 BARE SLAB REACTOR

We consider, as the simplest case, a bare slab reactor of width $2a$, extrapolated width $2a'$ (see Fig. 3.1), and consider further the conditions necessary to maintain criticality. We will take the reactor to be homogeneous so that all cross sections are position independent. Then the diffusion equation (3.4) becomes

$$\left(-D\frac{d^2}{dx^2} + \Sigma_a - \nu\Sigma_f\right)\phi(x) = \frac{\Lambda}{\bar{v}}\,\phi(x). \tag{3.8}$$

Assume a solution to this equation of the form[1]

$$\phi(x) = \phi_0 \cos Bx. \tag{3.9}$$

[1] The flux $\phi(x)$ must be symmetric about $x = 0$.

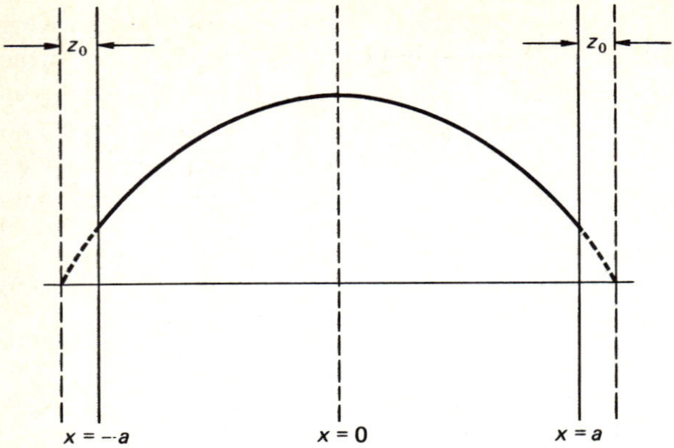

Fig. 3.1 Bare slab reactor.

Clearly not any value of B is acceptable. The neutron flux must vanish on the extrapolated boundary of the reactor. Since

$$\cos \frac{2n+1}{2} \pi = 0, \qquad n = 0, 1, 2, \ldots, \tag{3.10}$$

the only acceptable values of B are

$$B = B_n = \frac{2n+1}{2a'} \pi. \tag{3.11}$$

Then we find

$$(DB_n{}^2 + \Sigma_a - \nu\Sigma_f)\phi_0 = \frac{\Lambda_n}{\bar{v}} \phi_0, \tag{3.12}$$

or

$$\Lambda_n = \bar{v}\Sigma_a\left(L^2 B_n{}^2 + 1 - \frac{\nu\Sigma_f}{\Sigma_a}\right), \tag{3.13}$$

where we have used the thermal diffusion area $L^2 = D/\Sigma_a$ already discussed in Chap. 2.

From Eq. (3.11) we see that the B_n's increase with increasing n. Consequently, the Λ_n's also increase with increasing n. Thus Λ_0 is the smallest of the Λ_n's. We can write the neutron distribution in the reactor as a sum of the terms of the form of (3.3),

$$\phi(x, t) = \sum_{n=0}^{\infty} a_n e^{-\Lambda_n t} \cos B_n x. \tag{3.14}$$

Since Λ_0 is the smallest of the Λ_n's, after long enough times the first term in (3.14) will be the dominant contribution to $\phi(x,t)$,

$$\phi(x,t) \to a_0 e^{-\Lambda_0 t} \cos B_0 x. \tag{3.15}$$

We might note here that $\cos B_0 x$ is also the only term in (3.14) which is greater than zero for $-a < x < a$. It is physically reasonable that the persistent contribution to the flux is positive at every point in the reactor. B_0 is usually denoted by B_g and referred to as the *geometrical buckling.*

$$B_g = B_0 = \frac{\pi}{2a'}. \tag{3.16}$$

For a reactor to be critical, the persistent mode must be time independent; then all other contributions to the flux eventually die away. Thus for a critical reactor $\Lambda_0 = 0$ and (3.13) can be written, for $n = 0$,

$$\frac{\nu\Sigma_f/\Sigma_a}{1 + B_g{}^2 L^2} = 1. \tag{3.17}$$

This equation suggests that we define a quantity referred to as the *materials buckling* B_m by

$$B_m{}^2 = \frac{\nu\Sigma_f/\Sigma_a - 1}{L^2}. \tag{3.18}$$

With this definition of the materials buckling, the criticality condition (3.17) can be stated as

$$B_g = B_m. \tag{3.19}$$

The geometrical buckling is equal to the materials buckling for a critical reactor.

Let us consider another way of looking at the same problem. Return to Eq. (3.5), which can be written

$$\frac{d^2\phi(x)}{dx^2} + B_c{}^2\phi(x) = 0, \tag{3.20}$$

where

$$B_c{}^2 = \frac{\nu_c\Sigma_f - \Sigma_a}{D}. \tag{3.21}$$

This equation has solutions

$$\phi(x) = \phi_0 \cos B_c x, \tag{3.22}$$

where we require

$$B_c = \frac{2n + 1}{2a'} \tag{3.23}$$

in order that the extrapolated end-point condition

$$\phi(\pm a') = 0 \tag{3.24}$$

be satisfied. Because of the arguments given above, we must accept only the lowest eigenvalue $n = 0$. Clearly $B_g = B_c$. If also $B_g = B_m$, where B_m is given by Eq. (3.18), the reactor is critical. If $B_m > B_g$, clearly $\nu > \nu_c$ and the reactor is supercritical, and vice versa. We can calculate ν_c from Eq. (3.21),

$$\nu_c = \frac{DB_g^2 + \Sigma_a}{\Sigma_f}, \tag{3.25}$$

$$\nu_c = \frac{(1 + L^2B_g^2)\Sigma_a}{\Sigma_f}. \tag{3.26}$$

Thus, $k = \nu/\nu_c$ is given by

$$k = \frac{\nu\Sigma_f}{\Sigma_a(1 + L^2B_g^2)}. \tag{3.27}$$

To calculate the criticality of a bare slab reactor, then, the following steps are to be followed:

1. Find the geometrical buckling B_g^2 which is the lowest eigenvalue of the equation

$$\left(\frac{d^2}{dx^2} + B^2\right)\phi(x) = 0 \tag{3.28}$$

subject to the condition $\phi(\pm a') = 0$, where a' is the extrapolated boundary of the reactor.

2. Calculate k from (3.27) [ν_c from (3.25)] to determine if the reactor is supercritical, critical, or subcritical.

3.3 GENERAL REMARKS ON PRACTICAL CALCULATIONS

In practical reactor calculations, the procedure which is followed is basically that described at the end of the previous section. That is, a given geometrical configuration and material composition are assumed. Then ν_c is calculated by solving the flux diffusion equation, and this is compared with ν to determine if the reactor is or is not critical. If the reactor is found not to be critical, appropriate changes may be made, either in composition or in the geometry,[1] the changes depending upon whether k is greater or less than unity and by how much. Suppose one calculates $k < 1$. Then some of the poison

[1] But usually in the composition; more likely than not the geometry is fixed by practical engineering considerations.

material which is always present in a reactor, for example in the form of control rods, may be removed, and the calculation repeated (or the fuel loading might be increased). Then, after a few calculations of k have been carried out, a plot of k versus the variable parameter might be made, and the critical configuration determined in this way. A final calculation would be made in order to verify the critical condition.

As a practical matter, reactors are designed in such a way that k is *adjustable* to be greater or less than unity as the occasion demands. This adjustability is obtained through the use of control rods (or other means, e.g., soluble poison). Of course, while the reactor is operating at steady state, $k = 1$. However, as the reactor operates, certain effects come into play which decrease k. These include, for example, depletion of fuel (every fission means the loss of one uranium nucleus) and accumulation of fission-product poisons (the fission fragments left after the fission often absorb neutrons). On the positive side, i.e., tending to increase k, is the buildup of plutonium by neutron capture in U^{238} if there is a large inventory of U^{238} in the reactor. However, this factor is generally smaller than the factors which tend to decrease k. Then, some of the adjustibility, i.e., the amount by which k can by made to exceed unity, is used to compensate for the lost reactivity.

The excess reactivity at the beginning of life, i.e., the amount by which k potentially exceeds unity, is said to be *held down* by the control. Thus, the rods are inserted into the reactor at the beginning of life, so that the net value of k is equal to 1. As k decreases owing to the effects noted above, the rods are gradually withdrawn to keep $k = 1$ until, ultimately, the rods are fully withdrawn. Then as k decreases further, the reactor finally stops operating and the core must be refueled.[1]

Thus, when a reactor is designed, a value of excess reactivity is determined, which depends upon the operating lifetime requirements, the rate at which fuel is depleted, the rate at which plutonium buildup occurs, the amount of control available, etc. The problem of reactor design is very complicated because many of these important factors can only be estimated roughly in the beginning. Then when the reactor calculations are made, these factors must be recalculated, which leads to another calculation of the reactor, etc.

Another complicating feature is that practical reactors cannot be so simply calculated as the bare, homogeneous reactors we are dealing with in this chapter. Thus, the reactor core is actually a complicated array of fuel elements, coolant channels, etc., surrounded by a reflector or perhaps a

[1] In many reactors, particularly large water-moderated power reactors, k increases if temperature decreases (negative temperature coefficient). Thus even when a reactor has reached *end of life* it may be operated for some time at a reduced temperature (and hence a reduced power output). Practical considerations often require this "stretching" of the core life, e.g., if reloading would otherwise occur during a peak electrical-demand period in the case of a power reactor.

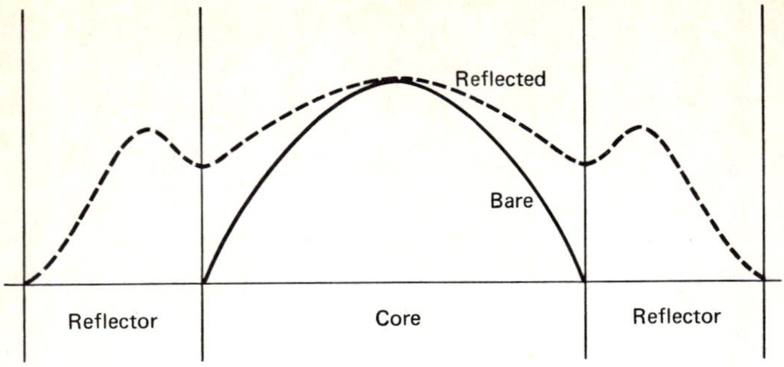

Fig. 3.2 Approximate thermal-neutron flux distributions in bare and reflected reactors. The peaking of the thermal flux in the reactor is described in Chapter 8.

blanket of natural uranium, a biological shield, etc. Even the fuel distribution in the core itself is rarely uniform because, as we have seen, a uniform fuel distribution leads to a flux which is highly peaked in the center of the reactor [$\phi \approx \cos(\pi x/2a')$]. Since the power generated is proportional to the fission rate, which is in turn proportional to the flux, such a reactor has a severe "hot spot" in and near the center. This introduces serious problems in heat removal, as can well be imagined. The reflector (or blanket) which surrounds most reactors improves this situation somewhat by returning neutrons to the outer edges of the core, but the situation is generally improved further by the use of nonuniform fuel loadings (see Fig. 3.2.)

In addition, it might be added that the one-group or one-speed approximation used in the present chapter is not used in practice. Most reactor calculations use an absolute minimum of two groups, and usually more—sometimes as many as twenty.

All these considerations mean that simple analytical solutions such as those obtained in this chapter cannot be used. Numerical methods are used to solve the diffusion equation(s). However, once the solution has been obtained, by whatever means, the procedures are basically those described here. All these problems, and some of the practical methods employed, are discussed in subsequent chapters of this book.

3.4 CRITICAL MASS AND CRITICAL SIZE

Bearing in mind the comments of the previous section, we present here some methods of determining the necessary size of a reactor once the composition is known, or the appropriate composition for a given size. These calculations are practical only for obtaining a general feel for the criticality conditions, because practical systems, as we have pointed out, require more complicated methods.

Suppose first that the reactor composition is given. Then Σ_f, Σ_a, and L^2 are known, and we can calculate $B_m{}^2$ from Eq. (3.18),

$$B_m{}^2 = \frac{\nu\Sigma_f/\Sigma_a - 1}{L^2}. \tag{3.18}$$

Then, the critical *extrapolated* width $2a'$ of a bare slab (see Fig. 3.1) is given by the condition

$$B_m{}^2 = B_g{}^2, \tag{3.29}$$

where $B_g{}^2$ is given by (3.16). Since $a' = a + z_0$, we have

$$a = \frac{\pi}{2B_m} - z_0. \tag{3.30}$$

Sometimes the reactor composition is such that no width, however large, will permit the attaining of a critical configuration. This corresponds to the condition $k_\infty < 1$, where k_∞ is defined in Eq. (1.11). However, for $k_\infty > 1$, a critical width always exists. (For $k_\infty = 1$, the critical width is infinite.)

As the inverse problem, suppose the dimensions are specified, and the necessary fuel loading (the critical mass) is required. Then $B_g{}^2$ is specified, and we have

$$\frac{\nu\Sigma_f/\Sigma_a - 1}{L^2} = B_g{}^2 = \left(\frac{\pi}{2a'}\right)^2. \tag{3.31}$$

For simplicity, assume that the reactor core contains only two materials, fuel (superscript F) and moderator (superscript M)[1]. Then

$$\Sigma_a = \Sigma_f + \Sigma_C{}^F + \Sigma_C{}^M, \tag{3.32}$$

where we recall that the subscript C means *capture*, i.e., absorption of a neutron without subsequent fission. Of course total absorption is the sum of fission *plus* capture.

[1] In the one-speed case the moderator serves as a sort of "internal reflector" tending to keep the neutrons within the core. In the practical energy-dependent case, of course, the moderator slows the neutrons down as well.

We have, further,

$$L^2 = \frac{D}{\Sigma_a} \tag{3.33a}$$

$$= \frac{1}{3\Sigma_{\text{tr}}\Sigma_a} \tag{3.33b}$$

$$= \frac{1}{3(\Sigma_{\text{tr}}{}^F + \Sigma_{\text{tr}}{}^M)(\Sigma_a{}^F + \Sigma_a{}^M)} . \tag{3.33c}$$

This expression can be written as

$$L^2 = \frac{L_M{}^2}{(1 + \Sigma_{\text{tr}}{}^F/\Sigma_{\text{tr}}{}^M)(1 + \Sigma_a{}^F/\Sigma_a{}^M)}, \tag{3.34a}$$

where

$$L_M{}^2 = \frac{1}{3\Sigma_{\text{tr}}{}^M\Sigma_a{}^M} . \tag{3.34b}$$

Because most reactors have relatively small fuel loadings,[1] it is customary to ignore the ratio $\Sigma_{\text{tr}}{}^F/\Sigma_{\text{tr}}{}^M$. Thus

$$L^2 = \frac{L_M{}^2}{1 + W}, \tag{3.35}$$

where

$$W = \frac{\Sigma_a{}^F}{\Sigma_a{}^M} . \tag{3.36}$$

Also, we can write

$$\frac{\nu\Sigma_f}{\Sigma_a} = \frac{\nu\Sigma_f}{\Sigma_a{}^F}\frac{\Sigma_a{}^F}{\Sigma_a{}^F + \Sigma_a{}^M} . \tag{3.37}$$

Note that the quantity

$$\frac{\nu\Sigma_f}{\Sigma_a{}^F} = \frac{\nu\sigma_f}{\sigma_a{}^F} \equiv \eta \tag{3.38}$$

does not depend upon the fuel concentration.[2] It is a nuclear constant depending only upon the fuel employed. Also

$$\frac{\Sigma_a{}^F}{\Sigma_a{}^F + \Sigma_a{}^M} = \frac{W}{1 + W} . \tag{3.39}$$

[1] This is not true in the case of fast reactors, to which the subsequent analysis is not even remotely applicable.

[2] The fuel efficiency $\eta = 2.08$ for thermal neutron fission of U^{235}.

Thus, Eq. (3.31) becomes

$$\frac{(\eta - 1)W - 1}{L_M^{\ 2}} = \left(\frac{\pi}{2a'}\right)^2, \tag{3.40}$$

from which W can be calculated (of course, $L_M^{\ 2}$ and η are known[1]). Having found W, we can find the critical mass in the following way. Equation (3.36) can be rewritten

$$W = \frac{N_F}{N_M}\frac{\sigma_a^{\ F}}{\sigma_a^{\ M}}. \tag{3.41}$$

Here N_F and N_M are the number densities of fuel and moderator nuclei, respectively, while the ratio of microscopic cross sections, $\sigma_a^{\ F}/\sigma_a^{\ M}$, is a known numerical factor. Again assuming the fuel occupies no volume, we have

$$N_M = \frac{n_M L_0 Z_M}{M_M}, \tag{3.42}$$

as given already in Eq. (2.11). Then N_F can be found, since all the other factors in Eq. (3.41) are known.

The critical mass m_F is the product of N_F times the mass of one fuel nucleus (A/L_0, where A is the *atomic* weight of the fuel and L_0 is Avogadro's number) times the volume of the reactor, V,

$$m_F = \frac{N_F A V}{L_0}. \tag{3.43}$$

For the slab reactor, the volume is infinite, and hence so is the critical mass. Then we might better speak of the critical mass per unit volume, m_F/V, or per unit area of the slab,

$$\frac{m_F}{\text{Unit area}} = \frac{N_F A 2a}{L_0}, \tag{3.44}$$

where $2a$ is the actual (*not* extrapolated) width of the slab.

In Table 3.1, some numerical factors are presented to permit actual criticality calculations to be carried out in this simple model.

[1] Perhaps one should point out that actually $L_M^{\ 2}$ depends upon the moderator concentration, which is not known until the fuel concentration is determined. However, the fuel concentration is assumed to be so low that it displaces no moderator, and the value of $L_M^{\ 2}$ for pure moderator is used. Note, the fuel may contain diluent species, such as oxygen in UO_2, but then the oxygen must be considered part of the fuel, not the moderator.

Table 3.1 Parameters for one-group critical calculations[a]

Substance	L^2 *cm*^{2}	σ_a *barns*	σ_{tr} *barns*	σ_f *barns*	η	ν	*n* *g/cm*^{3}
H_2O	7.45	0.58	69.5	—	—	—	1
D_2O	1.16×10^4	1×10^{-3}	25.9	—	—	—	1.1
Be^9	432	10×10^{-3}	5	—	—	—	1.85
C	2,900	3×10^{-3}	6	—	—	—	1.60
U^{233}	—	523	—	476	2.29	2.51	18.7
U^{235}	—	603	10	500	2.07	2.47	18.7
U^{238}	—	2.75	8.3	—	—	—	18.7
Pu^{239}	—	1,200	9.7	690	2.09	2.91	—
B^{10}	—	3,840	—	—	—	—	2.34
B	—	759	—	—	—	—	2.34
Cd^{113}	—	2.1×10^4	—	—	—	—	8.37
Xe^{135}	—	2.7×10^6	—	—	—	—	—

[a] Note that the microscopic cross sections for H_2O and D_2O refer to the entire molecule. Also, note that the water density given above is for 4°C; the density in an operating power reactor is vastly different. Similarly, the uranium density is for metal. Thus, these numbers are simply for illustrative calculations, although they could, with some validity, be applied to the analysis of critical (zero power) experiments.

3.5 CRITICALITY CALCULATIONS IN OTHER GEOMETRIES

In applying the methods described previously for a slab reactor to other geometrical shapes, the procedure is very similar. Thus, the Helmholtz equation

$$\nabla^2\phi + B_c^2\phi = 0 \tag{3.45}$$

is solved for the particular geometry under consideration, subject to the boundary condition that the flux vanish on the extrapolated boundary. Then, ν_c is calculated from the formula

$$\frac{\nu_c \Sigma_f/\Sigma_a}{1 + L^2 B_c^2} = 1. \tag{3.46}$$

Thus, the only difference between the infinite slab case and any other geometrical configuration lies in the solution of the Helmholtz equation (3.45) above. Actually, this equation, as has been mentioned earlier, is separable in 11 coordinate systems.[1]

[1] Philip M. Morse and Herman Feshbach: "Methods of Theoretical Physics," McGraw-Hill Book Company, New York, 1953.

We shall consider only some simple examples here.

a THE BARE SPHERICAL REACTOR

As in Sec. 2.9, ϕ is a function only of r. Equation (3.45) becomes

$$\frac{d^2}{dr^2} r\phi + B_c^2 r\phi = 0 \tag{3.47}$$

or

$$\phi = \phi_0 \frac{\sin B_c r}{r}. \tag{3.48}$$

Here ϕ_0 is a constant, and the term $\cos B_c r/r$ has been omitted because it approaches infinity as $r \to 0$.

The condition $\phi(R') = 0$ implies

$$\sin B_c R' = 0 \tag{3.49}$$

Thus the buckling of a spherical reactor (Fig. 3.3) is given by

$$B_c = \frac{\pi}{R'} \tag{3.50}$$

b THE FINITE CYLINDER

Here we must write Eq. (3.45) in cylindrical coordinates (Fig. 3.4). In this case ϕ depends on ρ and z. Thus,

$$\frac{\partial^2 \phi}{\partial \rho^2} + \frac{1}{\rho}\frac{\partial \phi}{\partial \rho} + \frac{\partial^2 \phi}{\partial z^2} + B_c^2 \phi = 0. \tag{3.51}$$

The partial differential equation is solved by the standard method, i.e., by separation of variables. The *ansatz*

$$\phi(\rho,z) = X(\rho)Z(z) \tag{3.52}$$

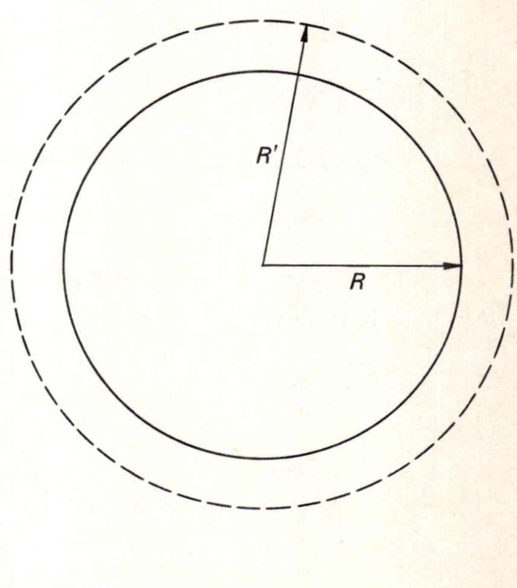

Fig. 3.3 Bare spherical reactor.

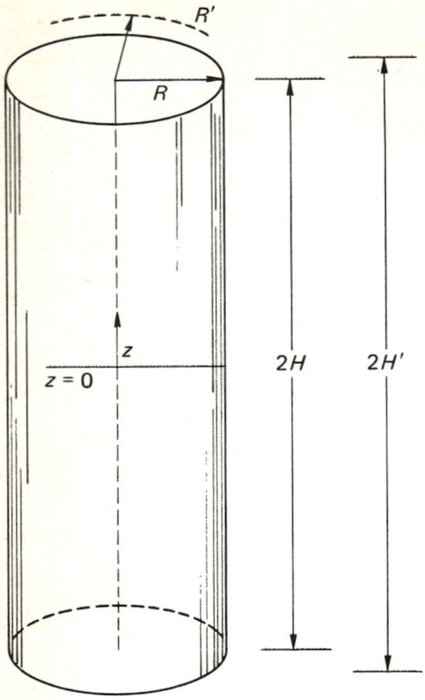

Fig. 3.4 Bare cylindrical reactor.

reduces Eq. (3.51) to [after division by $X(\rho)Z(z)$]

$$\frac{X''}{X} + \frac{1}{\rho}\frac{X'}{X} + \frac{Z''}{Z} = -B_c^2. \tag{3.53}$$

(Here a prime indicates differentiation with respect to the argument.) The variables are now separated. Thus,

$$X'' + \frac{X'}{\rho} + \alpha^2 X = 0 \tag{3.54a}$$

and

$$Z'' + \beta^2 Z = 0, \tag{3.54b}$$

where α and β are constants which must obey

$$\alpha^2 + \beta^2 = B_c^2. \tag{3.54c}$$

The appropriate solution of Eq. (3.54*b*), which vanishes at H' (H is the half-height of the cylinder), is

$$Z = Z_0 \cos \beta z \tag{3.55}$$

with

$$\beta = \frac{\pi}{2H'}. \tag{3.56}$$

Equation (3.54*a*) is Bessel's differential equation (cf. Appendix D). In fact, the substitution $x = \alpha\rho$ reduces (3.54*a*) to

$$X''(x) + X'(x) + \alpha^2 X(x) = 0, \tag{3.57}$$

the solution of which is

$$X = X_0 J_0(\alpha\rho) + X_0' Y_0(\alpha\rho). \tag{3.58}$$

Since $Y_0(\alpha\rho) \to \infty$ as $\rho \to 0$, $X_0' = 0$.[1] Furthermore, we require that

$$J_0(\alpha R') = 0. \tag{3.59}$$

The first zero of $J_0(x)$ occurs at $x = 2.405$ (cf. Appendix D). Thus

$$\alpha = \frac{2.405}{R'}. \tag{3.60}$$

The buckling of the finite cylinder then is given by

$$B_c^2 = \left(\frac{\pi}{2H'}\right)^2 + \left(\frac{2.405}{R'}\right)^2. \tag{3.61}$$

Also

$$\phi(\rho) = \phi_0 \cos\frac{\pi z}{2H'} J_0\left(\frac{2.405}{R'}\rho\right). \tag{3.62}$$

c THE HEMISPHERICAL REACTOR

In this case (see Fig. 3.5) the flux ϕ will be a function of r and θ, the polar angle, but not of the azimuthal angle. Thus

$$\nabla^2\phi = \left(\frac{\partial^2}{\partial r^2} + \frac{2}{r}\frac{\partial}{\partial r} + \frac{1}{r^2 \sin\theta}\frac{\partial}{\partial\theta}\sin\theta\frac{\partial}{\partial\theta}\right)\phi. \tag{3.63a}$$

[1] The terms X_0 and X_0' are arbitrary constants.

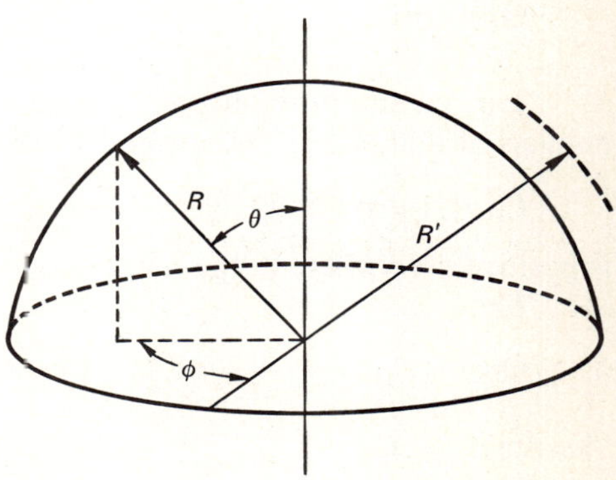

Fig. 3.5 Bare hemispherical reactor.

Again we attempt to separate variables,

$$\phi = X(r)\Theta(\theta), \tag{3.63b}$$

in which case we see that Eq. (3.45) becomes

$$\Theta X'' + \frac{2\Theta X'}{r} + \frac{X}{r^2 \sin\theta}\frac{\partial}{\partial\theta}(\sin\theta\Theta) + B_c{}^2 X\Theta = 0. \tag{3.64}$$

After division by Θ the variables become separated. We call the separation constant β^2 so that

$$X'' + \frac{2}{r}X' + \left(B_c{}^2 - \frac{\beta^2}{r^2}\right) X = 0 \tag{3.65}$$

and

$$\frac{1}{\sin\theta}\frac{\partial}{\partial\theta}(\sin\theta\Theta') + \beta^2\Theta = 0. \tag{3.66}$$

Equation (3.66) can be put in simpler form by utilizing the change of variable,

$$\mu = \cos\theta, \tag{3.67}$$

in which case we have

$$(1 - \mu^2)\Theta'' - 2\mu\Theta' + \beta^2\Theta = 0. \tag{3.68}$$

Equation (3.68) is Legendre's differential equation,[1] which has well-behaved solutions only if

$$\beta^2 = l(l+1), \qquad l = 0, 1, 2, 3, \ldots. \tag{3.69}$$

The corresponding solution of (3.68) is denoted by $P_l(\mu)$, where l is an integer, and is called a *Legendre polynomial of order l.*

In order that $\phi(r,\theta)$ vanish at $\theta = \pi/2$ for all r, we require that

$$\Theta\left(\frac{\pi}{2}\right) = 0. \tag{3.70}$$

This condition is obeyed by P_{2n+1}, that is, by all the Legendre polynomials of odd order. However, of these, only

$$P_1(\mu) = \mu = \cos\theta \tag{3.71}$$

is positive everywhere in the range $0 < \theta < \pi/2$.

Thus,

$$\Theta(\theta) = P_1(\cos\theta) = \cos\theta, \tag{3.72}$$

[1] See Appendix D.

and

$$\beta^2 = 2 \tag{3.73}$$

so that Eq. (3.65) becomes

$$X'' + \frac{2}{r} X' + \left(B_c^2 - \frac{2}{r^2}\right) X = 0. \tag{3.74}$$

The solution of this equation is a spherical Bessel function[1]

$$X = X_0 j_1(B_c r) = X_0 \left(\frac{\sin B_c r}{B_c^2 r^2} - \frac{\cos B_c r}{B_c r}\right), \tag{3.75}$$

Since $\phi(R')$ must vanish for all θ, we have

$$j_1(B_c R') = 0. \tag{3.76}$$

The first zero of $j_1(x)$ occurs at $x = 4.49$. Thus

$$B_c = \frac{4.49}{R'}, \tag{3.77}$$

while

$$\phi(r,\theta) = \phi_0 j_1\left(\frac{4.49r}{R'}\right) \cos\theta. \tag{3.78}$$

Note that the buckling is considerably larger for a hemisphere, $4.49/R'$, than for a sphere, $\pi/R' = 3.141/R'$, of the same radius. As we shall see in the next section, this is due to the fact that the hemisphere has a higher leakage.

3.6 PHYSICAL MEANING OF THE CRITICALITY CONDITION

Let us return now to the equation

$$k = \frac{\nu\Sigma_f/\Sigma_a}{1 + B_c^2 L^2} = 1 \tag{3.79}$$

for a critical reactor, and attempt to interpret it physically. The discussion here parallels much of the discussion in Sec. 1.2. We have already, in Sec. 3.4, broken the numerator into the product of two factors,

$$\frac{\nu\Sigma_f}{\Sigma_a} = \frac{\nu\sigma_f}{\sigma_a} \frac{\Sigma_a^F}{\Sigma_a^F + \Sigma_a^M} = \eta f, \tag{3.80}$$

[1] See Appendix D.

where η, the fuel efficiency, and f, the thermal utilization, have been discussed in detail (in Sec. 1.2).

If there were no leakage from the system, it is clear that the number ηf should be equal to unity for a critical system. Then the additional factor in Eq. (3.79) must represent somehow the probability that a neutron will not leak from the system (i.e., the probability that it will be absorbed). Then Eq. (3.79) tells us that

[Probability that a neutron will be absorbed, $(1 + L^2B^2)^{-1}$] $\times$
(probability that absorption will be by a fuel nucleus f) $\times$
(average number of neutrons emitted per neutron absorbed by fuel) (3.81)

must be equal *to unity* for a critical system. This result is indeed physically obvious.

It only remains to prove that the factor $(1 + L^2B_c{}^2)^{-1}$ does represent the nonleakage probability for a reactor. To see that it does, consider the bare spherical reactor, for which

$$\phi(\mathbf{r}) = \phi_0 \frac{\sin B_c r}{r}\,. \tag{3.82}$$

Then

$$J(\mathbf{r}) = -D\frac{\partial \phi}{\partial r}\,, \tag{3.83a}$$

$$= -\phi_0 DB_c\left(\frac{\cos B_c r}{r} - \frac{\sin B_c r}{r^2}\right). \tag{3.83b}$$

The nonleakable probability P_{nL} is simply given by the ratio of the number of neutrons absorbed in the reactor to the total number produced. The total number produced is clearly equal to the number which leak plus the number absorbed. Thus,

$$P_{nL} = \frac{\Sigma_a \int_{\text{reactor}} \phi(\mathbf{r})\, d^3r}{\int_{\text{surface}} \mathbf{J}\cdot d\mathbf{s} + \Sigma_a \int_{\text{reactor}} \phi(\mathbf{r})\, d^3r}\,. \tag{3.84}$$

Here let us make the slight approximation that the surface integration can be carried out over the extrapolated surface, rather than the actual surface, of the reactor. Then $\sin BR' = 0$ and $\cos BR' = \cos \pi = -1$, and using (3.83b), we have

$$\int_{\text{extr. surface}} \mathbf{J}\cdot d\mathbf{s} = 4\pi R'\phi_0 DB_0\,. \tag{3.85}$$

Similarly,

$$\int_{\text{extr. vol.}} \phi(\mathbf{r})\, d^3r = \frac{4\pi\phi_0 R'}{B_c}, \tag{3.86}$$

so that

$$P_{nL} = (1 + B_c^2 L^2)^{-1} \tag{3.87}$$

as predicted. Though this calculation has only been done here for a spherical reactor, it could easily be carried out for any type of bare reactor. The result is, of course, inaccurate in that the extrapolated boundaries of the reactor were used in calculating P_{nL}, rather than the actual physical boundaries.

3.7 POWER DISTRIBUTION

The diffusion equation, for example (3.51), obeyed by the neutron distribution in a reactor is *homogeneous*. Thus, the solution can be determined only to within an arbitrary multiplicative constant. In principle, then, the flux distribution in a reactor may have *any* amplitude between zero and infinity. In practice, the flux amplitude is limited by heat removal considerations and by temperature feedback effects.[1] But at any rate, the reactor will operate at critical, within a certain range of the power level.

To determine the amplitude of the neutron flux, the power level of the reactor must be specified. Since each fission liberates approximately 200 MeV of useful energy, we have

$$\text{Power} = \text{total fission rate} \times 200 \text{ MeV} \tag{3.88}$$

or, converting to watts ($1 \text{ MeV} = 1.6 \times 10^{-13}$ W-s) and introducing an expression for the fission rate,

$$P = \int_{\text{reactor}} \Sigma_f \phi(\mathbf{r})\, d^3r \times 3.20 \times 10^{-11} \qquad \text{W}. \tag{3.89}$$

For example, for a bare, homogeneous spherical reactor of radius R and, as in the previous section, setting $R \approx R'$ for simplicity, we have

$$\int \phi(\mathbf{r})\, d^3r = \frac{4\pi\phi_0 R}{B_c}, \tag{3.90}$$

[1] As the temperature changes, various changes in the reactor occur. For example, the size of the reactor changes owing to thermal expansion. In addition, the neutron spectrum changes, so that the average cross sections used in our one-group approximation will change.

so that

$$P = 3.20 \times 10^{-11}(4\phi_0 \Sigma_f R^2) \qquad \text{W}. \tag{3.91}$$

We might reasonably ask the question, how is the power level of a reactor changed? Here the control rods, already described in Sec. 3.3 as a means of compensating for fuel depletion, are used. Suppose a reactor is operating in steady state at a given power level P (or flux level ϕ). Then we have $k = 1$ and $\Lambda_0 = 0$, where, we recall, the flux varies as [cf. Eq. (3.3)]

$$\phi(t) \approx e^{-\Lambda_0 t}. \tag{3.92}$$

To increase the power level, it is necessary to increase k by withdrawing control rods. Noting Eq. (3.13), which may be written as

$$\Lambda_0 = v\Sigma_a(1 + B^2L^2)(1 - k), \tag{3.93}$$

we see that $k > 1$ implies $\Lambda_0 < 0$, and vice versa. After the power level has increased to the desired value, according to Eq. (3.92), the control rods are reinserted, which reduces k to 1 again. Similarly, the power level can be decreased by reducing k below unity. The actual dependence of Λ_0 on reactor parameters (taking into account delayed neutrons, which have so far not been considered) is discussed in Chap. 4 under reactor kinetics.

3.8 INTEGRAL EQUATION APPROACH

Another approach to the critical problem is based on an integral equation formulation for the neutron flux. This approach has the advantage of mathematical elegance. Even more important, it is readily generalized, for neutron flux that is described, not by diffusion theory, but by a more accurate approximation to the transport equation (or for the energy-dependent, rather than the one-speed, treatment in the present chapter).

The neutron flux $\phi(\mathbf{r})$ in a homogeneous reactor obeys the following integral equation:

$$\phi(\mathbf{r}) = v\Sigma_f \int_{\text{reactor}} \phi(\mathbf{r}')G(\mathbf{r}' \to \mathbf{r})\, d^3r', \tag{3.94}$$

where $G(\mathbf{r}' \to \mathbf{r})$ is the Green's function. That is, G is the solution of the equation

$$(-D\nabla^2 + \Sigma_a)G(\mathbf{r}' \to \mathbf{r}) = \delta(\mathbf{r}' - \mathbf{r}), \qquad G(\mathbf{r}' \to \mathbf{R}') = 0, \tag{3.95}$$

where $\mathbf{R}'$ is, of course, the extrapolated boundary of the reactor. Equation (3.95) is the generalization of the infinite-medium point-source Green's function Eq. (2.104) for finite systems.

We can see that (3.94) is equivalent to the critical equation (3.5), with $\nu = \nu_c$, by operating on (3.94) with $-D\nabla^2 + \Sigma_a$. This gives

$$(-D\nabla^2 + \Sigma_a)\phi(\mathbf{r}) = \nu\Sigma_f \int_{\text{reactor}} \phi(\mathbf{r}')(-D\nabla^2 + \Sigma_a)G(\mathbf{r}' \to \mathbf{r})\, d^3r' \tag{3.96a}$$

$$= \int_{\text{reactor}} \phi(\mathbf{r}')\nu\Sigma_f\, \delta(\mathbf{r}' - \mathbf{r})\, d^3r' \tag{3.96b}$$

$$= \nu\Sigma_f\, \phi(\mathbf{r}). \tag{3.96c}$$

We notice that the ϕ given by (3.94) automatically satisfies the boundary conditions

$$\phi(\mathbf{R}') = 0, \tag{3.97}$$

where $\mathbf{R}'$ is the extrapolated boundary.

At this point we make the approximation of asymptotic reactor theory.[1] This approximation results from replacing $G(\mathbf{r}' \to \mathbf{r})$ with $G_\infty(\mathbf{r}' \to \mathbf{r})$ as given in Eq. (2.104). Then in asymptotic reactor theory the integral equation is

$$\phi(\mathbf{r}) = \nu\Sigma_f \int \phi(\mathbf{r}')G_\infty(\mathbf{r}' - \mathbf{r})\, d^3r'. \tag{3.98}$$

It is convenient to examine the Fourier transform[2] of (3.98). We denote the transforms of $\phi(\mathbf{r})$ and $G_\infty(\mathbf{r})$ by

$$\bar{\phi}(\mathbf{k}) = \int d^3r\, e^{i\mathbf{k}\cdot\mathbf{r}}\phi(\mathbf{r}), \tag{3.99a}$$

$$\bar{G}_\infty(\mathbf{k}) = \int d^3r\, e^{i\mathbf{k}\cdot\mathbf{r}}G_\infty(\mathbf{r}). \tag{3.99b}$$

The Fourier integral theorem provides the inverse transform

$$\phi(\mathbf{r}) = \frac{1}{(2\pi)^3}\int d^3k\, \bar{\phi}(\mathbf{k})e^{-i\mathbf{k}\cdot\mathbf{r}}. \tag{3.99c}$$

The convolution theorem for Fourier integrals states that the Fourier transform of the convolution of f and g, written

$$f * g \equiv \int d^3r' f(\mathbf{r}')g(\mathbf{r}' - \mathbf{r}) \tag{3.100}$$

is simply the product of the transforms $\overline{fg}$.

[1] For a complete discussion of asymptotic reactor theory, see J. Ferziger and P. F. Zweifel: "The Theory of Neutron Slowing Down in Nuclear Reactors," Pergamon Press, New York, 1966.

[2] See Appendix C.

If we multiply (3.98) by $e^{i\mathbf{k}\cdot\mathbf{r}}$ and integrate over d^3r and use the convolution theorem, we obtain

$$\bar{\phi}(\mathbf{k}) = \nu\Sigma_f \bar{\phi}(\mathbf{k})\bar{G}_\infty(\mathbf{k}). \tag{3.101}$$

Equation (3.101) indicates that $\bar{\phi}(\mathbf{k})$ is zero unless

$$1 = \nu\Sigma_f \bar{G}_\infty(\mathbf{k}). \tag{3.102}$$

Equation (3.102) is the critical equation, while Eq. (3.98) can be solved to determine the flux. It is interesting to see how these results reduce to our previous results for the one-group case considered earlier. Then $G_\infty(\mathbf{r})$ is given by Eq. (2.104), and $G_\infty(\mathbf{k})$ can be found by taking the Fourier transform[1]

$$\bar{G}_\infty(\mathbf{k}) = \bar{G}_\infty(k) = \frac{1}{4\pi D}\int_0^\infty 4\pi r^2 \frac{e^{-Kr}}{r}\frac{\sin kr}{kr}\,dr \tag{3.103}$$

$$= \frac{1}{D(K^2 + k^2)}. \tag{3.104}$$

Inserting this result into (3.101), we arrive once more at the one-group critical equation

$$1 = \frac{\nu\Sigma f}{D(K^2 + k^2)}. \tag{3.105}$$

the solution of which is

$$k^2 = B_m{}^2. \tag{3.106}$$

Equation (3.98) can also be converted into a differential equation by noting, from Eq. (3.101), that $\bar{\phi}(\mathbf{k})$ vanishes unless $k^2 = B_m{}^2$. Thus, if we operate in Eq. (3.99*c*) with ∇^2, we obtain

$$\nabla^2\phi(\mathbf{r}) = \frac{1}{(2\pi)^3}\int d^3k\,\bar{\phi}(\mathbf{k})\nabla^2 e^{-i\mathbf{k}\cdot\mathbf{r}} \tag{3.107a}$$

$$= \frac{1}{(2\pi)^3}\int d^3k\,\bar{\phi}(\mathbf{k})(-k^2)e^{-i\mathbf{k}\cdot\mathbf{r}} \tag{3.107b}$$

$$= \frac{1}{(2\pi)^3}(-B_m{}^2)\int d^3k\,\bar{\phi}(\mathbf{k})e^{-i\mathbf{k}\cdot\mathbf{r}}. \tag{3.107c}$$

The quantity k^2 has been replaced by the constant $-B_m{}^2$ and removed from under the integral sign because the integral vanishes unless $k^2 = -B_m{}^2$. Thus

$$\nabla^2\phi(\mathbf{r}) = -B_m{}^2\phi(\mathbf{r}), \tag{3.108}$$

[1] See Appendix C.

i.e., the flux obeys the Helmholtz equation just as in our one-group simplified model! To find the flux shape, Eq. (3.108) is solved in any case. More accurate models (e.g., energy-dependent, transport-theoretical) may be used to compute a more accurate G_∞ than the one used here, and in this way, a more accurate criticality condition can be obtained.

3.9 NONHOMOGENEOUS REACTORS—REFLECTORS

As has been mentioned already in Sec. 3.3, reactors are not generally built in the form of bare, homogeneous assemblies. However, reactors are built as a combination of a number of homogeneous regions.[1] We say that in such cases the materials are uniform within a given region. Then the diffusion equation can be solved within each region, as was done in Sec. 2.8*c*, for example, with the solutions joined smoothly at the boundaries.

The simplest possible case, which we consider below, is slab reactor core surrounded by an infinite reflector (Fig. 3.6). There is fissionable material only in the core. We call the core *region 1* with properties D_1, Σ_{a1}, Σ_{f1}, and flux ϕ_1. Similarly, the reflector has properties D_2, Σ_{a2}, and flux ϕ_2.

Then we have two diffusion equations

$$-D_1\nabla^2\phi_1 + \Sigma_{a1}\phi_1 = \nu\Sigma_{f1}\phi_1 \tag{3.109a}$$

and

$$-D_2\nabla^2\phi_2 + \Sigma_{a2}\phi_2 = 0. \tag{3.109b}$$

We note that Eq. (3.109*a*) can be written in the form

$$\nabla^2\phi_1 + B_m{}^2\phi_1 = 0 \tag{3.110}$$

[1] For reactor design, a region is homogeneous if the physical inhomogeneities are small compared with a neutron mean free path. Even for inhomogeneities which are larger, various *homogenization* schemes exist (see Sec. 7.3) which may permit the application of these homogeneity techniques.

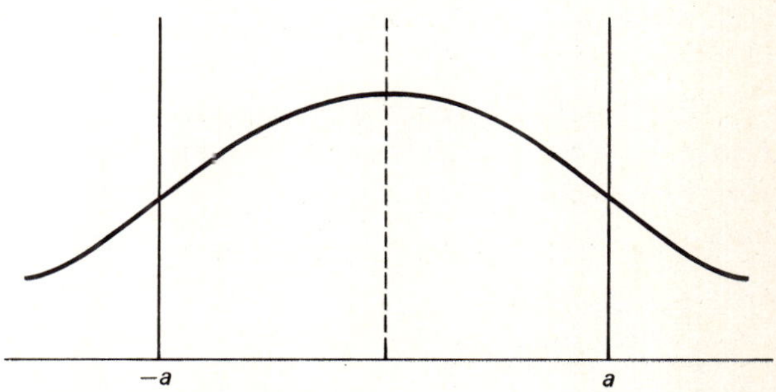

Fig. 3.6 Reflected slab reactor.

with

$$B_m^{\ 2} = \frac{\nu\Sigma_{f1} - \Sigma_{a1}}{D_1} \tag{3.111}$$

just as for the bare reactor. Thus, the materials buckling is the same for bare and reflected reactors; it is a property only of the composition of the core. Also, the solution for the core is the same as for the bare reactor,

$$\phi_1(x) = \phi_0 \cos B_m x, \tag{3.112}$$

but now the boundary condition is different. Instead of specifying that $\phi_1(a') = 0$, we now must join ϕ_1 smoothly (continuous flux and current) to ϕ_2, the solution for the reflector. This solution of Eq. (3.110) is given by

$$\phi_2(x) = \phi_0' e^{-x/L_2}, \tag{3.113}$$

where the positive exponential has been omitted because it diverges at infinity.[1]

Thus, from continuity of flux at the interface.

$$\phi_0 \cos B_m a - \phi_0' e^{-a/L_2} = 0, \tag{3.114}$$

while current continuity gives

$$\phi_0 D_1 B_m \sin B_m a - \phi_0' \frac{D_2}{L_2} e^{-a/L_2} = 0. \tag{3.115}$$

Equations (3.114) and (3.115) are two homogeneous equations for the unknown coefficients ϕ_0 and ϕ_0', A solution exists only if the determinant of the coefficients vanishes, which gives the result

$$\frac{D_2}{L_2} \cos B_m a = D_1 B_m \sin B_m a, \tag{3.116}$$

or

$$\tan B_m a = \frac{D_2}{D_1 L_2 B_m}. \tag{3.117}$$

It is interesting to see what happens in the limit that $L_2 \to 0$. Then no neutrons are returned to the core from the reflector, and we are again dealing with what is effectively a bare reactor. We see that Eq. (3.117) gives in this limit

$$\tan B_m a = \infty \tag{3.118a}$$

[1] Note that we are computing ϕ for $x > 0$. For $x < 0$, we have, of course, the symmetry condition $\phi(x) = \phi(-x)$. Again, ϕ_0 and ϕ_0' denote arbitrary constants.

or

$$B_m = \frac{\pi}{2a}, \tag{3.118b}$$

as for the bare reactor (except that the extrapolated end-point correction has been lost somewhere along the line).

Equation (3.117) is a relation between a and B_m. If the width of the core is given, B_m can be found and the critical mass obtained as described in Sec. 3.4. If, on the other hand, the composition is specified, then the critical thickness $2a$ can be found. Solving Eq. (3.117), we have

$$a = \frac{1}{B_m} \tan^{-1} \frac{D_2}{D_1 L_2 B_m}. \tag{3.119}$$

The "reflector-saving" δ is defined as the amount by which the half-width is decreased owing to the presence of the reflector. Since a bare reactor of the same composition (same B_m) would have a half-width of

$$a_B = \frac{\pi}{2B_m} - z_0, \tag{3.120}$$

we see that δ is given by

$$\delta = \frac{\pi}{2B_m} - z_0 - \frac{1}{B_m} \tan^{-1} \frac{D_2}{D_1 L_2 B_m}. \tag{3.121}$$

For a slab with a finite reflector, say of width b, the procedure for calculating the critical condition is similar. The major difference is that the boundary at infinity, which led to Eq. (3.113) for ϕ_2, is replaced by the condition that $\phi_2(a + b + z_0) = 0$, that is, the flux vanishes at the extrapolated boundary of the reflector. In such a case, of course, both the positive and negative exponential terms of ϕ_2 must be retained.

Another problem of some interest is one in which the core itself consists of two zones with different fuel loadings (done generally for power flattening, as discussed in Sec. 3.3). Consider a spherical reactor with three regions, two core regions plus a finite reflector (Fig. 3.7).

We have

$$\frac{d^2}{dr^2}(r\phi_1) + B_1{}^2(r\phi_1) = 0, \tag{3.122a}$$

$$\frac{d^2}{dr^2}(r\phi_2) + B_2{}^2(r\phi_2) = 0, \tag{3.122b}$$

and

$$\frac{d^2}{dr^2}(r\phi_3) - \frac{1}{L_3{}^2}(r\phi_3) = 0. \tag{3.123}$$

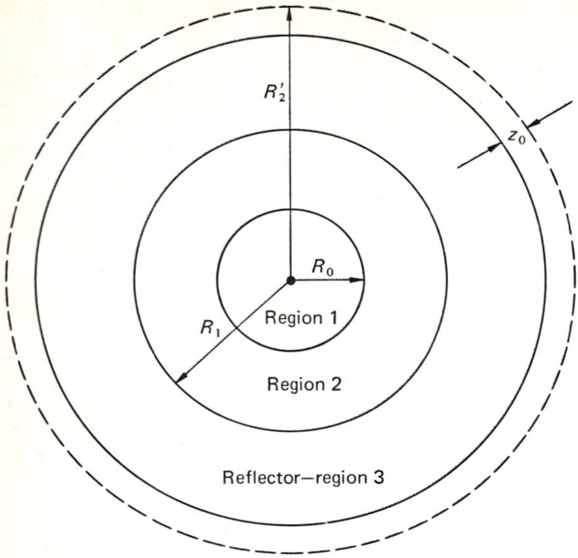

Fig. 3.7 Reflected spherical reactor with two-region core.

[$B_1{}^2$ and $B_2{}^2$ are defined as in Eq. (3.111).] For region 1, we have the solution

$$\phi_1 = A_1 \frac{\sin B_1 r}{B_1 r}. \tag{3.124a}$$

However, in region 2, the flux contains two terms

$$\phi_2 = A_2 \frac{\sin B_2 r}{B_2 r} + C_2 \frac{\cos B_2 r}{B_2 r} \tag{3.124b}$$

since there is no boundary condition at $r = 0$ to eliminate the cosine term (as there is for ϕ_1). Finally, in the reflector we have

$$\phi_3 = A_3 \frac{e^{-r/L_3}}{r/L_3} + C_3 \frac{e^{r/L_3}}{r/L_3}. \tag{3.124c}$$

There are three boundary conditions which may be applied here. They are:

1. Continuity of flux and current at $r = R_0$.
2. Continuity of flux and current at $r = R_1$.
3. $\phi_3(R_2') = 0$.

This leads to a critical condition which involves a 5×5 determinant. As the number of regions increases, the size of the determinant also increases.

The one-group treatment of reflected reactors described here does not give one of the important features of the reflectors, namely, the increased power production near the edge of the core. Physically, this effect is caused by fast neutrons which enter the reflector, are slowed down there, and return to the core as thermal neutrons. At least a two-group approach is necessary to describe this effect at all (see Chap. 8).

PROBLEMS

3.1 If $B_m{}^2 > B_g{}^2$, is a reactor supercritical or subcritical?

3.2 Calculate the critical mass per unit area of a bare slab reactor 20 cm wide, fueled with equal masses of U^{235} and Pu^{239}, and moderated with equal masses of H_2O and D_2O.

3.3 Calculate the mass of U^{233} in a bare slab reactor required to give $k = 1.1$. Then compute the amount of B^{10} which must be uniformly added to make the reactor critical.

3.4 Calculate the dimensions of a cylindrical reactor of a given composition which has the smallest critical mass.

3.5 Calculate the geometrical buckling of a bare orthogonal parallelpiped of sides a, b, and c.

3.6 Calculate the buckling of a bare quadrant of a sphere.

3.7 Calculate the fuel depletion in a reactor per kilowatthour of energy produced.

3.8 Calculate δ for a reflected slab reactor, with core width $2a$ surrounded by a reflector of width b.

3.9 Complete the solution of the critical condition for a reflected spherical reactor with a two-region core in Sec. 3.9.

4
Time-dependent Problems

4.1 INTRODUCTION

We now turn our attention to the time dependence of the neutron flux in a nuclear reactor and to the related problems of reactor control. The effects which cause a critical reactor to change from the critical configuration are of two types, long term and short term. The long-term reactivity changes are caused by such effects as fuel depletion and fission product buildup, while the short-term changes are caused by temperature effects, random fluctuations of one sort or another, or deliberate changes in order to increase or decrease the power level in a reactor. Both of these effects have already been mentioned in Sec. 3.3, and we shall consider these events in the present chapter. First we consider short-term changes, which are generally classed under the heading *Reactor Kinetics*.

4.2 ASYMPTOTIC REACTOR KINETICS[1]

We begin with the time-dependent diffusion equation (2.50). We shall, for simplicity, consider cross sections that are constant.[2] Then Eq. (2.50) can be rewritten in the form

$$\frac{1}{\bar{v}}\frac{\partial}{\partial t}\phi(\mathbf{r},t) = (D\nabla^2 - \Sigma_a + \nu\Sigma_f)\phi(\mathbf{r},t), \tag{4.1}$$

where we have of course replaced the source term $S(\mathbf{r},t)$ with the fission source $\nu\Sigma_f\phi(\mathbf{r},t)$. To solve this equation, we expand $\phi(\mathbf{r},t)$ in the complete set of functions $x_n(\mathbf{r})$, which satisfy the Helmholtz equation subject to the condition that they vanish on the extrapolated boundary of the reactor,

$$(\nabla^2 + B_n{}^2)x_n(\mathbf{r}) = 0, \tag{4.2a}$$

$$x_n(\mathbf{R}') = 0. \tag{4.2b}$$

Then we write

$$\phi(\mathbf{r},t) = \sum_n A_n(t)x_n(\mathbf{r}). \tag{4.3}$$

Inserting Eq. (4.3) into (4.1) gives a differential equation for $A_n(t)$,

$$\sum_n x_n(\mathbf{r})\left[\frac{1}{\bar{v}}\dot{A}_n(t) + (DB_n{}^2 - \nu\Sigma_f + \Sigma_a)A_n(t)\right] = 0. \tag{4.4}$$

Because of the linear independence of the x_n, the bracket must vanish for every n so that

$$\frac{1}{\bar{v}}\dot{A}_n(t) + (DB_n{}^2 - \nu\Sigma_f + \Sigma_a)A_n(t) = 0. \tag{4.5}$$

Thus

$$A_n(t) = A_n(0)e^{-\bar{v}(DB_n{}^2 - \nu\Sigma_f + \Sigma_a)t}. \tag{4.6}$$

We note that

$$l_a = \frac{\lambda_a}{\bar{v}} = \frac{1}{\bar{v}\Sigma_a} \tag{4.7}$$

is the mean lifetime for a neutron before it is absorbed. Since we have already interpreted DB^2 as a leakage cross section,[3] we see that

$$l_{\text{eff}} = \frac{1}{\bar{v}(\Sigma_a + DB^2)} = \frac{1}{\bar{v}\Sigma_a(1 + L^2B^2)} \tag{4.8}$$

[1] This treatment is often called, for mysterious reasons, the *point kinetics model.*

[2] As in the earlier chapters of the book, the one-speed approximation is used here. This assumption is relaxed in Sec. 8.8. Also, we do not consider the effect of delayed neutrons until Sec. 4.3.

[3] We now have three different symbols for the geometrical buckling, B_g, B_c, and B_0. We shall combine all these into one, namely, B.

represents the mean lifetime of a neutron before it is absorbed *or* leaked. By adding the superscript (n) to l_{eff} associated with B_n^2, Eq. (4.6) may be rewritten in the form

$$A_n(t) = A_n(0)e^{(k_n - 1)t/l_{\text{eff}}^{(n)}}. \tag{4.9a}$$

Here we have defined, in analogy with Eq. (3.27),

$$k_n = \frac{\nu\Sigma_f/\Sigma_a}{1 + L^2B_n^{\,2}}, \tag{4.9b}$$

$$l_{\text{eff}}^{(n)} = \frac{1}{\bar{v}\Sigma_a(1 + L^2B_n^{\,2})}. \tag{4.9c}$$

One frequently finds the kinetic equation written in terms of $l_{\text{eff}}^{(n)}$ and $\delta k_n = k_n - 1$. Thus, Eq. (4.5) becomes

$$\dot{A}_n(t) - \frac{\delta k_n}{l_{\text{eff}}^{(n)}} A_n(t) = 0. \tag{4.10}$$

We note that the k_n form a monotonic sequence

$$k = k_0 > k_1 > k_2 \cdots \tag{4.11}$$

because the $B_n^{\,2}$ increase with increasing n. [Cf., for example, Eq. (3.11). Note also that for two- and three-dimensional systems, the index n really is an abbreviation for two or three indices, one associated with each dimension. But still the $B_n^{\,2}$ can be arranged in an increasing sequence, so that (4.11) will apply.] It is easiest to see from Eq. (4.6) that $n = 0$ is the dominant term, for long times. Therefore, we ignore all the n except $n = 0$, and write

$$\phi(\mathbf{r},t) = A_0(0)x_0(\mathbf{r})e^{(k-1)t/l_{\text{eff}}}. \tag{4.12}$$

If k is very different from unity, this assumption is not valid and a different approach is necessary.

At this point, we should examine the magnitude of the higher-order terms in (4.3). Suppose, for example, we are considering an exactly critical reactor. Then the condition $k = 1$ can be written

$$DB^2 + \nu\Sigma_f + \Sigma_a = 0. \tag{4.13}$$

Using this result, we can write (4.6) as

$$A_n(t) = A_n(0)e^{-\bar{v}D(B_n^2 - B^2)t}. \tag{4.14}$$

Typically, $D(B_n^{\,2} - B^2) \approx n^2 \text{ cm}^{-1}$ for thermal neutrons (as, e.g., in Chap. 3). Then, since $\bar{v} \approx 10^5$ cm/s, we have

$$A_n(t) \approx A_n(0)e^{-10^5tn^2}. \tag{4.15}$$

Clearly the decay of the higher modes ($n \geq 1$) is very fast. In fact, the decay is usually so fast that it violates the conditions under which Fick's rule (2.49*a*) is applicable.

Thus we may say that the higher modes ($n \geq 1$) in (4.3) die out very rapidly. Also, we must conclude that the terms for $n \geq 1$ in (4.3) do *not* adequately represent the very short time behavior of the flux.[1] If we expect to treat the very short time variations of the neutron flux, we must use a better approximation than diffusion theory. It usually is not sensible to compute the higher-order terms in (4.3).

Returning to Eq. (4.12), assume $k = 1.001$. This is only a very small deviation from criticality, but we observe that for $l_{\text{eff}} \approx 10^{-3}$ seconds (a typical value of D_2O or graphite thermal reactors) that

$$\phi(\mathbf{r},t) = A_0(0)x_0(\mathbf{r})e^t, \tag{4.16}$$

i.e., the reactor power rises as e does in a 1-s period.[2] This is a very rapid rise, particularly for such a small deviation from criticality, and it would make reactors very difficult to hold in the steady state if it were not for the mitigating effects of *delayed neutrons*. In turns out that of the 2.5 neutrons emitted in fission, about 99.3 percent are emitted at the time of fission (*prompt neutrons*), but the other 0.7 percent are emitted with delay times ranging from a fraction of a second to almost a minute later. This is of no concern in steady-state calculations, such as the criticality calculations described in an earlier chapter of this book. The reason is that the 0.7 percent deficit at any given time is compensated for by the delayed neutrons emitted by fission at an earlier time (in the steady state). We see, of course, that in a situation where the flux is changing with time, say increasing, the delayed neutrons from earlier fission are fewer than the loss due to the delay at the present time; so delayed neutrons must be included if an adequate kinetic equation is to be derived.

We give in Table 4.1 the six *groups* of delayed neutrons along with the fraction of fission neutrons which fall in each group and the decay constant (lifetime $= \lambda_i^{-1}$).[3] Then in Eq. (4.12) the quantity l_{eff} should not be the lifetime of the prompt neutrons, $\sim 10^{-3}$, but the average lifetime of all the

[1] Here *very short time* means simply the variations in ϕ which cannot be described by asymptotic theory ($n = 0$). Also, for *very* large reactors, $D(B_n^2 - B^2) \ll n^2$ cm, and the higher modes can be treated in diffusion theory.

[2] The situation is even worse in enriched reactors, for which l_{eff} may be of order of 10^{-5} s.

[3] G. R. Keepin: "Physics of Nuclear Kinetics," Addison-Wesley Publishing Company, Inc., Reading, Mass., 1965. The average energy of a delayed neutron is less than the average energy of a prompt neutron. This fact means that when we treat the energy-dependent case, we shall have to take into account this difference in the two energy spectra.

Table 4.1 Characteristics of delayed groups for thermal fission of U^{235}; $\beta = 0.0065$, $\nu = 2.43$

Delayed group	*Decay constant* λ_i, s^{-1}	*Relative abundance* β_i/β	*Mean energy of delayed neutron* keV
1	0.0124	0.033	250 ± 50
2	0.0305	0.219	550 ± 100
3	0.111	0.196	550 ± 100
4	0.301	0.395	600 ± 200
5	1.14	0.115	400 ± 70
6	3.01	0.042	

neutrons, prompt and delayed, i.e.,

$$\bar{l} \approx (1 - \beta) l_{\text{eff}} + \sum_i \beta_i l_i, \tag{4.17}$$

where $\beta = \sum_i \beta_i$. Although β is very small, the l_i are so long that one calculates

$$\bar{l} \approx 0.1 \text{ s}. \tag{4.18}$$

Thus, for $k = 1.001$, we find, instead of (4.16),

$$\phi(\mathbf{r},t) = A_0(0) x_0(\mathbf{r}) e^{t/100}, \tag{4.19}$$

i.e., the reactor power increases by a factor e in 100 s. This rate of change is obviously much easier to control, and in fact, makes the nuclear reactor a practical possibility.

4.3 KINETICS EQUATIONS WITH DELAYED NEUTRONS[1]

Each of the delayed neutron groups listed in Table 4.1 is presumably attached to a β radioactive nucleus, called a *precursor nucleus*, which after decay instantaneously emits the neutron. We denote the precursor concentration associated with the ith group of delayed neutrons as $C_i(\mathbf{r},t)$, and it obeys the equation

$$\frac{\partial}{\partial t} C_i(\mathbf{r},t) = -\lambda_i C_i(\mathbf{r},t) + \beta_i \nu \Sigma_f \phi(\mathbf{r},t), \qquad i = 1, \ldots, 6. \tag{4.20]}$$

[1] This material must be modified if one wishes to treat a particular type of reactor, namely, the *circulating-fuel* reactor. In such a reactor, a suspension or molten form of fuel is circulated through the core and an external heat exchanger. Thus, many of the delayed neutrons are emitted outside the core. We do not consider the special problems introduced by this type of reactor.

Here $\lambda_i C_i$ represents the usual radioactive decay law, while of course $\beta_i \nu\Sigma_f \phi$ represents the rate at which the precursor nuclei are created in fission. Clearly $C_i(\mathbf{r},t)$ will have the same spatial behavior as $\phi(\mathbf{r},t)$. In fact, this can be seen by considering the steady state $\partial C_i/\partial t = 0$ and denoting the steady-state quantities by subscript 0,

$$C_{i0}(\mathbf{r}) = \frac{\beta_i}{\lambda_i} \nu\Sigma_f \phi_0(\mathbf{r}). \tag{4.21}$$

In our point-kinetics model, the flux is separable in space and time, so the spatial proportionality of $C_i(\mathbf{r},t)$ and $\phi(\mathbf{r},t)$ will still hold.

Coupled to Eq. (4.20) is the flux equation

$$\frac{1}{\bar{v}}\frac{\partial}{\partial t}\phi(\mathbf{r},t) = [D\nabla^2 - \Sigma_a + (1-\beta)\nu\Sigma_f]\phi(\mathbf{r},t) + \sum_i \lambda_i C_i(\mathbf{r},t) \tag{4.22a}$$

$$= [-DB^2 - \Sigma_a + (1-\beta)\nu\Sigma_f]\phi(\mathbf{r},t) + \sum_i \lambda_i C_i(\mathbf{r},t) \tag{4.22b}$$

which has been written in accord with the assumption, explained in the previous section, that only the mode with the smallest buckling contributes.

This equation is frequently seen written in terms of δk and l_{eff} [see Eqs. (4.8) and (3.27)] and the neutron density $n = \phi/\bar{v}$,[1]

$$\frac{\partial n}{\partial t} = \frac{\delta k}{l_{\text{eff}}} n - \beta\frac{k}{l_{\text{eff}}} n + \sum_i \lambda_i C_i. \tag{4.22c}$$

In the same notation, Eq. (4.20) becomes

$$\frac{\partial C_i}{\partial t} = -\lambda_i C_i + \beta_i \frac{k}{l_{\text{eff}}} n. \tag{4.20'}$$

To solve Eqs. (4.20′) and (4.22*c*), let us assume a time dependence of the densities,

$$C_i(\mathbf{r},t) = \tilde{C}_i e^{\omega t}, \tag{4.23a}$$

and

$$\phi(\mathbf{r},t) = \tilde{\phi} e^{\omega t}, \tag{4.23b}$$

substitute in Eqs. (4.20′) and (4.22*c*), and obtain an equation for $\tilde{\phi}$. From (4.20′) we obtain

$$(\omega + \lambda_i)\tilde{C}_i = \beta_i \nu\Sigma_f \tilde{\phi}, \tag{4.24a}$$

[1] We have previously used the symbol ρ to refer to neutron density, as is common in transport theory. Here we switch to n in order to avoid confusion with the reactivity ρ.

and from (4.22)

$$\left\{\frac{\omega}{\bar{v}} + \Sigma_a(1 + L^2B^2)[1 - (1 - \beta)k]\right\}\tilde{\phi} = \sum_i \lambda_i \tilde{C}_i. \tag{4.24b}$$

Eliminating $\tilde{C}_i$ between (4.24*a*) and (4.24*b*) gives

$$\left\{\frac{\omega}{\bar{v}} + \Sigma_a(1 + L^2B^2)[1 - (1 - \beta)k]\right\}\tilde{\phi} = \nu\Sigma_f \sum_i \frac{\beta_i \lambda_i}{\omega + \lambda_i} \tilde{\phi}, \tag{4.25}$$

or by dividing through by $\Sigma_a(1 + L^2B^2)$ and using Eq. (4.8) and canceling $\tilde{\phi}$ from both sides,

$$[\omega l_{\text{eff}} + (1 - k) + \beta k] = k \sum_i \frac{\beta_i \lambda_i}{\omega + \lambda_i}. \tag{4.26a}$$

Thus

$$\frac{\omega l_{\text{eff}}}{k} - \rho = \sum_i \beta_i \left(\frac{\lambda_i}{\omega + \lambda_i} - 1\right) \tag{4.26b}$$

$$= -\omega \sum_i \frac{\beta_i}{\omega + \lambda_i}. \tag{4.26c}$$

We have used the *reactivity* ρ defined in Chap. 3 by

$$\rho = \frac{k - 1}{k}, \tag{4.27}$$

and in going from (4.26*a*) to (4.26*b*), we have divided by k and then taken the term β from the left-hand side and placed it inside the sum as $-\beta_i$. Equation (4.26*c*) can be written in somewhat more conventional form as

$$\rho = \frac{\omega l_{\text{eff}}}{k} + \omega \sum_i \frac{\beta_i}{\omega + \lambda_i}, \tag{4.28}$$

or alternatively, since from (4.27) we have

$$k = \frac{1}{1 - \rho}, \tag{4.29}$$

$$\rho = \frac{\omega l_{\text{eff}}}{1 + \omega l_{\text{eff}}} + \frac{\omega}{1 + \omega l_{\text{eff}}} \sum_{i=1}^{N} \frac{\beta_i}{\omega + \lambda_i}. \tag{4.30}$$

Here we have explicitly assumed that we have N delayed groups. We might note that k should be close to unity,[1] so that for the branch nearest $\omega = 0$,

[1] In fact, the point model used here (i.e., the assumption that only the term $n = 0$ in Eq. (4.3) is important) can be shown to break down if k is appreciably different from unity.

Eq. (4.28) might be rewritten approximately as

$$\rho = \omega l_{\text{eff}} + \omega \sum_{i=1}^{N} \frac{\beta_i}{\omega + \lambda_i}. \tag{4.31}$$

Returning now to Eq. (4.30), we might calculate ω as a function of ρ. We see that (4.30) is of $(N + 1)$th degree in ω, so that we can expect $N + 1$ solutions in ω. The one which is algebraically largest of course corresponds to the asymptotic solution which prevails at long times. (Here, long times generally correspond only to fractions of a second, so it is only this asymptotic term which is usually of any practical importance.)

From Eq. (4.30) we see that ρ versus ω is a multibranched curve which has infinite discontinuities at the points $\omega = -\lambda_i$, $i = 1, \ldots, N$, and $\omega = -1/l_{\text{eff}}$. Also, as $\omega \to \infty$, so $\rho \to 1$ asymptotically. This leads to a curve of the type in Fig. 4.1.

Thus we see that Eq. (4.30) has $N + 1$ solutions; call them ω_i, which may be ordered

$$\omega_0 > \omega_1 > \omega_2 > \cdots > \omega_N, \tag{4.32}$$

so that the time-dependent flux has the form

$$\phi(\mathbf{r},t) = \phi_0 e^{\omega_0 t} + \phi_1 e^{\omega_1 t} + \cdots + \phi_N e^{\omega_N t} \tag{4.33a}$$

$$\to \phi_0 e^{\omega_0 t}, \tag{4.33b}$$

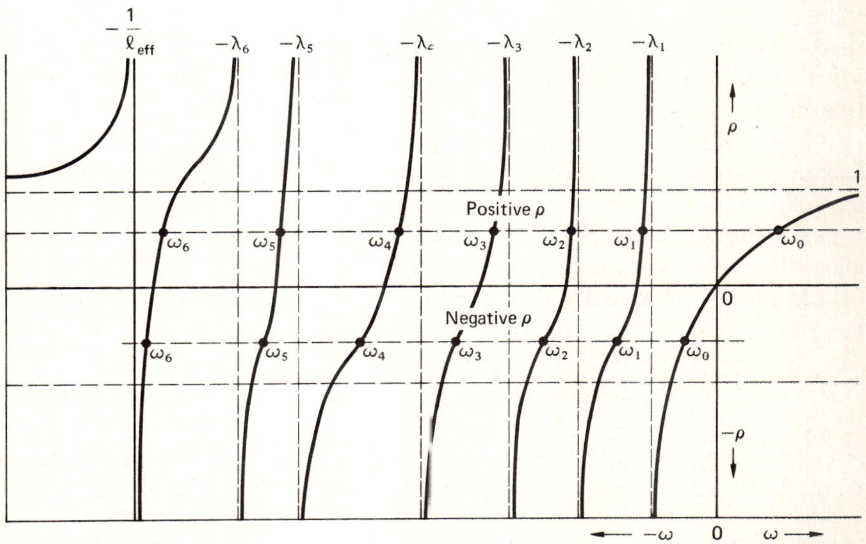

Fig. 4.1 General plot of ω as a function of ρ.

where the arrow indicates the asymptotic (i.e. long-time) limit. We note that by virtue of the assumptions made earlier, all the ϕ_i have the same spatial dependence (that is, $\sim\phi_0$) differing only in amplitude.

The quantity $\omega_0{}^{-1} = T$ is called the *reactor period* (or sometimes, the *stable period*). If we substitute $\omega = 1/T$ in Eq. (4.28), we find

$$\rho = \frac{l_{\text{eff}}}{kT} + \sum_i \frac{\beta_i}{1 + \lambda_i T}, \tag{4.34}$$

a relation generally called the *inhour formula*. The *inhour* is the unit of reactivity ρ for which the stable period T is equal to 1 h [that is, let $T = 3{,}600$ s in Eq. (4.34), and solve for ρ.] Note that a reactivity of *2* inhours does not give a stable period of 30 min, however. This strange, nonlinear unit is not in common use today.[1]

Equations (4.33) should be interpreted in the following way. A reactor is presumed to be operating at critical, so that $\phi(\mathbf{r},t) = x_0(\mathbf{r})$, that is, $T = \infty$. At a certain time, say, $t = 0$, the reactivity is changed by a certain amount. The flux now begins to change according to Eq. (4.33*a*) and, after a short time,[2] approaches (4.33*b*). From the rate of change of ϕ, the period $T = 1/\omega$ can be determined from Eq. (4.34).

4.4 MULTISTAGE PROCESSES PRODUCING DELAYED NEUTRONS

In Sec. 4.3, we considered the delayed neutrons emitted instantaneously at the end of a single β decay of a fission fragment. Suppose the delayed neutron is emitted at the end of a radioactive series of β decays of a fission fragment. This can also be treated by a slight modification of the preceding treatment.

For example, let us assume that the delayed neutrons are emitted instantaneously after a series of two β decays. That is, we assume that a fission fragment A β-decays with a time constant λ' into the precursor nuclide C, which emits a neutron with a time constant λ. If this is the only process giving rise to delayed neutrons, we can write the kinetic equations for the concentrations of A and C (we denote the concentrations by A and C respectively) as

$$\frac{\partial A}{\partial t} = -\lambda' A + \beta\nu\Sigma_f \phi, \tag{4.35a}$$

$$\frac{\partial C}{\partial t} = -\lambda C + \lambda' A. \tag{4.35b}$$

[1] Thank God!

[2] As we shall see in Sec. 4.6, this time is of the order of a fraction of a second.

Of course, we also have the flux equation

$$\frac{1}{\bar{v}}\frac{\partial \phi}{\partial t} = [-DB^2 - \Sigma_a + (1-\beta)\nu\Sigma_f]\phi + \lambda C. \tag{4.35c}$$

As we did in (4.23), we assume an exponential time dependence.

$$A = \tilde{A}e^{\omega t}, \tag{4.36a}$$

$$C = \tilde{C}e^{\omega t}, \tag{4.36b}$$

$$\phi = \tilde{\phi}e^{\omega t}. \tag{4.36c}$$

We use (4.36) in (4.35) and follow the procedures leading to (4.24) to obtain

$$(\omega + \lambda')\tilde{A} = \beta\nu\Sigma_f\tilde{\phi}, \tag{4.37a}$$

$$(\omega + \lambda)\tilde{C} = \lambda'\tilde{A}, \tag{4.37b}$$

$$\left\{\frac{\omega}{\bar{v}} + \Sigma_a(1 + L^2B^2)[1 - (1-\beta)k]\right\}\tilde{\phi} = \lambda\tilde{C}. \tag{4.37c}$$

Now using (4.37*a*) and (4.37*b*) to eliminate $\tilde{C}$ from (4.37*c*), we have

$$\left\{\frac{\omega}{v} + \Sigma_a(1 + L^2B^2)[1 - (1-\beta)k]\right\}\tilde{\phi} = \frac{\beta\nu\Sigma_f\lambda\lambda'}{\lambda' - \lambda}\,\tilde{\phi}\left(\frac{1}{\omega + \lambda} - \frac{1}{\omega + \lambda'}\right). \tag{4.38}$$

Clearly (4.38) has the same mathematical form as (4.25) with two delayed neutron groups. However, in (4.38) the quantities appearing in place of delayed-neutron fractions are not generally positive. If either of the lifetimes is short (that is, $\lambda \gg \lambda'$ or $\lambda' \gg \lambda$), then (4.38) reduces to (4.25) for one delayed-neutron group.

We can generalize (4.38) to arbitrary numbers of decays in a radioactive chain leading to delayed neutrons and arbitrary numbers of different chains. This generalization gives rise to an equation of the form of (4.25) with again the possibility of negative effective delayed-neutron fractions.

Experimental studies of delayed-neutron effects do not usually include a detailed study of the precursor nuclides or the radioactive decays leading to delayed neutrons. Usually a sample of fissionable material is irradiated by a known flux in a reactor and then removed, and the neutron activity is fitted by using (4.25) and treating the β's and λ's and number of groups as parameters. For this reason, negative delayed-neutron fractions have not been observed in ordinary fission. However, in one specific case, they have been observed, namely, a delayed photoneutron emission in D_2O and Be reactors.[1]

[1] G. R. Keepin, *op. cit.*

Since both deuterium and beryllium have rather substantial (γ,n) cross sections, delayed gamma-ray emission by fission fragments has the effect of introducing additional delayed-neutron groups into the reactor system. The parameters of the various additional delayed groups can be found in the Keepin reference cited previously. One notes, in fact, that with beryllium negative β's do occur, which is explained by the analysis given above.

4.5 SPECIFIC RESULTS FOR ONE DELAYED-NEUTRON GROUP[1]

To illustrate the numerical calculations which might be carried out to analyze a typical kinetics experiment, let us make the simplifying approximation that there is only a single group of delayed neutrons with parameters

$$\beta = \sum_i \beta_i = 0.0065 \tag{4.39a}$$

and

$$\lambda = \left(\frac{1}{\beta}\sum_i \frac{\beta_i}{\lambda_i}\right)^{-1} = 0.077 \text{ s}^{-1}. \tag{4.39b}$$

While this is of course a gross approximation, it does permit us to obtain a qualitative feeling for the kinetic behavior of a nuclear reactor.

Thus we rewrite Eq. (4.31) in the form

$$\rho = \omega l_{\text{eff}} + \frac{\beta\omega}{\omega + \lambda}, \tag{4.40}$$

which is a quadratic equation for ω. Specifically,

$$\omega_\pm = -\frac{\beta - \rho + \lambda l_{\text{eff}}}{2l_{\text{eff}}}\left[1 \mp \sqrt{1 + \frac{4\lambda\rho l_{\text{eff}}}{(\beta - \rho + \lambda l_{\text{eff}})^2}}\right]. \tag{4.41}$$

To simplify this expression, we recall that $l_{\text{eff}} \approx 10^{-3}$ s, $\lambda \approx 0.08 \text{ s}^{-1}$, and ρ should be of the order of 10^{-3} (if $\rho \geq \beta$, we see that the reactor is at *prompt* critical, i.e., critical on prompt neutrons alone). In any case, we see that even for $\rho - \beta \approx 10^{-2}$ we still have

$$(\beta - \rho + \lambda l_{\text{eff}})^2 \gg 4\lambda\rho l_{\text{eff}}, \tag{4.42}$$

so that Eq. (4.41) may be written approximately,

$$\omega_\pm = -\frac{\beta - \rho + \lambda l_{\text{eff}}}{2l_{\text{eff}}}\left[1 \mp \frac{2l_{\text{eff}}\lambda\rho}{(\beta - \rho + \lambda l_{\text{eff}})^2}\right] \tag{4.43}$$

[1] Clearly the results of this section will apply only for times shorter than the shortest delayed group.

or

$$\omega_+ = \frac{\lambda\rho}{\beta - \rho + \lambda l_{\text{eff}}} \tag{4.44a}$$

and

$$\omega_- = -\frac{\beta - \rho + \lambda l_{\text{eff}}}{l_{\text{eff}}}. \tag{4.44b}$$

Again, in $\omega_\pm$ we may ignore λl_{eff} compared with $\beta - \rho$ so long as we are not too close to prompt critical, in which case

$$\omega_+ = \frac{\lambda\rho}{\beta - \rho}, \tag{4.45a}$$

$$\omega_+ = \frac{\lambda\$}{1 - \$}, \tag{4.45b}$$

and

$$\omega_- = -\frac{1 - \$}{l_{\text{eff}}/\beta}. \tag{4.45c}$$

Here we have defined the \$ (dollars)[1] as ρ/β. The cent (¢) is, of course, one-hundredth of a dollar.

Thus the time-dependent flux in this approximation obeys the equation

$$\phi(t) = \phi_+ e^{[\lambda\$/(1-\$)]t} + \phi_- e^{-[(1-\$)/(l_{\text{eff}}/\beta)]t}. \tag{4.46}$$

We note that $\omega_+ > \omega_-$, so that the stable period is given by

$$T = \frac{1}{\omega_+} = \frac{1 - \$}{\lambda\$}. \tag{4.47}$$

Similarly

$$C(t) = C_+ e^{\lambda\$/(1-\$)t} + C_- e^{[(1-\$)/(l_{\text{eff}}/\beta)]t}. \tag{4.48}$$

We can see the range of validity of our approximate calculations in Fig. 4.2 is $\$ \leq 1$.

[1] A reactor \$1 supercritical would be prompt critical if the $\lambda l_{\text{eff}} \ll |\beta - \rho|$ held. Of course, it does not, and we must refer to (4.41). For example, see Fig. 4.2.

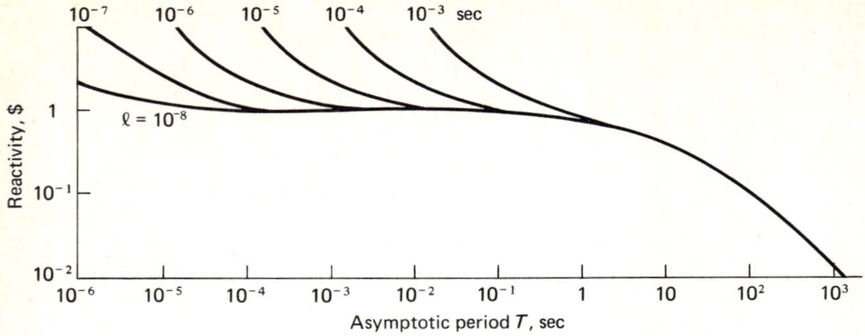

Fig. 4.2 Reactivity vs. asymptotic period for various neutron lifetimes (l) for U^{235} as computed from Eq. (4.41).

4.6 MEASUREMENT OF REACTIVITY: STABLE PERIOD

We assume the reactor to be operating at steady state until, at time $t = 0$, a reactivity change is made. Then a simple measurement of the flux, or power, as a function of time yields the stable period (we continue to assume one delayed-neutron group),

$$T = \frac{1 - \$}{\lambda \$}, \tag{4.49}$$

so that the reactivity can be defined since λ is known. Suppose we are considering a small reactivity, say, 1¢. Let us determine how rapidly the transient flux dies out. We have

$$\omega_- = -\frac{0.99 \times 0.0065}{0.001} = -6.43\ \text{s}^{-1}, \tag{4.50a}$$

$$T_- = -\frac{1}{\omega_-} = 0.156\ \text{s}, \tag{4.50b}$$

so that the decay of the transient term is very rapid indeed. The stable period is about

$$T = \frac{0.99}{0.08 \times 0.01} = 20\ \text{min} \tag{4.51}$$

in this case. As the reactivity increases, the situation is not quite so favorable, but it is still true that the transient term dies out very rapidly as long as we are somewhat below \$1 supercritical. Thus, consider a reactivity of 90¢.

$$\omega_- = -\frac{0.1 \times 0.0065}{0.001} = -0.75\ \text{s}^{-1}, \tag{4.52}$$

while

$$T \approx \frac{0.1}{0.08 \times 0.9} = 1.4 \text{ s.} \tag{4.53}$$

In this case, $\omega_\pm$ are approximately equal in magnitude, but because ω_- is negative, after a couple of seconds the reactor will still be on a stable period.

For negative reactivities, the situation is similar, as can easily be seen by substituting representative numbers in the expressions for $\omega_\pm$. One can, in this way, always evaluate how long he must wait after reactivity change has taken place to measure the stable period. Of course, experimentally there is no question. When a pure exponential decay or increase of the power is observed, the stable period has been achieved. Note that for $\rho > 0$, the $\omega_\pm$ are of different sign except for positive reactivities in excess of $\rho = 1$, which cannot be obtained anyway [see (4.27)]. Thus, when the reactor is on a positive period, the transient is a decreasing exponential. For $\rho < 0$, both terms are negative, but $|\omega_0| < |\omega_1|$. We can easily see from Fig. 4.1 that the general equation (4.31) has one positive root for $\rho > 0$ and no positive root for $\rho < 0$.

It is interesting to calculate the value of 1 inhour in \$. Since for a reactivity of 1 inhour the stable period is 3,600 s, we have, from (4.49),

$$\frac{1 - \$}{\$} = 3{,}600 \times 0.08 \tag{4.54}$$

or

$$\frac{\$}{\text{inhour}} = \frac{1}{300} \tag{4.55}$$

or 1/3¢. Thus we readily see that the inhour is a very small unit of reactivity indeed!

4.7 ROD-DROP EXPERIMENTS

A direct measurement of the reactivity in dollars can be made without measuring the period if the change in reactivity is so sudden that it can be considered instantaneous. Since these measurements are generally made by dropping a control rod into a critical reactor, they are referred to as *rod-drop experiments*, although they could equally well be performed by introducing a sudden positive reactivity change (for example, by withdrawing a rod from a critical reactor). The positive reactivity experiments are obviously not quite safe as the rod-drop variety.

To analyze these experiments, we return to Eqs. (4.20) and (4.22), rewriting them below for convenience (using only one group of delayed neutrons, however).[1]

$$\frac{\partial}{\partial t} C = -\lambda C + \beta \nu \Sigma_f \phi, \tag{4.56a}$$

$$\frac{1}{\bar{v}} \frac{\partial}{\partial t} \phi = [-DB^2 - \Sigma_a + (1-\beta)\nu\Sigma_f]\phi + \lambda C. \tag{4.56b}$$

We now take the Laplace transform[2] of these equations, obtaining

$$\omega \hat{C} - C(0) = -\lambda \hat{C} + \beta \nu \Sigma_f \hat{\phi}, \tag{4.57a}$$

$$\frac{1}{\bar{v}} [\omega \hat{\phi} - \phi(0)] = [-DB^2 - \Sigma_a + (1-\beta)\nu\Sigma_f]\hat{\phi} + \lambda \hat{C}. \tag{4.57b}$$

Since the reactor was assumed to be operating at steady state prior to the rod drop, $C(0)$ and $\phi(0)$ are related by Eq. (4.21), which, combined with (4.57a), gives

$$\hat{C} = \frac{\beta \nu \Sigma_f}{\omega + \lambda} \left[\hat{\phi} + \frac{1}{\lambda} \phi(0) \right]. \tag{4.58}$$

Substituting in (4.57b) gives for $\hat{\phi}$, after some rearrangement,

$$\phi(\omega) = \phi(0) \frac{\omega + \lambda + \bar{v}\beta\nu\Sigma_f}{\omega(\omega+\lambda) - [(\omega+\lambda)/l_{\text{eff}}][(1-\beta)k - 1] - \bar{v}\beta\nu\Sigma_f \lambda}. \tag{4.59}$$

The Laplace inverse is given as

$$\phi(t) = \phi(0)(R_+ e^{\omega_+ t} + R_- e^{\omega_- t}), \tag{4.60}$$

where $\omega_\pm$ are the poles of the right-hand side, and $R_\pm$ the corresponding residues. Of course, the poles $\omega_\pm$ are precisely the roots $\omega_\pm$ already found by another method [see Eq. (4.41) for the exact, or (4.45) for the approximate, forms]. The residues are readily evaluated in the same (4.45) approximation. We find

$$\phi(t) = \phi(0)\left(\frac{1}{1-\$} e^{[\lambda\$/(1-\$)]t} - \frac{\$}{1-\$} e^{-[(1-\$)t/(l_{\text{eff}}/\beta)]} \right). \tag{4.61}$$

[1] Although the assumption of one delayed-neutron group makes the physical ideas behind the rod-drop measurement clear, practical application requires that all the groups be taken into account. This leads to a much more complicated result, but the analysis can be carried through by the interested student in direct analogy with that given here.

[2] See Appendix C.

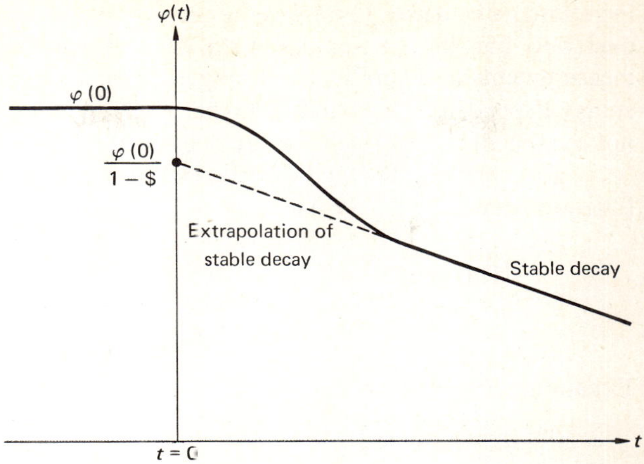

Fig. 4.3 Neutron flux as a function of time after a rod drop.

As before, the second term, i.e., the one involving $e^{\omega_- t}$, dies out within a second or so. Thus the asymptotic flux is measured as

$$\phi(t) \to \frac{\phi(0)}{1-\$} e^{[\lambda\$/(1-\$)]t}. \tag{4.62}$$

When this is extrapolated back to $t = 0$ (see Fig. 4.3), we see that the prompt drop in flux is given by the factor $1/(1-\$)$.[1] In this way, the reactivity in \$ can be measured directly, with no need to know any delayed neutron or reactor parameters (as β, λ, or l_{eff}). By combining a measurement such as that described above with a measurement of the period, we see that λ can also be determined, since the rod reactivity in \$ is known as

$$T = \frac{1-\$}{\lambda\$}. \tag{4.63}$$

It is clear that a positive reactivity can be analyzed in precisely the same fashion, except that one would have an increasing rather than a decreasing exponential.

4.8 PULSED-NEUTRON EXPERIMENTS

Another method of measuring ρ, particularly for large negative values, is to use the pulsed-neutron technique. In this method, a burst of neutrons is injected into a reactor at time $t = 0$ (by a neutron generator of some type), and

[1] The problem of extrapolating backward to zero is not so simple in practice because of the existence of more than one delayed group.

the subsequent decay is observed at times so long that the transients have died out, but short compared with delayed-neutron lifetimes (this means measurements in the millisecond range). These transients are introduced by energy-dependent effects and by higher buckling modes. We shall of course not see the energy transient in our one-speed analysis.

The appropriate equation[1] then is Eq. (4.22) with an additional delta function source $S_0\,\delta(\mathbf{r}-\mathbf{r}_0)\,\delta(t)$,

$$\frac{1}{\bar{v}}\frac{\partial}{\partial t}\phi(\mathbf{r},t) = D\,\nabla^2\,\phi(\mathbf{r},t) - \Sigma_a\phi(\mathbf{r},t) + (1-\beta)\nu\Sigma_f\phi(\mathbf{r},t) + S_0\,\delta(\mathbf{r}-\mathbf{r}_0)\,\delta(t). \tag{4.64}$$

Expanding, as in Sec. 4.2,

$$\phi(\mathbf{r},t) = \sum_n \alpha_n(t)x_n(\mathbf{r}) \tag{4.65}$$

and

$$\delta(\mathbf{r}-\mathbf{r}_0) = \sum_n x_n(\mathbf{r}_0)x_n(\mathbf{r}), \tag{4.66}$$

we obtain an equation for $\alpha_n(t)$,

$$\frac{1}{\bar{v}}\,\dot{\alpha}_n(t) = -[DB_n{}^2 + \Sigma_a - (1-\beta)\nu\Sigma_f]\alpha_n(t) + S_0\,x_n(\mathbf{r}_0)\,\delta(t). \tag{4.67}$$

By introducing the usual notation, this can be rewritten

$$\dot{\alpha}_n(t) = -\frac{1}{l_{\text{eff}}^{(n)}}\left(1 - \frac{1-\beta}{1-\rho_n}\right)\alpha_n(t) + \bar{v}S_0\,x_n(\mathbf{r}_0)\,\delta(t). \tag{4.68}$$

Here, we have defined ρ_n by

$$k_n = \frac{1}{1-\rho_n} \tag{4.69}$$

[recall (4.9*b*)]. The solution of this equation may be obtained by the method of Laplace transforms,

$$\omega\hat{\alpha}_n - \alpha_n(0) + \frac{1}{l_{\text{eff}}^{(n)}}\left(1 - \frac{1-\beta}{1-\rho_n}\right)\hat{\alpha}_n = \bar{v}S_0\,x_n(\mathbf{r}_0), \tag{4.70}$$

$$\alpha_n(t) = [\alpha_n(0) + \bar{v}S_0x_n(\mathbf{r}_0)]e^{(t/l_{\text{eff}}^{(n)})\{[(1-\beta)/(1-\rho_n)]-1\}}. \tag{4.71}$$

The α's for $n > 0$ decay more rapidly than the fundamental $n = 0$ in analogy with the arguments given in Sec. 4.2.[2] Here, though, we merely

[1] Since we are interested in times short compared with $1/\lambda_i$, we can ignore the terms $\lambda_i C_i$.
[2] It is still true that the terms $n \geq 1$ do not accurately describe the transient behavior.

wait sufficiently long that only the fundamental decay is left (this can easily be checked experimentally by waiting for a pure exponential decay). By inserting representative numbers into Eq. (4.71), it can be seen that this will still happen in a time short compared with the delayed-neutron lifetime. Then the exponential decay constant is given by

$$\lambda = \frac{1}{l_{\text{eff}}}\left(\frac{1-\beta}{1-\rho} - 1\right) \approx \frac{-\beta}{l_{\text{eff}}}(\$ - 1), \tag{4.72}$$

where, as previously, \$ represents ρ/β. Note that for $\rho < \beta$, that is, the reactor below prompt critical, λ is negative, corresponding to a decreasing flux. Of course, this is true only for short times. If the reactor is above delayed critical but below prompt critical, that is, $0 < \rho < \beta$, then the initial decrease will of course be followed by a rise as soon as the delayed neutrons begin to take effect.

To measure \$, we note that the ratio ρ/l_{eff} can be measured by pulsing a reactor which is exactly critical, $\rho = 0$, since then

$$\lambda \to -\frac{\beta}{l_{\text{eff}}}. \tag{4.73}$$

This technique is most often used to determine the *shutdown* margin of control rods in a reactor.[1]

Some interesting information can also be obtained by pulsing a nonmultiplying system ($k = 0$, hence $\rho = -\infty$). In this case,

$$\lambda = -\frac{1}{l_{\text{eff}}} = -\bar{v}(\Sigma_a + DB^2). \tag{4.74}$$

The measurements are generally made as a function of B^2 by pulsing assemblies of identical composition but of different size. When $\lambda(B^2)$ is plotted (see Fig. 4.4) versus B^2, the intercept extrapolated to $B^2 = 0$ yields $\bar{v}\Sigma_a$, while

[1] See B. E. Simmons and J. S. King: *Nuc. Sci. Eng.* **3**, 595 (1958). In this paper, refinements of the experimental techniques are described.

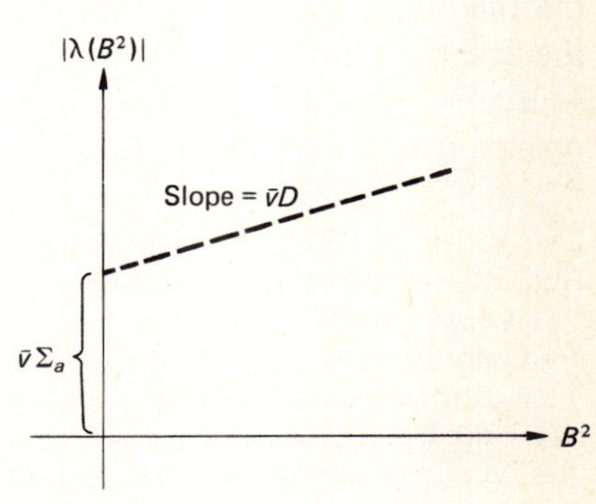

Fig. 4.4 Decay constant versus B^2 for a pulsed-neutron experiment.

the slope gives $\bar{v}D$. Actually, the simple analysis given here is not completely accurate. When energy-dependent effects are taken into account, it turns out that λ contains terms proportional to B^4, B^6, and so forth. This effect, known as *diffusion cooling*, is discussed in various advanced texts.[1]

4.9 INVERSE MULTIPLICATION

Consider a subcritical assembly in which a stationary neutron distribution is maintained by the introduction of a neutron source of strength, say S. From the discussion in Sec. 3.6 we know that this source will give rise to a number kS of fission neutrons in the assembly. These, in turn, will give rise to $kkS = k^2S$, and in fact, the total neutron population in the assembly will be

$$N = S(1 + k + k^2 + \cdots +) \tag{4.75a}$$

$$= \frac{S}{1-k}. \tag{4.75b}$$

This immediately suggests that k might be measured by measuring the quantity called the *multiplication*,

$$M = \frac{N}{S} = \frac{1}{1-k}. \tag{4.76}$$

Actually, the inverse multiplication $1/M$ is a more useful quantity.

Suppose one is attempting to start up a reactor which is shut down, say, by the withdrawal of control rods. As long as a source is present in the reactor, the quantity $1/M$ can be measured as a function of control-rod position as the rods are gradually withdrawn (see Fig. 4.5). By extrapolating the curve to $1/M = 0$, that is, to $k = 1$, the critical rod position can be estimated in advance and approached safely. The same technique can be used when a reactor is being fueled, in which case $1/M$ is plotted as a function of the fuel loading, and the critical mass estimated. Note that M, and hence the flux level in a reactor, approach infinity as $k \to 1$. For this reason, if the source is not removed from the *critical reactor*, the reactor will actually be operating in a slightly subcritical condition.[2] The value of k can of course be

[1] M. M. R. Williams: "The Slowing Down and Thermalization of Neutrons," North-Holland Publishing Company, Amsterdam, 1966.

[2] There are almost always extraneous sources of neutrons in a reactor, such as those produced by (γ,n) reactions, the γ's coming from "hot" fuel elements. In some cases, these extraneous sources are in time strong enough that an extra source, such as Pu(α,n)Be, is not needed. This is particularly true of D_2O or Be reactors since both deuterium and beryllium have large (γ,n) cross sections.

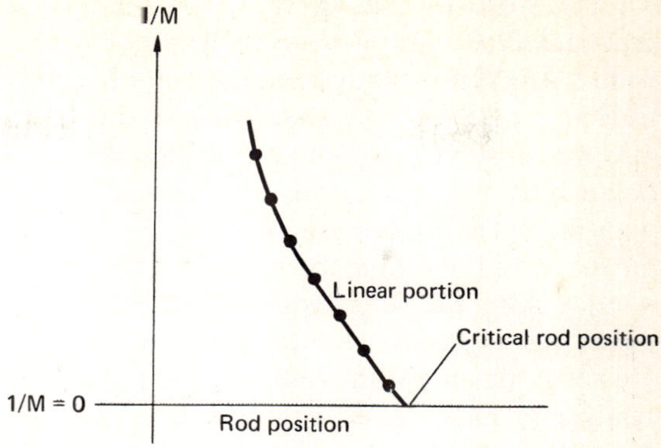

Fig. 4.5 Inverse multiplication.

calculated if the neutron population in the reactor is known,

$$N = \frac{1}{V} \int_{\text{reactor}} \phi(\mathbf{r})\, d^3r, \tag{4.77}$$

and from (4.75*b*),

$$k = 1 - \frac{S}{N}. \tag{4.78}$$

Actually, as always, the situation is considerably more complicated than outlined above. The factor k multiplies only those neutrons with the same energy and spatial distribution as the fundamental mode flux (and the same energy spectrum as the fission spectrum). Thus, the factors k_i in each *generation* are different, although, indeed, they converge fairly rapidly to k in practice (after three or four generations). A more accurate calculation can be made by the source iteration methods described in Chapter 6. The initial *guess* is the actual extraneous source distribution S. Then the first *iterate* S_1 actually represents the first generation of fission neutrons. S_2 represents the second generation. Eventually S_{j+1} will be found to be proportional to S_j for all $j > j_0$, and this proportionality constant is of course k. Then

$$N = S_0 + S_1 + \cdots + S_{j_0} + \frac{S_{j_0+1}}{1-k}. \tag{4.79}$$

By combining calculations with experiments in this way, a very accurate value of k can be determined. However, we note that for k close to unity, the last term in Eq. (4.79) may be expected to dominate, and so the simple formula (4.75*b*) will be valid after all. This implies that a plot of $1/M$ will be linear near the critical configuration; that configuration thus can be estimated by linear extrapolation.

A word or two of caution should be inserted here with respect to practical experimental measurements of $1/M$. Suppose a change is made from a subcritical configuration at multiplication of M_1 to another configuration M_2, perhaps by withdrawing a control rod. As we have seen in Sec. 4.2, the flux will contain transient terms corresponding to higher modes (k_n, $n > 0$), which die away and so leave the fundamental mode associated with k_0. The multiplication cannot be measured until the transient modes have decayed, the time for which can be estimated from Eq. (4.6), by setting $n = 1$. Also we note, from Eq. (4.76), that the total neutron population enters the expression for M; thus absolute counting seems to be required. However, this is not the case. If every multiplication measurement is known relative to an absolute counting rate, then each point in Fig. 4.5 will be multiplied by the same constant factor. The $1/M$ curve will have a different slope, but will intersect the x axis at the same point. From a practical standpoint, this means counters can be placed at various locations adjacent to the subcritical assembly as convenience dictates.

4.10 THE REACTOR TRANSFER FUNCTION

A concept particularly popular among engineers interested in the dynamic behavior of nuclear reactors is that of the *transfer function*. To understand this concept, assume a system is described by a linear operator H, a *response function* or *output* $O(t)$, and an *input* or source $I(t)$. That is,

$$HO(t) = I(t). \tag{4.80}$$

If we solve this problem by the method of Fourier transforms[1] (cf. Appendix C), we have

$$\bar{O}(\omega) = R(\omega)\bar{I}(\omega) \tag{4.81}$$

or

$$O(t) = \frac{1}{2\pi}\int_{-\infty}^{\infty} e^{i\omega t}R(\omega)\bar{I}(\omega)\,d\omega. \tag{4.82}$$

The function $R(\omega)$ is the transfer function. We note that if $R(\omega)$ has poles in the lower-half complex ω plane, $O(t)$ will have an increasing exponential time dependence. Such a system is said to be *unstable*. Thus, the poles of the transfer function determine the stability of a (linear) system. In particular, it is often possible to measure the transfer function experimentally in a reasonably simple manner, in which case stability criteria can be obtained easily.

[1] Laplace transform works as well.

Let us study an explicit example, namely, the reactor system described by the kinetic equations (4.22c) and (4.20′). The input in this case is generally a sinusoidal δk (for example, by a control rod oscillating into and out of a reactor). Thus, assuming $k = 1$ initially, we have[1] $\delta k(t) = \delta k_0\, e^{i\omega_0 t}$, and

$$\frac{\partial n}{\partial t} = \frac{n}{l_{\text{eff}}}\, \delta k_0\, e^{i\omega_0 t} - \frac{\beta n}{l_{\text{eff}}} (1 + \delta k_0\, e^{i\omega_0 t}) + \sum_i \lambda_i C_i , \tag{4.83a}$$

$$\frac{\partial C_i}{\partial t} = -\lambda_i C_i + \frac{\beta_i n}{l_{\text{eff}}} (1 + \delta k_0\, e^{i\omega_0 t}). \tag{4.83b}$$

We note that $n(t)$ will be given by

$$n(t) = n_0 + n_1(t), \tag{4.84a}$$

$$C_i(t) = C_{i0} + C_{i1}(t), \tag{4.84b}$$

where n_0 and C_{i0} are the steady-state values of $n(t)$ and $C_i(t)$ associated with the critical reactor, while $n_1(t)$ is the response to the input $\delta k(t)$. We assume that $n_1(t)$ is small so that products like $n_1\, \delta k$ can be ignored [which linearizes Eqs. (4.83)]. This gives, in place of (4.83),

$$\frac{\partial n_1}{\partial t} = \frac{\delta k_0}{l_{\text{eff}}}\, n_0\, e^{i\omega_0 t} - \frac{\beta}{l_{\text{eff}}} [(1 + \delta k_0\, e^{i\omega_0 t}) n_0 + n_1(t)] + \sum_i \lambda_i (C_{i0} + C_{i1}(t)) \tag{4.85a}$$

and

$$\frac{\partial C_{i1}(t)}{\partial t} = -\lambda_i (C_{i0} + C_{i1}(t)) + \frac{\beta_i}{l_{\text{eff}}} [(1 + \delta k_0\, e^{i\omega_0 t}) n_0 + n_1(t)]. \tag{4.85b}$$

These equations can be simplified further by use of the steady-state relations[2]

$$\frac{\beta}{l_{\text{eff}}}\, n_0 = \sum_i \lambda_i C_{i0} , \tag{4.86a}$$

$$\lambda_i C_{i0} = \frac{\beta_i n_0}{l_{\text{eff}}} , \tag{4.86b}$$

to yield

$$\frac{\partial n_1}{\partial t} = \frac{\delta k_0\, n_0}{l_{\text{eff}}}\, e^{i\omega_0 t} - \frac{\beta}{l_{\text{eff}}} (\delta k_0\, n_0\, e^{i\omega_0 t} + n_1) + \sum_i \lambda_i C_{i1}, \tag{4.87a}$$

$$\frac{\partial C_{i1}}{\partial t} = -\lambda_i C_{i1} + \frac{\beta_i}{l_{\text{eff}}} (\delta k_0\, n_0\, e^{i\omega_0 t} + n_1). \tag{4.87b}$$

[1] Of course δk cannot be complex. Thus to make comparison of experiments, we must take the real or the imaginary part of (4.83a) and (4.83b).

[2] Use (4.22a), and set $\partial n/\partial t$ and δk equal to zero; use (4.20′) and set $\partial C_i/\partial t$ equal to zero and k equal to unity.

The Fourier transformed equations are

$$-i\omega\bar{n}_1(\omega) = \frac{2\pi\,\delta k_0\,n_0}{l_{\text{eff}}}\,\delta(\omega+\omega_0) + \sum_i \lambda_i\,\bar{C}_{i1}(\omega) - \frac{\beta}{l_{\text{eff}}}\,[2\pi\,\delta k_0\,n_0\,\delta(\omega+\omega_0) + \bar{n}_1(\omega), \tag{4.88a}$$

and

$$\bar{C}_{i1}(\omega)(-i\omega+\lambda_i) = \frac{\beta_i}{l_{\text{eff}}}\,[2\pi\,\delta k_0\,n_0\,\delta(\omega+\omega_0) + \bar{n}_1(\omega)]. \tag{4.88b}$$

Solving for $\bar{n}_1(\omega)$ gives, after a little algebra,

$$\frac{\bar{n}_1(\omega)}{n_0} = 2\pi\,\delta k_0\,\delta(\omega+\omega_0)\,\frac{1-\beta+\sum_i(\lambda_i\beta_i/l_{\text{eff}})}{-i\omega\{l_{\text{eff}}+\sum_i[\beta_i/(i\omega+\lambda_i)]\}} \tag{4.89a}$$

$$= 2\pi\,\delta k_0\,\delta(\omega+\omega_0)\,R(\omega). \tag{4.89b}$$

The quantity $R(\omega)$ is called the *zero power* reactor transfer function. We note the close similarity to the inhour equation (4.34), $-i\omega = T^{-1}$. This transfer function is not very interesting, because of the assumptions made in deriving it. In fact, the reactor is automatically stable for $k(0) \leq 1$ and unstable for $k(0) > 1$. Of more interest is the power reactor transfer function. For a power reactor, the coefficients (cross sections, etc.) are functions of the power, or neutron density, because of temperature feedback effects, etc.[1] Then the kinetics equation becomes highly nonlinear, and it is obvious because of the feedback that stability cannot be determined by as simple a criterion as $k(0) \leq 1$. It is in such cases that measurements of a transfer function are most important.

4.11 NEUTRON WAVE EXPERIMENTS[2]

We described, in Sec. 2.11, a method for measuring thermal diffusion lengths by a simple exponential attenuation of thermal neutrons. Then in Sec. 4.9, pulsed-neutron experiments were described. Neutron wave experiments can be considered, in a sense, a combination of these two methods. For example, the experimental arrangement might be similar to that described in Sec. 2.11, except that the source is modulated, perhaps by oscillating a cadmium shutter

[1] That is, as the power increases, the temperature increases and the reactor material expands. This obviously affects the macroscopic cross sections. There are many other effects as well.
[2] R. E. Uhrig: *Trans. Amer. Nucl. Soc.*, **2**, 2 (1959).

in front of the thermal column. Then, Eq. (2.130a) applies, except that there must be an additional term present, $(1/\bar{v})\,(\partial/\partial t)\,\chi$, to take account of the fact that the flux will be time dependent. Also, the source will be modulated by a factor $1 + S_1 e^{i\omega_0 t}$. The appropriate equation describing the time-dependent flux in the z direction is therefore[1]

$$\frac{1}{\bar{v}}\frac{\partial \chi}{\partial t} - D\left[-\left(\frac{\pi}{2a'}\right)^2 - \left(\frac{\pi}{2b'}\right)^2 + \frac{\partial^2}{\partial z^2}\right]\chi(z,t) + \Sigma_a \chi(z,t)$$
$$= S_0\,\delta(z)\,(1 + S_1 e^{i\omega_0 t}). \tag{4.90}$$

Assuming a solution of the form

$$\chi(z,t) = \chi_0(z) + \chi_1(z) e^{i\omega_0 t} \tag{4.91a}$$

gives

$$\frac{i\omega_0}{\bar{v}}\chi_1 e^{i\omega_0 t} + \Sigma_b(\chi_0 + \chi_1 e^{i\omega_0 t}) - D(\chi_0'' + \chi_1'' e^{i\omega_0 t}) = S_0\,\delta(z)\,(1 + S_1 e^{i\omega_0 t}). \tag{4.91b}$$

Here we have written, for simplicity,

$$\Sigma_b = \Sigma_a + D\left[\left(\frac{\pi}{2a'}\right)^2 + \left(\frac{\pi}{2b'}\right)^2\right], \tag{4.92a}$$

$$\chi_j'' = \frac{\partial^2}{\partial z^2}\chi_j(z). \tag{4.92b}$$

(That is, Σ_b is the sum of the true absorption and the leakage absorption.)

If we now equate, in Eq. (4.91b), the terms on each side of the equation which are time independent and those proportional to $e^{i\omega_0 t}$, we find

$$\Sigma_b\chi_0 - D\chi_0'' = S_0\,\delta(z) \tag{4.93a}$$

and

$$\left(\Sigma_b + \frac{i\omega_0}{\bar{v}}\right)\chi_1 - D\chi_1'' = S_0 S_1\,\delta(z). \tag{4.93b}$$

Thus, χ_0 is precisely the same exponentially decreasing function which one measures in the diffusion length experiments of Sec. 2.11. Superimposed on this solution, however, is χ_1.

To solve Eq. (4.93b) for χ_1, we take the Fourier transform (k = transform variable) and find

$$\bar{\chi}_1(k) = \frac{S_0 S_1/D}{k^2 + (\Sigma_b + i\omega_0/\bar{v})/D}. \tag{4.94}$$

[1] As in (4.83a), we must take the real or the imaginary part of (4.90) to describe an experiment.

Thus

$$\chi_1(z,t) = R_+ \exp\left(-z\sqrt{\frac{\Sigma_b + i\omega_0/\bar{v}}{D}} + i\omega_0 t\right), \tag{4.95}$$

where R_+ is $2\pi i$ times the residue,

$$R_+ = \pi \frac{S_0 S_1/D}{\sqrt{(\Sigma_b + i\omega_0/\bar{v})/D}}. \tag{4.96}$$

The square root is easily calculated,

$$\sqrt{\frac{\Sigma_b + i\omega_0/\bar{v}}{D}} = a + ib \tag{4.97a}$$

or

$$\frac{\Sigma_b + i\omega_0/\bar{v}}{D} = a^2 - b^2 + 2iab. \tag{4.97b}$$

Equating real and imaginary parts, we find

$$a_\pm = \pm\sqrt{\frac{1}{2D}(\Sigma_b + \sqrt{\Sigma_b{}^2 + \omega_0{}^2/\bar{v}^2})}, \tag{4.98a}$$

$$b_\pm = \pm\frac{\omega_0}{\bar{v}\sqrt{2D(\Sigma_b + \sqrt{\Sigma_b{}^2 + \omega_0{}^2/\bar{v}^2})}}. \tag{4.98b}$$

Then

$$\chi(z,t) = R_+ e^{i(-b_+ z + \omega_0 t) - a_+ z}. \tag{4.99}$$

We must choose $a > 0$ [the + sign in Eq. (4.98)] so that (4.99) represents a damped wave traveling to the right. The phase velocity is given by

$$C_p \equiv \frac{\omega_0}{b_+} = \bar{v}\sqrt{2D(\Sigma_b + \sqrt{\Sigma_b{}^2 + \omega_0{}^2/\bar{v}^2})}. \tag{4.100}$$

By measuring the phase velocity of the wave and the damping factor a_+ [Eq. (4.98*a*)], two relations are obtained, from which Σ_b and D are obtained. There are some experimentalists who prefer this technique to the pulsed-neutron technique, which essentially measures the same thing. In fact, by the nature of the Fourier expansion of the delta function,

$$\delta(t) = \frac{1}{2\pi}\int_{-\infty}^{\infty} e^{i\omega t}\, d\omega, \tag{4.101}$$

we see that a *wave* experiment corresponds to a single Fourier component of a *pulsed* experiment.

PROBLEMS

4.1 For one delayed group, compute the period for a reactivity of 2 inhours.

4.2 For one delayed group, compute the period for a reactivity of \$1.

4.3 Show explicitly, by use of the transfer function formalism, that a reactor with $k(0) \leq 1$ (> 1) is stable (unstable).

4.4 Integrate the response for a neutron wave experiment all over ω_0 to obtain the pulsed-neutron response.

5 Calculation of Reactivity

5.1 INTRODUCTION

In the previous chapter we discusssd how a reactor will respond to a given reactivity ρ. Suppose now we wish to calculate the reactivity. We might go back to the methods of Chap. 3 and, particularly, use Eq. (3.25) in the form

$$\nu_c = \frac{DB^2 + \Sigma_a}{\Sigma_f}. \tag{5.1}$$

Then, of course,

$$\rho = \frac{\nu - \nu_c}{\nu}. \tag{5.2}$$

If $\nu \neq \nu_c$, the reactor is not critical. We can also compute the changes in the various parameters which would make the reactor critical, simply by differentiating (5.1),

$$\delta \nu_c = \frac{[(2DB\,\delta B + B^2\,\delta D + \delta\Sigma_a)\Sigma_f - (DB^2 + \Sigma_a)\,\delta\Sigma_f]}{{\Sigma_f}^2}. \tag{5.3}$$

The trouble with this approach is that the changes which are represented as differentials in Eq. (5.3) must be made *uniformly* throughout the whole reactor, because Eq. (5.1) applies only to the bare, homogeneous system. Equation (5.3) does not tell us how to make a small, localized change in any of the reactor parameters (for example, by changing the position of a control element) so as to achieve criticality.

Another approach would be to carry out a numerical calculation, as described in Chap. 6, and arbitrarily change parameters until criticality is achieved. However, this method is cumbersome since the necessary changes can only be guesssed at beforehand and, besides, the numerical calculations are generally long and expensive (in machine time). Another difficulty in this approach is that one is really calculating k (or v_c) directly, not ρ. Thus, since

$$\rho = \frac{1}{1-k}, \tag{5.4a}$$

we have

$$\frac{\delta\rho}{\rho} = -\frac{\delta k}{1-k} \tag{5.4b}$$

Since we may expect k to be close to unity, we see that an accurate calculation of ρ requires accuracy in the calculation of k, which is generally not available.

In this chapter we discuss methods for calculating ρ directly for nonuniform changes of reactor composition. As a by-product, we obtain methods which permit the determination of critical conditions for nonuniform reactors.[1]

5.2 LINEAR OPERATORS

In order to develop the methods mentioned in the previous section, we must first give a brief description of the theory of linear operators.[2] An operator θ will be defined as a mapping of a function space $\{f\}$ into itself. In plain English, this means that if f is some function in the space, θf is another function of the same space, call it g. Mathematically we write

$$\theta f = g. \tag{5.5}$$

[1] These methods cannot be applied very well if a reactor is grossly heterogeneous. Then the numerical methods are the only recourse. But for more or less continuous changes of material properties throughout the reactor, these methods suffice very well.

[2] For a more detailed discussion, see, e.g., N. I. Akheizer and I. M. Glazman: "Theory of Linear Operators in Hilbert Space," vols. I and II, Frederick Ungar Publishing Co., New York, 1961.

By the term *same space* we mean, for example, functions which have certain properties which we know should hold physically for the systems we wish to consider.

It is customary to consider functions which are square integrable (in the Lebesque sense) on some interval, i.e., for which $\int_a^b |f|^2\,dx < \infty$. We refer to this as the space $L_2(a,b)$. The operator θ will not operate on all functions f in the space, however. First, it is required that θf also be square integrable. Second, θ will operate only on functions which obey certain boundary conditions. The set of functions on which an operator θ operates is called the *domain of* θ and is represented by the symbol $D(\theta)$. We shall usually be concerned with the domain of functions which vanish at the (extrapolated) boundaries of the reactor under consideration and which may have certain other properties that are necessary for the functions to be acceptable as neutron fluxes (e.g., positivity).

An example of an operator is d/dx. It changes a function f into its derivative f'. Another example is *cos*. Now, a *linear* operator has the property that if f and g are two functions and c_1 and c_2 are two (complex) constants, then

$$\theta(c_1 f + c_2 g) = c_1\theta f + c_2\,\theta g. \tag{5.6}$$

We see that d/dx is indeed a linear operator, while cos is not [since cos $(f+g) \neq \cos f + \cos g$].

In the domain of θ there will in general be a *subspace* of functions $\{\xi_n\}$ with the property that operation on one of the ξ_n by θ reproduces the same function multiplied by a (complex) constant. Mathematically we write

$$\theta\xi_n = a_n\,\xi_n, \tag{5.7}$$

where, as stated above, a_n is a complex number. The set of functions $\{\xi_n\}$ are called *eigenfunctions* of the operator θ, and the number a_n associated with a given ξ_n is called the *eigenvalue*.[1]

We have given an example above of a linear operator d/dx. This is, we note, a *differential* operator. If we replace θ in Eq. (5.7) with d/dx, we obtain a differential equation for the eigenfunctions,

$$\frac{d}{dx}\,\xi_n(x) = a_n\,\xi_n(x). \tag{5.8}$$

As is well known, the solution of an nth-order differential equation requires the specification of *n boundary conditions*. We must always be certain, in dealing with differential operators, that a sufficient number of boundary

[1] In some cases, several of the ξ_n will have the same eigenvalue. We then say this eigenvalue is *N-fold degenerate*, where N is the number of eigenfunctions with the same eigenvalue.

conditions is specified so that the domain is defined. It is customary, in dealing with the simple operator d/dx, to specify what is called the *periodic boundary condition*, which states simply that ξ_n is a periodic function with a given periodicity;[1] for example, we could say

$$\xi_n(x) = \xi_n(x + d), \tag{5.9}$$

where d is some constant of our choosing.

The general solution of Eq. (5.3) is

$$\xi_n = C_n e^{a_n x}, \tag{5.10}$$

where C_n is an arbitrary constant. The eigenvalue a_n can be determined from the boundary condition (5.9),

$$e^{a_n(x+d)} = e^{a_n x}, \tag{5.11a}$$

so that

$$e^{a_n d} = 1 \tag{5.11b}$$

or

$$a_n d = 2n\pi i, \qquad n = 0, \pm 1, \pm 2, \ldots. \tag{5.11c}$$

Thus, the eigenfunctions are given by

$$\xi_n(x) = C_n e^{2n\pi i x/d}, \tag{5.12a}$$

and the eigenvalue associated with ξ_n is

$$a_n = \frac{2n\pi i}{d}, \tag{5.12b}$$

with $n = 0, \pm 1, \pm 2, \ldots.$

We note the presence of the arbitrary constant C_n. This comes about because the eigenvalue equation (5.7) is *homogeneous*; any multiple of a solution is again a solution. The constant C_n can be determined if a *normalization* condition is imposed. For example, we might set

$$\int_0^d |\xi_n(x)|^2 \, dx = 1, \tag{5.13a}$$

from which we see

$$C_n = \sqrt{\frac{1}{d}} \tag{5.13b}$$

independent of n.

[1] Other choices of boundary conditions are of course possible.

Another differential eigenvalue equation has already been encountered in Chap. 3, namely,

$$\nabla^2\phi + B^2\phi = 0. \tag{5.14}$$

Consider the bare slab differential $\nabla^2 = d^2/dx^2$. Let us also affix subscripts to ϕ and B^2 to remind us that (5.14) is really an eigenvalue equation,

$$\frac{d^2}{dx^2}\,\phi_n(x) = -B_n{}^2\phi_n(x). \tag{5.15}$$

The appropriate boundary conditions [two, because (5.15) is a *second*-order equation] are already known:

$$\phi_n(x) = \phi_n(-x), \tag{5.16a}$$

$$\phi_n(a') = 0. \tag{5.16b}$$

Condition (5.16*a*) states that the flux is symmetric about the center of the reactor ($x = 0$), while condition (5.16*b*) is the statement that the flux vanishes at the extrapolated boundary.

The general solution of Eq. (5.15) is

$$\phi_n(x) = A_n \cos B_n x + C_n \sin B_n x; \tag{5.17}$$

by virtue of Eq. (5.16*a*), $C_n = 0$. From (5.16*b*), we see that

$$B_n = \frac{2n+1}{2a'}\,\pi, \qquad n = 0, 1, 2, \ldots \tag{5.18a}$$

and

$$\phi_n(x) = A_n \cos\left(\frac{2n+1}{2a'}\,\pi x\right). \tag{5.18b}$$

These eigenfunctions have already been met in Chap. 3. There, only ϕ_0 was considered, since only it obeyed the physical requirement $\phi > 0$, with $-a \leq x \leq a$, that is, the neutron flux should be everywhere positive. However, the complete set of eigenfunctions may be needed for certain mathematical purposes, for example, in dealing with time-dependent problems (see Chap. 4).

The constants A_n may also be determined from an arbitrary normalization assumption. For example, in Chap. 3 we normalized A_0 such that the integral of $\Sigma_f\,\phi_0$ over the reactor volume was equal to the power divided by 2.88×10^{-11} W [see (3.89)].

In many cases, the operator *inverse* to Θ, denoted by Θ^{-1}, will exist. This operator has the property that $\Theta\Theta^{-1} = \Theta^{-1}\Theta = 1$.

Let us return now to our general discussion of eigenvalues and eigenfunctions. We shall in general assume every set $\{\xi_i\}$ to be *complete*, which means that any function in the space in which θ operates can be expanded in terms of the $\{\xi_i\}$,

$$f = \sum_i f_i \xi_i, \tag{5.19}$$

where f_i are (complex) constants called *expansion coefficients*.

Next we define a *scalar product* in our space. This is defined as a correspondence of a complex number with every pair of functions f and g in the space, and is denoted by the symbol (f,g).[1] This complex number has the following properties:

$$(f,g) = (g,f)^* \tag{5.20a}$$

(the asterisk means complex conjugate),

$$(f + h,g) = (f,g) + (h,g), \tag{5.20b}$$

$$(f,Cg) = C(f,g), \tag{5.20c}$$

where C is any complex constant. From (5.20*a*) and (5.20*c*) it follows that

$$(Cf,g) = C^*(f,g). \tag{5.21}$$

If $(f,g) = 0$, we say f and g are *orthogonal*. Note $(f,f) \geq 0$, with equality holding only if $f = 0$.

Next, we define the adjoint $\theta^\dagger$ of the operator θ by the relation

$$(f,\theta g) = (\theta^\dagger f,g). \tag{5.22}$$

Just as θ had associated with it certain boundary conditions, so will $\theta^\dagger$ have associated with it (possibly) different boundary conditions (called the *adjoint boundary conditions*). In the definition, Eq. (5.22), it is assumed that the function g obeys the boundary conditions associated with θ, while f obeys the adjoint boundary conditions.[2]

To consider a specific example, let us introduce a definition of scalar product

$$(f,g) = \int_{\text{reactor}} f^*(\mathbf{r})g(\mathbf{r})\, d^3r. \tag{5.23}$$

[1] A space in which a scalar product is defined is called a *Hilbert space* if it is *complete* and *separable*. A space is complete if every convergent sequence converges to an element of the space (i.e., the space contains all its limit points). A space is *separable* if it contains a *basis*, i.e., a denumerable set of functions like the ξ_i defined earlier which form a complete set. The quantity $(f,f)^{1/2}$ is called the *norm* of f and may be thought of as the "length" of the function f. In our Hilbert space it is required that the norm of every function be a finite number less than infinity. Note that a Hilbert space is linear in that if f and g are elements of the space, so is $c_1 f + c_2 g$ [cf. Eq. (5.20)]. A function space with all the properties of a Hilbert space except the scalar product is called a *Banach space*.

[2] We shall in the future call θ the *direct operator*, and its boundary conditions the *direct conditions*.

This definition is, we note, consistent with the properties (5.20). Next, consider the operator d/dx with the direct boundary conditions given by Eq. (5.9), and attempt to determine the adjoint operator $(d/dx)^\dagger$ and the adjoint boundary condition.[1] Then

$$(f,\theta g) = \left(f, \frac{d}{dx} g\right) \tag{5.24a}$$

$$= \int_0^d f^*(x) \frac{d}{dx} g(x)\, dx. \tag{5.24b}$$

Integration by parts yields

$$(f,\theta g) = -\int_0^d \frac{df^*(x)}{dx} g(x)\, dx + f^*(x)g(x)\Big|_0^d. \tag{5.24c}$$

The boundary term vanishes if we set

$$f^*(d) = f^*(0). \tag{5.25}$$

Then

$$\left(f, \frac{d}{dx} g\right) = \left(-\frac{df}{dx}, g\right), \tag{5.26a}$$

so

$$\left(\frac{d}{dx}\right)^\dagger = -\frac{d}{dx}, \tag{5.26b}$$

with the adjoint condition given by (5.25). They are the same as the direct conditions (5.9); $D(d/dx) = D(d/dx)^\dagger$.

In a similar manner, the adjoint of d^2/dx^2 with conditions (5.16) can be found. In fact,[2] it is readily shown that

$$\left(\frac{d^2}{dx^2}\right)^\dagger = \frac{d^2}{dx^2}, \tag{5.27}$$

with both direct and adjoint boundary conditions given by Eqs. (5.16). Thus the operator d^2/dx^2 is said to be *self-adjoint*.[3]

Note also for any operator θ that

$$\{\theta^\dagger\}^\dagger = \theta. \tag{5.28}$$

[1] We consider the reactor boundaries in this case to be $x = 0$ and $x = d$.

[2] See Prob. 5.9. In this case it is more convenient to take the reactor boundaries to be $x = -d$ and $x = d$. Then $(f,g) = \int_{-d}^{d} f^*g\, dx$.

[3] If $\theta^\dagger = \theta$, but $D(\theta) \neq D(\theta^\dagger)$, the operator is *not* self-adjoint. Rather it is called *symmetric* or *hermitian*.

Actually, Eq. (5.28) does not hold for operators which are not *closed.* However, we shall not meet such operators in diffusion theory.

The adjoint operator $\theta^\dagger$ has eigenfunctions and eigenvalues just as θ does:

$$\theta^\dagger \zeta_i = b_i \zeta_i . \tag{5.29}$$

Next we prove an important theorem.

Theorem 5.1 *Let ξ_k be an eigenfunction of θ corresponding to eigenvalue a_k, and let ζ_l be an eigenfunction of $\theta^\dagger$ with eigenvalue b_l. Either*

$$a_k = b_l^* \tag{5.30a}$$

or

$$(\zeta_l, \xi_k) = 0. \tag{5.30b}$$

Proof The proof is simple. Write the two eigenvalue equations

$$\theta \xi_k = a_k \xi_k , \tag{5.31a}$$

$$\theta^\dagger \zeta_l = b_l \zeta_l . \tag{5.31b}$$

Now take the scalar product, with ζ_l from the left, of Eq. (5.31*a*); then take the scalar product, with ξ_k from the right, of (5.31*b*). We obtain

$$(\zeta_l, \theta\xi_k) = a_k(\zeta_l, \xi_k), \tag{5.32a}$$

$$(\theta^\dagger \zeta_l, \xi_k) = b_l^*(\zeta_l, \xi_k). \tag{5.32b}$$

(Here we have used (5.20*a*) and (5.21) on the right-hand side of these two equations.) We note from (5.22) that the left sides of Eq. (5.32) are equal. Subtracting these, we find

$$(a_k - b_l^*)(\zeta_l, \xi_k) = 0, \tag{5.33}$$

which proves the theorem.

The practical consequence of this theorem is the following: One can associate and label the eigenfunctions of an operator and its adjoint in such a way that the eigenvalues associated with ξ_i are the complex conjugates of the eigenvalues associated with ζ_i, the eigenfunctions.[1] Then ξ_i and ζ_l are orthogonal for $i \neq l$, that is,

$$(\zeta_l, \xi_i) = 0, \qquad i \neq l. \tag{5.34}$$

[1] For the purist, we note that the operators of reactor physics do not, in general, possess a residual spectrum.

In particular, we shall assume that the eigenfunction can always be normalized, so that we may set[1]

$$(\zeta_l,\xi_i) = \delta_{il}. \tag{5.35}$$

Of great interest are *self-adjoint* operators θ, defined by

$$\theta = \theta^\dagger, \tag{5.36}$$

with the boundary conditions on θ and on $\theta^\dagger$ identical. We see from the theorem proof (5.33) that self-adjoint operators always have real eigenvalues, although, of course, non-self-adjoint operators may have real eigenvalues also. Then from (5.35) we see that self-adjoint operators generate an orthogonal set of eigenfunctions. Unfortunately, the operators of nuclear reactor theory are not self-adjoint. However, in the one-speed approximation, they are.

The adjoint operator has the following physical interpretation: Let $G(\boldsymbol{\tau}' \to \boldsymbol{\tau})$ be the Green's function of an operator θ, that is,

$$\theta G(\boldsymbol{\tau}_1 \to \boldsymbol{\tau}) = \delta(\boldsymbol{\tau} - \boldsymbol{\tau}_1); \tag{5.37a}$$

and let $G^\dagger$ be the Green's function of the adjoint,

$$\theta^\dagger G^\dagger(\boldsymbol{\tau}_2 \to \boldsymbol{\tau}) = \delta(\boldsymbol{\tau} - \boldsymbol{\tau}_2). \tag{5.37b}$$

Then we shall prove that

$$G(\boldsymbol{\tau}_1 \to \boldsymbol{\tau}_2) = G^\dagger(\boldsymbol{\tau}_2 \to \boldsymbol{\tau}_1), \tag{5.38}$$

i.e., the adjoint Green's function is obtained from the direct Green's function by interchanging source and field variables. Here $\boldsymbol{\tau}$ must be interpreted to include all coordinates upon which the operator θ may depend. In one-speed diffusion theory there are only spatial coordinates, but in energy-dependent diffusion, the energy coordinates enter as well. This explains why the energy-dependent neutron-diffusion operators are not self-adjoint. If they were, Eq. (5.38) would imply that the insertion of a higher-energy neutron into a system would yield the same distribution of thermal neutrons, say, as the distribution of fast neutrons caused by a thermal neutron. This we know is certainly not the case.[2]

Now to prove Eq. (5.38), multiply Eq. (5.37*a*) by $G^\dagger(\boldsymbol{\tau}_2 \to \boldsymbol{\tau})$, and multiply Eq. (5.37*b*) by $G(\boldsymbol{\tau}_1 \to \boldsymbol{\tau})$, integrate over $\boldsymbol{\tau}$, and subtract. If (5.22) is used, the theorem proof (5.38) follows immediately.

[1] The Kronecker delta symbol is defined by $\delta_{ik} = 0$, $i \neq k$, $\delta_{ii} = 1$.

[2] In fancy language, the general equation of neutron transport is not time-reversal invariant, although the one-speed transport equation generally is.

5.3 VARIATIONAL PRINCIPLE

Let us now turn to a consideration of calculating the eigenvalues of an operator θ,

$$\theta\xi_i = a_i\xi_i. \tag{5.39}$$

We might solve for a_i directly by taking the scalar product of Eq. (5.39) with any function, say, f, in which case we should write

$$a_i = \frac{(f,\theta\xi_i)}{(f,\xi_i)} \tag{5.40}$$

[provided, of course, that $(f,\xi_i) \neq 0$]. An equally valid expression may be obtained from the eigenvalue equation for $\theta^\dagger$,

$$\theta^\dagger\zeta_i = b_i\zeta_i, \tag{5.41}$$

whereupon, after scalar multiplication by any function g,

$$b_i = \frac{(g,\theta^\dagger\zeta_i)}{(g,\zeta_i)}, \tag{5.42a}$$

and, of course,

$$b_i = a_i^*. \tag{5.42b}$$

If, in particular, we chose the function ζ_i for f in Eq. (5.40) [or, equivalently, the function ξ_i for g in Eq. (5.41a)], we obtain an expression

$$a_i = \frac{(\zeta_i,\theta\xi_i)}{(\zeta_i,\xi_i)}, \tag{5.43}$$

which has very interesting properties. Suppose neither ζ_i nor ξ_i is known, but that we make guesses at them; call these guesses ζ_T and ξ_T (they are called *trial functions*). In particular, these trial functions differ from the correct eigenfunctions by some errors $\delta\zeta_i$ and $\delta\xi_i$,

$$\zeta_T = \zeta_i + \delta\zeta_i, \tag{5.44a}$$

$$\xi_T = \xi_i + \delta\xi_i. \tag{5.44b}$$

Then we shall calculate, not a_i, but

$$a_T = a_i + \delta a_i; \tag{5.45}$$

from Eq. (5.43)

$$a_T = \frac{(\zeta_T,\theta\xi_T)}{(\zeta_T,\xi_T)}. \tag{5.46}$$

The problem is to estimate the magnitude of δa_i, that is, to answer the question, how well can one approximate the calculation of an eigenvalue by Eq. (5.43), in which one only *guesses* at eigenfunctions?

The answer to this question is somewhat surprising. We insert Eqs. (5.44) and (5.45) into (5.46) to obtain

$$(a_i + \delta a_i)(\zeta_i + \delta\zeta_i, \xi_i + \delta\xi_i) = [\zeta_i + \delta\zeta_i, \theta(\xi_i + \delta\xi_i)]. \tag{5.47}$$

Expanding, we find

$$\begin{aligned}(a_i + \delta a_i)[(\zeta_i,\xi_i) + (\delta\zeta_i,\xi_i) + (\zeta_i,\delta\xi_i) + (\delta\zeta_i,\delta\xi_i)]\\ = [(\zeta_i,\theta\xi_i) + (\delta\zeta_i,\theta\xi_i) + (\zeta_i, \theta\,\delta\xi_i) + (\delta\zeta_i, \theta\,\delta\xi_i)].\end{aligned} \tag{5.48}$$

This expression simplifies when we note that

$$(\zeta_i,\theta\xi_i) = a_i(\zeta_i,\xi_i), \tag{5.49a}$$

$$(\delta\zeta_i,\theta\xi_i) = a_i(\delta\zeta_i,\xi_i), \tag{5.49b}$$

and

$$(\zeta_i,\theta\,\delta\xi_i) = (\theta^\dagger\zeta_i,\delta\xi_i) \tag{5.50a}$$

$$= (b_i\zeta_i,\delta\xi_i) \tag{5.50b}$$

$$= b_i^*(\zeta_i,\delta\xi_i) \tag{5.50c}$$

$$= a_i(\zeta_i,\delta\xi_i). \tag{5.50d}$$

Canceling common terms then, from the right- and left-hand sides of Eq. (5.48), we find

$$\delta a_i = \frac{(\delta\zeta_i,\theta\,\delta\xi_i) - a_i(\delta\zeta_i,\delta\xi_i)}{(\zeta_T,\xi_T)}\,. \tag{5.51}$$

Thus, the error in a_i, δa_i, is proportional to the product of the errors in the eigenfunctions. Speaking very roughly, if ζ_i and ξ_i are each known to within 10 percent, we might expect to be able to calculate a_i to within 1 percent. That is the great virtue of this so-called variational method. Frequently, a very rough guess at the eigenfunction will yield a very accurate value of the eigenvalue. Of course, one must choose the trial functions with some care; for example, they should obey any symmetry properties and boundary conditions which we know, in advance, the exact solution must obey.

Suppose one chooses two sets of trial functions, and calculates two different values for a_T. He might naturally be interested in knowing which value is closer to reality. Unfortunately, this question cannot be answered in general. However, it can be in the special case that θ is a self-adjoint operator, and one is seeking either the largest or the smallest eigenvalue.[1] Then, Eq. (5.46) becomes

$$a_T = \frac{(\xi_T, \theta\xi_T)}{(\xi_T, \xi_T)}. \tag{5.52}$$

Let us expand ξ_T in terms of the ξ_i,

$$\xi_T = \sum_i \alpha_i \xi_i. \tag{5.53}$$

Then,

$$a_T = \frac{\sum_{i,j} \alpha_i^* \alpha_j (\xi_i, \theta\xi_j)}{\sum_{i,j} \alpha_i^* \alpha_j (\xi_i, \xi_j)} \tag{5.54a}$$

$$= \frac{\sum_{i,j} \alpha_i^* \alpha_j (\xi_i, \xi_j) a_i}{\sum_{i,j} \alpha_i^* \alpha_j (\xi_i, \xi_j)} \tag{5.54b}$$

$$= \frac{\sum_i |\alpha_i|^2 a_i}{\sum_i |\alpha_i|^2}. \tag{5.54c}$$

Here we have used Eq. (5.35). From (5.54*c*) we see immediately that

$$a_{\min} \leq a_T \leq a_{\max}. \tag{5.55}$$

Thus, suppose one is looking for the lowest eigenvalue, and has calculated two values of a_T from two different trial functions. The lower value of a_T is closer to the correct value. It is common to parameterize the trial function, in which case the minimum (or maximum, as the case may be) may be found from differentiation with respect to the parameters.

As an example, let us calculate the eigenvalue ν_c of a bare, homogeneous slab reactor. The equation (cf. Sec. 3.2) obeyed by the flux is

$$\theta\phi = \nu_c \phi, \tag{5.56a}$$

where

$$\theta = -\frac{D}{\Sigma_f}\frac{d^2}{dx^2} + \frac{\Sigma_a}{\Sigma_f}. \tag{5.56b}$$

We shall use the trial function in Fig. 5.1. Note that we have used some judgment in choosing the trial function to be an even function of x and to vanish at the extrapolated end points. For simplicity in the calculation,

[1] As one generally is.

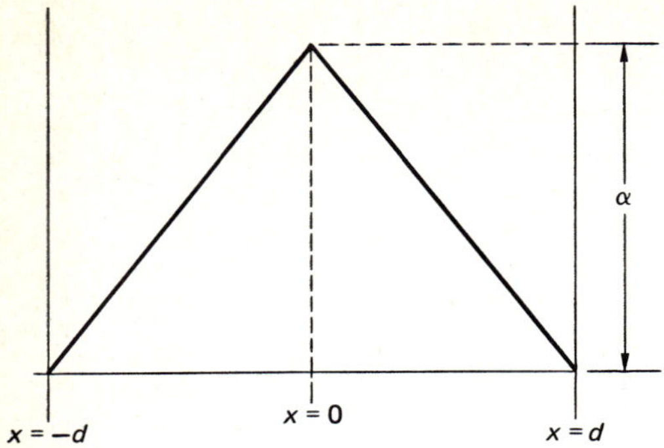

Fig. 5.1 Trial function for bare slab reactor.

however, we shall ignore the extrapolation distance correction and set $d' = d$. Then we see

$$\phi_T(x) = \begin{cases} \alpha\left(1 + \dfrac{x}{d}\right), & x < 0, \qquad (5.57a) \\ \alpha\left(1 - \dfrac{x}{d}\right), & x > 0, \qquad (5.57b) \end{cases}$$

and using the results derived in Appendix B,

$$\frac{d^2}{dx^2}\,\phi_T(x) = -\,\frac{2\alpha}{d}\,\delta(x). \tag{5.58}$$

With these functions, the variational expression

$$\nu_{cT} = \frac{\int_{-d}^{d} dx\, \phi_T(x)\theta\phi_T(x)}{\int_{-d}^{d} dx\, \phi_T{}^2\,(x)} \tag{5.59}$$

may be easily evaluated. The result is

$$\nu_{cT} = \frac{\dfrac{3}{d^2}\,D + \Sigma_a}{\Sigma_f}, \tag{5.60a}$$

which is to be compared with the exact value [Eq. (3.36)],

$$\nu_c = \frac{\dfrac{\pi^2}{4d^2}\,D + \Sigma_a}{\Sigma_f}. \tag{5.60b}$$

We see that the agreement is quite good. Note that in this case the parameter α did not enter the final expression for ν_c. If it had, we should have had an expression $\nu_c(\alpha)$. Setting the derivative $d\nu_c/d\alpha = 0$ and solving for α would have given the best value of v_c with a trial function of this type. If the extrapolated end-point correction is *not* ignored, incidentally, then d should be replaced by d' in Eqs. (5.57) and (5.58), although the integrations in (5.59) would still be carried out from $-d$ to $+d$. Also note that if the correct form of $\phi(x)$, that is, $\cos \pi x/2d'$, is used in Eq. (5.59), the correct result, that is, Eq. (5.60*b*), is obtained.

Another interesting application of the variational method is the calculation of the buckling of a reactor, i.e., the largest eigenvalue of the equation

$$\nabla^2\phi = -B^2\phi. \tag{5.61}$$

We consider a spherical reactor, in which case we have for $\chi = r\phi$ (cf. Chap. 3)

$$\frac{d^2}{dr^2}\chi = -B^2\chi. \tag{5.62}$$

Take a trial function of the form shown in Fig. 5.2. (Note that this trial function has been chosen in a somewhat intelligent fashion, since the correct eigenfunction $\chi(r) = (1/r) \sin \pi r$ vanishes at $r = 0$ and $r = R$, with, as before, the extrapolated end-point correction ignored. The value of α is as yet undetermined.)

We see that

$$\chi_T(r) = \frac{r}{\alpha}, \qquad 0 \leq r \leq \alpha, \tag{5.63a}$$

$$\chi_T(r) = 1, \qquad \alpha \leq r \leq R - \alpha, \tag{5.63b}$$

and

$$\chi_T(r) = \frac{R - r}{\alpha}, \qquad R - \alpha \leq r \leq R. \tag{5.63c}$$

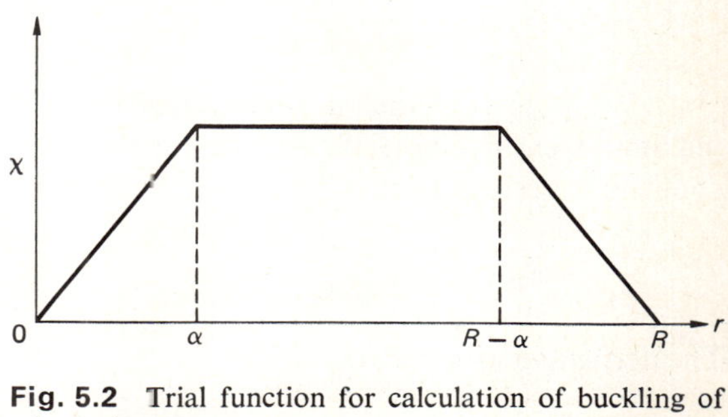

Fig. 5.2 Trial function for calculation of buckling of spherical reactor.

Since

$$\chi_T'(r) = \frac{1}{\alpha}, \qquad 0 \le r \le \alpha, \tag{5.64a}$$

$$\chi_T'(r) = 0, \qquad \alpha \le r \le R - \alpha, \tag{5.64b}$$

$$\chi_T'(r) = -\frac{1}{\alpha}, \qquad R - \alpha \le r \le R, \tag{5.64c}$$

we see that

$$\chi_T''(r) = [\chi_T'(\alpha + \varepsilon) - \chi_T'(\alpha - \varepsilon)]\, \delta(r - \alpha) + [\chi_T'(R - \alpha + \varepsilon) - \chi_T'(R - \alpha - \varepsilon)]\, \delta(r - R + \alpha). \tag{5.65}$$

(In (5.65), ε is small and positive, and the limit $\varepsilon \to 0$ is to be taken.) Thus

$$\chi_T''(r) = -\frac{1}{\alpha} [\delta(r - \alpha) + \delta(r - R + \alpha)], \tag{5.66}$$

or

$$\int_0^R \chi_T(r)\chi_T''(r)\, dr = -\frac{2}{\alpha}. \tag{5.67}$$

One easily calculates

$$\int_0^R [\chi_T(r)]^2\, dr = \int_0^\alpha \frac{r^2}{\alpha^2}\, dr + \int_\alpha^{R-\alpha} dr + \int_{R-\alpha}^R \frac{(R - r)^2}{\alpha^2}\, dr \tag{5.68}$$

$$= R - \frac{4\alpha}{3}. \tag{5.69}$$

Thus

$$-B_T{}^2 = \frac{\int_0^R \chi_T(r)\chi_T''(r)\, dr}{\int_0^R [\chi_T(r)]^2\, dr} \tag{5.70}$$

or

$$-B_T{}^2(\alpha) = -\frac{2}{R\alpha - 4/3\alpha^2}. \tag{5.71}$$

We note that the result depends upon α. Since we are seeking the maximum eigenvalue (i.e., the minimum value of $B_T{}^2$), we maximize this result with respect to α.

$$\frac{\partial}{\partial \alpha} [-B_T{}^2(\alpha)] = \frac{2(R - 8\alpha/3)}{(R\alpha - 4\alpha^2/3)^2}. \tag{5.72}$$

The derivative vanishes for

$$\alpha = \alpha_m = \tfrac{3}{8}R. \tag{5.73}$$

Then

$$-B_T^{\,2}(\alpha_m) = -\frac{32}{3R^2} = -\frac{10.67}{R^2}, \tag{5.74}$$

which is to be compared with the exact value

$$-B^2 = -\frac{\pi^2}{R^2} = -\frac{9.86}{R^2}. \tag{5.75}$$

The variational principle may be used conveniently to compute the criticality of a reactor with nonuniform material properties. Suppose, for example, the moderator and poison concentrations are constant, but fuel loading is modified in some way in an attempt to flatten the power. Then we have, approximately,

$$D = \text{constant}, \tag{5.76a}$$

$$\Sigma_f = \Sigma_f(x), \tag{5.76b}$$

$$\Sigma_a = \Sigma_{am} + \alpha\,\Sigma_f(x). \tag{5.76c}$$

(The contribution of the uranium absorption to D is being neglected.) Then

$$\nu_{cT} = \frac{\int dx\, \phi_T(x)\left[-\dfrac{D}{\Sigma_f(x)}\dfrac{d^2}{dx^2} + \dfrac{\alpha\Sigma_f(x) + \Sigma_{am}}{\Sigma_f(x)}\right]\phi_T(x)}{\int dx\, [\phi_T(x)]^2} \tag{5.77}$$

and the integrations can be carried out for any choice of the functions $\Sigma_f(x)$ and for any choice of trial function. In this case, if the fuel loading is not too different from uniform, one might intelligently choose the uniform flux distribution $\cos(\pi x/2d')$. On the other hand, if the fuel loading increases rapidly toward the outer edge of the core, common sense would dictate a trial function more flattened in the center. In any event, it should be chosen to be symmetric about $x = 0$ and to vanish at the (extrapolated) boundary of the system.[1]

It must be obvious how this method would be applied in other geometries and to other material distributions, so we shall not pursue the matter any further.

5.4 PERTURBATION THEORY

A most convenient application of the variational principle derived in the last section may be made when the operator to be treated differs only slightly from another operator whose eigenfunctions and eigenvalues are known.

[1] Assuming the fuel distribution is symmetric.

For example, this might apply when a bare, homogeneous reactor is *perturbed* by the addition of a small amount of absorber material at one position (or more) in the reactor. We then write the operator θ as the sum of the operator θ_0, whose eigenfunctions and eigenvalues, as mentioned above, are known, and the perturbation $\delta\theta$:

$$\theta = \theta_0 + \delta\theta, \tag{5.78a}$$

where

$$\theta_0 \xi_k^{(0)} = a_k^{(0)} \xi_k^{(0)}. \tag{5.78b}$$

In addition, of course, one knows the eigenfunctions and eigenvalues of $\theta_0^\dagger$,

$$\theta_0^\dagger \zeta_k^{(0)} = a_k^{*(0)} \zeta_k^{(0)}. \tag{5.78c}$$

We wish to find the eigenvalues a_k of θ,

$$\theta \xi_k = a_k \xi_k. \tag{5.79}$$

Using the variational principle developed in the previous section to solve for a_k, we have

$$a_k \approx \frac{(\zeta_k^{(0)}, \theta \xi_k^{(0)})}{(\zeta_k^{(0)}, \xi_k^{(0)})}; \tag{5.80}$$

in other words, we are using the eigenfunctions and adjoint eigenfunctions of the unperturbed problem as trial functions in the calculation of a_k.

If we use Eqs. (5.78) and (5.80), we obtain

$$a_k - a_k^{(0)} \approx \frac{(\zeta_k^{(0)}, \delta\theta \xi_k^{(0)})}{(\zeta_k^{(0)}, \xi_k^{(0)})}. \tag{5.81}$$

Equation (5.81) is the basic formula of perturbation theory. It describes the change in an eigenvalue due to a perturbation of the system, i.e., a change in the operator which gave rise to the original eigenvalue. If we write $a_k - a_k^{(0)} = \delta a_k$, we have

$$\delta a_k = \frac{(\zeta_k^{(0)}, \delta\theta \xi_k^{(0)})}{(\zeta_k^{(0)}, \xi_k^{(0)})}. \tag{5.82}$$

Since the first-order perturbation theory, as described here, is derived from a variational principle, it is quite accurate even for fairly "large" perturbations $\delta\theta$.

In applying perturbation theory, it is convenient to consider, not ν_c as the eigenvalue, but rather Λ in Eq. (3.4).[1]

$$\theta\phi = \Lambda\phi, \tag{5.83}$$

[1] This approach may also be used in the variational method. The advantage is that one need not form differentials of quantities like D/Σ_f or Σ_a/Σ_f.

where

$$\theta = \bar{v}(-\nabla \cdot D\nabla + \Sigma_a - \nu\Sigma_f). \tag{5.84}$$

If we interpret Λ as the smallest eigenvalue of (5.83), then Λ^{-1} represents the stable period, and it is related to ρ through the inhour equation (4.34) or the approximate formula (4.47). If one wants to avoid solving the kinetic equations, then ν_c must be used as the eigenvalue, at the expense of a somewhat more complicated variational expression (see footnote, page 120).

From (5.84) we obtain

$$\delta\theta = \frac{\delta\bar{v}}{\bar{v}}\,\theta^{(0)} + \bar{v}(-\nabla \cdot \delta D\nabla + \delta\Sigma_a - \nu\,\delta\Sigma_f). \tag{5.85}$$

We must allow for the possibility that $\bar{v}$, the average neutron speed, might change for example, by a change in reactor temperature.

Then Eq. (5.82) gives (recall that $\theta = \theta^{\dagger}$ in the one-speed form)

$$\delta\Lambda = \Big[(\delta\bar{v}/\bar{v}) \int \phi^{(0)}(\mathbf{r})\,\theta^{(0)}\phi^{(0)}(\mathbf{r})\,d^3r + \bar{v}\int \phi^{(0)}(\mathbf{r})\,(-\nabla \cdot \delta D\nabla + \delta\Sigma_a - \nu\,\delta\Sigma_f)\phi^{(0)}(\mathbf{r})\,d^3r\Big]\Big/\int[\phi^{(0)}(\mathbf{r})]^2\,d^3r. \tag{5.86}$$

The first two terms can be simplified since

$$\int \phi^{(0)}\theta\phi^{(0)}\,d^3r = \Lambda^{(0)}\int (\phi^{(0)})^2\phi^{(0)}\,d^3r, \tag{5.87}$$

and after a partial integration in which the surface terms are seen to vanish,

$$\int \phi^{(0)}\nabla \cdot \delta D\nabla\phi^{(0)}\,d^3r = -\int (\nabla\phi^{(0)})^2\,\delta D\,d^3r. \tag{5.88}$$

Then

$$\delta\Lambda = \frac{\Lambda^{(0)}\,\delta\bar{v}}{\bar{v}} + \bar{v}\,\frac{\int[(\bar{v}\phi^{(0)})^2\,\delta D + (\phi^{(0)})^2(\delta\Sigma_a - \nu\,\delta\Sigma_f)]\,d^3r}{\int[\phi^{(0)}(\mathbf{r})]^2\,d^3r}. \tag{5.89}$$

If we have perturbed a *critical* reactor, $\Lambda^{(0)} = 0$, in which case the first term on the right-hand side of (5.89) will be absent, of course.

Actually, $\delta\Lambda$ may be related to ρ in a more convenient manner through the use of the inhour formula if $\delta\Lambda$ is indeed small (so that T is large). In this case $k \approx 1$, and from Eq. (4.34) we see that

$$\rho \approx \frac{1}{T}\left(l_{\text{eff}} + \sum_i \frac{\beta_i}{\lambda_i}\right), \tag{5.90}$$

an expression sometimes called the *small reactivity inhour formula*. The quantity

$$l_{\text{eff}} + \sum_i \frac{\beta_i}{\lambda_i},$$

is approximately the mean neutron lifetime including the effect of delayed neutrons, and as we have already seen in Chap. 4, the first term is generally negligible compared with the second; so Eq. (5.90) becomes, to a quite good approximation,

$$\rho = \sum_i \frac{\beta_i}{\lambda_i T}. \tag{5.91}$$

If our perturbation formula (5.89) refers to a reactor initially critical, then

$$\delta\Lambda = \Lambda = -\frac{1}{T} \tag{5.92}$$

and we have, finally,[1]

$$\frac{\rho}{\beta} = -\frac{\bar{v}}{\lambda}\frac{\int[(\nabla\phi^{(0)})^2\,\delta D + (\phi^{(0)})^2(\delta\Sigma_a - v\,\delta\Sigma_f)]\,d^3r}{\int(\phi^{(0)})^2\,d^3r}, \tag{5.93}$$

where

$$\frac{1}{\lambda} = \frac{1}{\beta}\sum_i \frac{\beta_i}{\lambda_i}.$$

Sometimes the term *reactivity coefficient* is used. (This was formerly called the "danger coefficient," but in the interest of good public relations the former term is preferable.) This merely refers to the reactivity induced in a critical reactor, generally by a small sample of material placed at a given point. The units are usually cents per mole. We can see this as follows: If the sample, of volume δV, is inserted at $\mathbf{r}_0$ and we assume ϕ_0 is constant over δV, we have

$$\frac{\rho}{\beta} = \$ = \frac{\bar{v}}{\lambda}\frac{\{-\delta D\,[\nabla\phi^{(0)}(\mathbf{r}_0)]^2 + [v\bar{\Sigma}_f - \bar{\Sigma}_a)(\phi^{(0)}(\mathbf{r}_0)]^2\}\delta V}{\int(\phi^{(0)})^2\,d^3r}, \tag{5.94}$$

where we are using δV, $\bar{D}$, $\bar{\Sigma}_f$, and $\bar{\Sigma}_a$ now to represent the quantities associated with the sample. If the scattering of the sample is negligible, the first term can be ignored. Then,

$$\bar{\Sigma} = N\bar{\sigma} \tag{5.95}$$

$$= \frac{L_0 m}{\delta V}\bar{\sigma}, \tag{5.96}$$

[1] In the non-self-adjoint case, $(\nabla\phi^{(0)})^2 \to \nabla\phi^{(0)\dagger} \cdot \nabla\phi^{(0)}$ and $(\phi^{(0)})^2 \to \phi^{(0)\dagger}\phi^{(0)}$. Otherwise the formula is the same except that, of course, in the energy-dependent case additional perturbations are possible.

where, as before, L_0 is Avogadro's number, and m represents the number of moles. Thus

$$RC = \frac{\$}{m} = \frac{\bar{v}L_0}{\lambda} \frac{(\nu\bar{\sigma}_f - \bar{\sigma}_a)[\phi^{(0)}(\mathbf{r}_0)]^2}{\int [\phi^{(0)}(\mathbf{r})]^2 \, d^3r} . \tag{5.97}$$

Sometimes the RC is calculated by performing two calculations of ν_c or k, one with the sample in place and one without it. The method described here is preferable, however, since it does not require finding small differences of large numbers.

We note from Eq. (5.97) that, as expected, an increase in fission cross section leads to a positive reactivity, while an increase in absorption cross section has the opposite effect. Also, we see that an increase in D *always* decreases reactivity, since it increases leakage.[1]

5.5 ITERATED FISSION PROBABILITY

It is of some interest to ask the question, if a neutron is inserted into a critical system at a particular point, what is the subsequent behavior of the reactor. Clearly, the material properties of the reactor do not change, so the system is still critical. Thus, the additional neutron introduces some time transients into the system, which eventually decay away and so leave a steady-state situation again at a higher flux (or power) level.

Actually, we can answer the question of how high this rise in power level is in a somewhat more general context than that of one-speed reactor theory. Thus, suppose any linear system is described by a steady-state equation of the form

$$M\psi = 0, \tag{5.98}$$

where the linear operator M and its adjoint $M^\dagger$ have eigenfunctions ξ_n and ζ_n,

$$M\xi_n = \Lambda_n \xi_n, \tag{5.99a}$$

$$M^\dagger \zeta_n = \Lambda_n^* \zeta_n . \tag{5.99b}$$

We now consider the response of this system to a unit impulse,

$$\left(\frac{1}{\bar{v}} \frac{\partial}{\partial t} + M\right)\psi(t) = \delta(t)\,\delta(\mathbf{w} - \mathbf{w}_0), \tag{5.100}$$

where the vector $\mathbf{w}$ represents all the variables upon which ψ may depend.[2]

[1] This statement is not true for the energy-dependent case.

[2] In one-speed reactor theory, $\mathbf{w} = \mathbf{r}$. However, in energy-dependent theory $\mathbf{w} = (\mathbf{r}, E)$, and so forth.

We expand the solutions to this equation in terms of the eigenfunctions,

$$\psi(t) = \sum_n \alpha_n(t)\xi_n(\mathbf{w}), \tag{5.101}$$

so that

$$\sum_n \xi_n \left(\frac{\dot{\alpha}_n}{\bar{v}} + \Lambda_n \alpha_n\right) = \delta(t)\, \delta(\mathbf{w} - \mathbf{w}_0). \tag{5.102}$$

Multiplying by ζ_m and integrating over $d\mathbf{w}$ gives [with $(\zeta_m, \xi_n) = \delta_{mn}$]

$$\frac{\dot{\alpha}_m}{\bar{v}} + \Lambda_m \alpha_m = \delta(t)\, \zeta_n(\mathbf{w}_0). \tag{5.103}$$

The Laplace transform of this equation is (transform variable $= p$)

$$\left(\frac{p}{v} + \Lambda_m\right) \alpha_m(p) = \zeta_n(\mathbf{w}_0) + \frac{1}{v}\alpha_m(0), \tag{5.104}$$

so that

$$\alpha_m(t) = [\bar{v}\zeta_n(\mathbf{w}_0) + \alpha_m(0)]e^{-\Lambda_m \bar{v} t} \tag{5.105}$$

and, as $t \to \infty$,

$$\psi(t) \to [\bar{v}\zeta_0(\mathbf{w}_0) + \alpha_0(0)]e^{-\Lambda_0 \bar{v} t}\xi_0(\mathbf{w}), \tag{5.106}$$

where Λ_0 is the smallest eigenvalue. Clearly, since the system is initially (and after a long time) in a steady state, $\Lambda_0 = 0$ and

$$\phi(t) = \psi_0 + \bar{v}\zeta_0(\mathbf{w}_0)\xi_0(\mathbf{w}), \tag{5.107}$$

where we have replaced $\alpha_0(0)\xi_0$ with ψ_0. The increase in this is then seen to be proportional to the adjoint $\zeta_0(\mathbf{w}_0)$ evaluated at the (phase-space) point at which the impulse was introduced.

The one-speed reactor equation is, of course, self-adjoint. Thus, the increase in the flux due to the insertion of a neutron at point $\mathbf{r}_0$ is proportional to $\phi^{(0)}(\mathbf{r}_0)$.[1]

Because of this property, the adjoint is sometimes referred to as the *importance* function. If a neutron is inserted into a reactor in a place where the adjoint is large (small), the corresponding power increase will be large (small) so that the neutron is of greater (lesser) *importance*. (Needless to say, the above analysis could be carried out for the removal of a neutron from a

[1] In the (non-self-adjoint) energy-dependent reactivity, the increase we see is proportional to the adjoint flux $\phi^{(0)\dagger}(\mathbf{w}_0)$. In the most general transport theoretical case, if a neutron is inserted with a velocity $\mathbf{v}_0$ at position $\mathbf{r}_0$, the increase in the angular density ψ will be proportional to the adjoint angular density $\psi^\dagger(\mathbf{r}_0, \mathbf{v}_0)$. This quantity obeys the adjoint transport equation.

reactor, in which case there would be a net *decrease* in power level. Also, note that delayed-neutron effects have not been included in the above analysis. This has been for simplicity only. The analysis could include delayed neutrons and be carried through with identical conclusions.)

Hurwitz[1] has normalized the adjoint in a specific way and labeled it the *iterated fission probability*. The fission rate N_F in a reactor (which is of course proportional to the power) is given by

$$N_F = \int \Sigma_f(\mathbf{r})\phi(\mathbf{r})\, d^3r. \tag{5.108}$$

Before the neutron is inserted, we have

$$N_{F0} = \alpha_0(0) \int \Sigma_f(\mathbf{r})\xi_0(\mathbf{r})\, d^3r, \tag{5.109a}$$

while much later

$$N_{F\infty} = [\bar{v}\zeta_0(\mathbf{r}_0) + \alpha_0(0)] \int \Sigma_f(\mathbf{r})\xi_0(\mathbf{r})\, d^3r. \tag{5.109b}$$

The *change* in fission rate, ΔN_F, is given by

$$\Delta N_F = \bar{v}\zeta_0(\mathbf{r}_0) \int \Sigma_f(\mathbf{r})\xi_0(\mathbf{r})\, d^3r, \tag{5.110}$$

or

$$\frac{\Delta N_F}{N_F} = \frac{\bar{v}\zeta_0(\mathbf{r}_0)}{\alpha_0(0)}. \tag{5.111}$$

These ideas were used by Hurwitz[2] to develop perturbation theory in a particularly elegant fashion.

In a critical reactor, the adjoint ζ_0 should be normalized such that

$$\int \nu\Sigma_f \phi\zeta_0\, d^3r = \int \Sigma_f \phi\, d^3r \tag{5.112}$$

since every fission should give rise to *one* subsequent fission. When ζ_0 is normalized such that Eq. (5.112) is obeyed, it is called by Hurwitz the iterated fission probability, and is denoted by $F(\mathbf{r})$. Explicitly

$$F(\mathbf{r}) = \frac{\zeta_0(\mathbf{r}) \int \Sigma_f(\mathbf{r}')\phi(\mathbf{r}')\, d^3r}{\nu \int \Sigma_f(\mathbf{r}')\phi(\mathbf{r}')\zeta_0(\mathbf{r}')\, d^3r'}. \tag{5.113}$$

In terms of $F(\mathbf{r})$, a perturbation formula can be written down immediately. For example, suppose an absorber is inserted into a reactor at position $\mathbf{r}_0$. This is equivalent to a negative source of strength $\Sigma_a(\mathbf{r}_0)\phi(\mathbf{r}_0)$, and the

[1] H. Hurwitz, Jr.: Note on the Theory of Danger Coefficient, *KAPL* 98, September, 1948.
[2] *Ibid.*

reduction in the fission rate due to this absorber will be $\Sigma_a(\mathbf{r}_0)\phi(\mathbf{r}_0)F(\mathbf{r}_0)$. Thus

$$\Delta k = -\Sigma_a(\mathbf{r}_0)\phi(\mathbf{r}_0)F(\mathbf{r}_0), \tag{5.114}$$

and so forth, for various other perturbations.

5.6 SUPERVARIATIONAL PRINCIPLE

In Secs. 5.3 and 5.4 we considered computation of eigenvalues, first using a variational principle and then using perturbation theory to obtain appropriate trial functions to use in the variational scalar products. It is possible to obtain very accurate estimates of all the eigenvalues of an operator by combining these techniques.

Consider the separation of θ as given in (5.78*a*). Let us assume that the "small" operator $\delta\theta$ can be written

$$\delta\theta = \varepsilon\theta_1, \tag{5.115}$$

where ε is small and a constant, and the operator θ_1 is comparable to θ_0 [that is, the size of $(\zeta,\theta_0\xi)$ is comparable to $(\zeta,\theta_1\xi)$]. We can expand the eigenfunctions of θ and $\theta^\dagger$ in terms of the eigenfunctions of θ_0 and $\theta_0^\dagger$ [cf. (5.87) and (5.79)],

$$\xi_k = \sum_{k'} b_{kk'}\,\xi_{k'}{}^{(0)}, \tag{5.116a}$$

$$\xi_k = \sum_{k'} c_{kk'}\,\zeta_{k'}{}^{(0)}. \tag{5.116b}$$

Assume we can expand $b_{kk'}$ and $c_{kk'}$ in a power series in the small parameter ε,

$$b_{kk'} = \sum_{n=0}^{\infty} \varepsilon^n b_{kk'}^{(n)}, \tag{5.117a}$$

$$c_{kk'} = \sum_{n=0}^{\infty} \varepsilon^n c_{kk'}^{(n)}. \tag{5.117b}$$

The first terms in (5.117*a*) and (5.117*b*) must give the eigenvectors of θ_0 and $\theta_0^\dagger$ so that

$$b_{kk'}^{(0)} = c_{kk'}^{(0)} \tag{5.118a}$$

$$= \delta_{kk'}. \tag{5.118b}$$

Suppose we have computed ξ_k and ζ_k correctly to order ε^N. That is, we have approximate eigenfunctions

$$\xi_k{}^{(N)} = \sum_{n=0}^{N}\sum_{k'} b_{kk'}^{(n)}\varepsilon^n\xi_{k'}{}^{(0)} + \sum_{n=N+1}^{\infty}\sum_{k'} B_{kk'}^{(n)}\varepsilon^n\xi_{k'}{}^{(0)}, \tag{5.119a}$$

$$\zeta_k{}^{(N)} = \sum_{n=0}^{N}\sum_{k'} c_{kk'}^{(n)}\varepsilon^n\zeta_{k'}{}^{(0)} + \sum_{n=N+1}^{\infty}\sum_{k'} C_{kk'}^{(n)}\varepsilon^n\zeta_{k'}{}^{(0)}. \tag{5.119b}$$

In Eqs. (5.119*a*) and (5.119*b*), we have assumed that we know the eigenvalues and eigenfunctions *correctly* to order ε^N. Contributions of order ε^{N+1} and higher can be present ($B_{kk'}^{(n)}$ and $C_{kk'}^{(n)}$ are not generally zero); however, they are not correct,

$$(B_{kk'}^{(n)} \neq b_{kk'}^{(n)}, \qquad C_{kk'}^{(n)} \neq c_{kk'}^{(n)}).$$

Comparing (5.117) and (5.119), we see that we can write the eigenvectors and eigenvalues as the $\zeta_k^{(n)}$ and $\xi_k^{(n)}$ plus corrections for the higher orders in ε,

$$\xi_k = \xi_k^{(N)} + \sum_{n=N+1}^{\infty} \sum_{k'} b_{kk'}^{\prime(n)} \varepsilon^n \xi_{k'}^{(0)}, \tag{5.120a}$$

$$\zeta_k = \zeta_k^{(N)} + \sum_{n=N+1}^{\infty} \sum_{k'} c_{kk'}^{\prime(n)} \varepsilon^n \zeta_{k'}^{(0)}, \tag{5.120b}$$

where

$$b_{kk'}^{\prime(n)} = b_{kk'}^{(n)} - B_{kk'}^{(n)}, \tag{5.121a}$$

$$c_{kk'}^{\prime(n)} = c_{kk'}^{(n)} - C_{kk'}^{(n)}. \tag{5.121b}$$

The eigenvalue equation (5.79) for ξ_k can be written

$$(\theta_0 + \varepsilon\theta_1)\left(\xi_k^{(N)} + \sum_{n=N+1}^{\infty} \sum_{k'} \varepsilon^n b_{kk'}^{\prime(n)} \xi_{k'}^{(0)}\right)$$
$$= a_k\left(\xi_k^{(N)} + \sum_{n=N+1}^{\infty} \sum_{k'} \varepsilon^n b_{kk'}^{\prime(n)} \xi_{k'}^{(0)}\right). \tag{5.122a}$$

Similarly the eigenvalue equation for ζ_k can be written

$$(\theta_0^\dagger + \varepsilon\theta_1^\dagger)\left(\zeta_k^{(N)} + \sum_{n=N+1} \sum_{k'} \varepsilon^n c_{kk'}^{\prime(n)} \zeta_{k'}^{(0)}\right)$$
$$= a_k^*\left(\zeta_k^{(N)} + \sum_{n=N+1}^{\infty} \sum_{k'} c_{kk'}^{\prime(n)} \zeta_{k'}^{(0)}\right). \tag{5.122b}$$

Suppose we wish to improve our approximation for the eigenvectors. We can do this by writing equations which are the same as (5.122*a*) and (5.122*b*) to order ε^{N+1}. Then, to order ε^{N+1}, we can write (5.122*a*) as

$$(\theta_0 + \varepsilon\theta_1)\left(\xi_k^{(N)} + \varepsilon^{N+1} \sum_{k'} b_{kk'}^{\prime(N+1)} \xi_{k'}^{(N)}\right)$$
$$= a_k\left(\xi_k^{(N)} + \varepsilon^{N+1} \sum_{k'} b_{kk'}^{\prime(N+1)} \xi_{k'}^{(N)}\right). \tag{5.123a}$$

Here we have thrown away a number of terms of order greater than ε^{N+1} and used the fact that $\varepsilon^{N+1}\xi_{k'}^{(0)}$ is equal to $\varepsilon^{N+1}\xi_{k'}^{(n)}$ to order $N + 1$. Thus (5.123*a*) is the same as (5.122*a*) to terms of order ε^{N+1}. Both equations contain terms of higher order in ε, and these terms are different for each equation. The important point to note here is that the solution of (5.123*a*) will be correct to order ε^{N+1}.

Similarly we can approximate (5.122*b*) to order ε^{N+1} as

$$(\theta_0^\dagger + \varepsilon\theta_1^\dagger)\left(\zeta_k{}^{(N)} + \varepsilon^{N+1}\sum_{k'} c_{kk'}^{\prime(N+1)}\zeta_{k'}{}^{(N)}\right)$$

$$= a_k^*\left(\zeta_k{}^{(N)} + \varepsilon^{N+1}\sum_{k'} c_{kk'}^{\prime(N+1)}\zeta_{k'}{}^{(N)}\right). \qquad (5.123b)$$

We are going to take the scalar product of both sides of (5.123*a*) with $\zeta_{k_0}{}^{(N)}$. There are a number of points we must consider first. We have shown in Sec. 5.3 that the variational eigenvalue estimate is correct to second order in the errors in the eigenfunctions. Thus the eigenvalue estimate

$$a_{kN} \equiv \frac{(\zeta_k{}^{(N)},\theta\xi_k{}^{(N)})}{(\zeta_k{}^{(N)},\xi_k{}^{(N)})} \qquad (5.124)$$

is correct to order ε^{2N}. Also consider the scalar product $(\zeta_{k''}{}^{(N)},\theta\xi_{k'}{}^{(N)})$. We shall need this quantity to order ε^0 for $k' \neq k''$,

$$(\zeta_{k''}{}^{(N)},\theta\zeta_{k'}{}^{(N)}) = (\zeta_{k''}{}^{(0)},\theta_0\,\xi_{k'}{}^{(0)}) + 0(\varepsilon) \qquad (5.125a)$$

$$= a_{k'}{}^{(0)}\,\delta_{k'k''}(\zeta_{k'}{}^{(0)},\,\xi_{k'}{}^{(0)}) + 0(\varepsilon) \qquad (5.125b)$$

$$= a_{k'N}\,\delta_{k'k''}(\zeta_{k'}{}^{(N)},\xi_{k'}{}^{(N)}) + 0(\varepsilon). \qquad (5.125c)$$

Now we can take the scalar product of (5.123*a*) with $\zeta_{k_0}{}^{(N)}$ $(k_0 \neq k)$,

$$(\zeta_{k_0}{}^{(N)},\theta\xi_k{}^{(N)}) + \varepsilon^{N+1}\sum_{k'} b_{kk'}^{\prime(N+1)}(\zeta_{k_0}{}^{(N)},\theta\xi_{k'}{}^{(N)})$$

$$= a_k[(\zeta_{k_0}{}^{(N)},\xi_k{}^{(N)}) + \varepsilon^{N+1}\sum_{k'} b_{kk'}^{\prime(N+1)}(\zeta_{k_0}{}^{(N)},\xi_{k'}{}^{(N)})]. \qquad (5.126)$$

Of course the $\zeta_{k_0}{}^{(N)}$ are orthogonal to $\xi_k{}^{(N)}$ for $k_0 \neq k$.

To order ε^{2N}, we can replace a_k with a_{kN} [see (5.124)]. In the term containing the factor ε^{N+1}, we only need to compute the scalar product to order ε^0, so that we can use (5.125*c*). With these, (5.126) can be written, to order ε^{N+1},

$$(\zeta_{k_0}{}^{(N)},\theta\xi_k{}^{(N)}) + \varepsilon^{N+1}b_{kk_0}^{\prime(N+1)}a_{k_0n}(\zeta_{k_0}{}^{(N)},\xi_{k_0}{}^{(N)})$$

$$= a_{kN}\,\varepsilon^{N+1}b_{kk_0}^{\prime(N+1)}(\zeta_{k_0}{}^{(N)},\xi_{k_0}{}^{(N)}). \qquad (5.127)$$

A solution of (5.127) is

$$\varepsilon^{N+1}b_{kk_0}^{\prime(N+1)} = \frac{(\zeta_{k_0}{}^{(N)},\theta\xi_k{}^{(N)})}{a_{kN}(\zeta_k{}^{(N)},\xi_k{}^{(N)}) - a_{k_0N}(\zeta_{k_0}{}^{(N)},\xi_{k_0}{}^{(N)})}, \qquad k \neq k_0. \qquad (5.128)$$

Any value for $b'^{(N+1)}_{kk}$ will give a solution correct to order ε^{N+1}. Zero is the simplest choice. Of course, we must be careful; Eq. (5.128) is a solution only if

$$a_{kN} \neq a_{k_0N}. \tag{5.129}$$

Thus when (5.129) is satisfied, we can use (5.128) and (5.124*a*) to find the eigenvector accurate to ε^{N+1},

$$\xi_k^{(N+1)} = \xi_k^{(N)} + \sum_{k'\neq k} \frac{(\zeta_{k'}^{(N)},\theta\xi_k^{(N)})\xi_{k'}^{(N)}}{a_{kN}(\zeta_{k'}^{(N)},\xi_{k'}^{(N)}) - a_{k'N}(\zeta_{k'N}^{(N)},\xi_{k'}^{(N)})}. \tag{5.130a}$$

In the last term in (5.130*a*) we have used the fact that $\varepsilon^{N+1}\xi_{k'}^{(0)} = \varepsilon^{N+1}\xi_{k'}^{(N)}$ to order ε^{N+1}.

In an identical calculation we can use (5.122*b*) to obtain

$$\zeta_k^{(N+1)} = \zeta_k^{(N)} + \sum_{k'\neq k} \frac{(\xi_{k'}^{(N)},\theta^\dagger\zeta_k^{(N)})\zeta_{k'}^{(N)}}{a_{kN}^*(\xi_k^{(N)},\zeta_k^{(N)}) - a_{k'N}^*(\xi_{k'}^{(N)},\zeta_{k'}^{(N)})}. \tag{5.130b}$$

To apply this technique, we need the initial eigenfunction estimates, say, the eigenfunctions of θ_0 and $\theta_0^\dagger$. Then by using the variational estimate, the eigenvalue of θ can be determined (5.124); then improved estimates of the eigenfunction can be determined from (5.130*a*) and (5.130*b*), and the iteration can be carried out to arbitrary order. Convergence should be fairly rapid since for the *N*th step in the iteration the eigenvalue estimate is accurate to order ε^{2N}.

5.7 VARIATIONAL PRINCIPLES FOR INHOMOGENEOUS EQUATIONS[1]

The variational principle, as described in Sec. 5.3, is used to estimate eigenvalues. Putting it in another way, it is applied to homogeneous equations. It is also possible to develop a variational principle for *inhomogeneous* equations. Such variational principles are increasingly being applied to problems in reactor physics.

Consider, then, an inhomogeneous equation of the form

$$\theta\phi(\omega) = f(\omega), \tag{5.131}$$

where, as before, θ is a linear operator, and ω represents all the variables on which the flux ϕ may depend. The function $f(\omega)$ is known (for example, it may represent neutron sources). The solution $\phi(\omega)$ may be obtained if the inverse operator θ^{-1} can be determined, for example, by the numerical methods

[1] The authors are indebted to Dr. A. G. Bullard for considerable help in the preparation of this section. The variational was first published by P. Roussopolous: *Compt. Rend.*, **236**, 1858 (1953).

described in Chap. 6. On the other hand, a reactor designer may be interested in obtaining certain information, for example, the total fission rate in the reactor, to a good approximation, and at the same time wish to avoid additional flux calculations. The procedure is to consider the adjoint inhomogeneous equation

$$\theta^\dagger \phi^\dagger(\omega) = g(\omega), \tag{5.132}$$

where g is, at the moment, an arbitrary source. Later we shall see how to choose it to suit our needs.

Consider a quantity J defined to be

$$J = -(\phi^\dagger,\theta\phi) + (\phi^\dagger,f) + (g,\phi). \tag{5.133}$$

J is referred to as a *functional of the flux and adjoint flux* ϕ and $\phi^\dagger$. At this point, it is not at all clear what would be the use of (5.133). To see this we can use (5.131) in the first term of (5.133) to obtain

$$J = -\,(\phi^\dagger,f) + (\phi^\dagger,f) + (g,\phi) \tag{5.134a}$$

$$= (g,\phi). \tag{5.134b}$$

We recall that g was the source term for $\phi^\dagger$. Thus, we can see the utility of J and, at the same time, provide a choice for g. For example, suppose we wish to know the absorption rate in U^{238}. Then by taking g to be the absorption cross section for U^{238}, J would be the absorption rate in U^{238}

Also we could write (5.133), by using the definition of $\theta^\dagger$, as

$$J = (\theta^\dagger\phi^\dagger,\phi) + (\phi^\dagger,f) + (g,\phi). \tag{5.135a}$$

Then we can use (5.132) to obtain

$$J = (\phi^\dagger,f). \tag{5.135b}$$

We have seen from (5.134*b*) that J can represent physically meaningful quantities. However, we might well ask why we have used this rather devious method for introducing expressions for these quantities. The reason is that J is variationally stationary to second order. That is, we can compute J by using approximations for $\phi^\dagger$ and ϕ in Eq. (5.133), and the error in J will be second order in the error in $\phi^\dagger$ and ϕ.

To prove this, we can introduce trial functions ϕ_T and $\phi_T^\dagger$ such that

$$\phi_T = \phi + \delta\phi \tag{5.136a}$$

and

$$\phi_T^\dagger = \phi^\dagger + \delta\phi^\dagger. \tag{5.136b}$$

Then the error in J, namely, δJ, will be proportional to the product $\delta\phi\ \delta\phi^\dagger$.

To prove this, we calculate

$$J + \delta J = -(\phi_T^\dagger, \theta\phi_T) + (\phi_T^\dagger, f) + (g, \phi_T) \tag{5.137a}$$

$$= -(\phi^\dagger + \delta\phi^\dagger, \theta[\phi + \delta\phi]) + (\phi^\dagger + \delta\phi^\dagger, f) + (g, \phi + \delta\phi). \tag{5.137b}$$

Multiplying and using (5.134*a*), we see that

$$\delta J = -(\phi^\dagger, \theta\,\delta\phi) - (\delta\phi^\dagger, \theta\phi) - (\delta\phi^\dagger, \theta\,\delta\phi) + (\delta\phi^\dagger, f) + (g, \delta\phi). \tag{5.137c}$$

In the first term above, we note that

$$(\phi^\dagger, \theta\,\delta\phi) = (\theta^\dagger\phi^\dagger, \delta\phi), \tag{5.138a}$$

$$(\phi^\dagger, \theta\,\delta\phi) = (g, \delta\phi), \tag{5.138b}$$

where the definitions of the adjoint, (5.22) and (5.132), have been used. Thus, the first term in (5.137*c*) cancels the last. Similarly, the second term cancels the next to last, and we write

$$\delta J = -(\delta\phi^\dagger, \theta\,\delta\phi), \tag{5.139}$$

which proves that J is variationally stationary.

An example of the application of this method has already been mentioned, the estimation of the total fission rate in the reactor. We note, for Eq. (5.134*b*), that (introducing an integration as the definition of scalar product)

$$J = \int d\omega g(\omega)\phi(\omega), \tag{5.140}$$

where we have assumed g to be real.

Suppose now we specify the adjoint source in Eq. (5.132) to be the fission cross section. Then $\phi^\dagger$ is a solution to the adjoint diffusion (or transport) equation whose source has the same spatial (and energy) dependence as the fission cross section. If the solution to this equation can be obtained in some approximate and rapid manner, and if the solution to the flux equation (5.131) can be estimated in a similar manner, then an evaluation of the expression for J, (5.133), will yield, to second-order accuracy, the fission rate as given in Eq. (5.135), that is,

$$J = \int d\omega\, \Sigma_f(\omega)\phi(\omega). \tag{5.141}$$

As a simple application of the method, consider the calculation of the absorption rate. We are given a slab of width $2a$ (region 1) that is inserted in a medium (region 2) and has a constant neutron source of amplitude q (see Fig. 5.3).

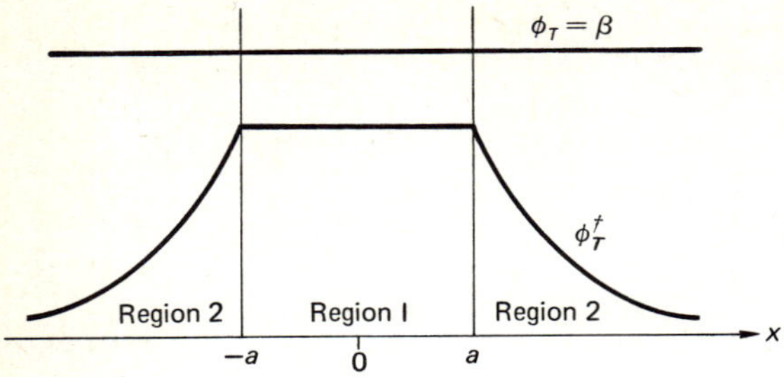

Fig. 5.3 Trial functions $\phi_T{}^\dagger$ and ϕ_T.

The problem is to calculate the absorption rate in region 1. Thus we take for our two basic equations

$$\theta\phi = -D\frac{d^2}{dx^2}\phi(x) + \Sigma_a\phi(x) = q \tag{5.142a}$$

and

$$\theta^\dagger\phi^\dagger = -D\frac{d^2}{dx^2}\phi^\dagger(x) + \Sigma_a\phi^\dagger(x) = \Sigma_{a1}H(x), \tag{5.142b}$$

where

$$H(x) = \begin{cases} 1, & -a \le x \le a, \quad (5.143a) \\ 0, & |x| > a. \quad (5.143b) \end{cases}$$

Let us take for the trial functions

$$\phi_T^\dagger = \begin{cases} \alpha, & |x| < a, \quad (5.144a) \\ \alpha e^{-|x-a|/L}, & |x| > a. \quad (5.144b) \end{cases}$$

Note that $\phi_T^\dagger$ is continuous and symmetric about $x = 0$ and vanishes at infinity. For ϕ_T we take a constant

$$\phi_T = \beta. \tag{5.145}$$

The functional

$$J = -2\left[\int_0^\infty \phi_T^\dagger\theta\phi_T\,dx - \int_0^\infty \phi_T^\dagger q\,dx - \int_0^\infty \Sigma_{a1}H(x)\phi_T(x)\,dx\right] \tag{5.146}$$

gives, as we have shown earlier in this section, an estimate of the total absorption rate in region 1,

$$N_1 = 2\int_0^\infty \phi(x)\Sigma_{a1}(x)H(x)\,dx \tag{5.147a}$$

$$= 2\int_0^a \phi(x)\Sigma_{a1}(x)\,dx. \tag{5.147b}$$

The integrals in Eq. (5.146) are readily carried out:

$$J = 2(-\alpha\beta a\Sigma_{a1} - \alpha\beta L_2 \Sigma_{a2} + \alpha q L_2 + \beta\Sigma_{a1} a). \tag{5.148}$$

To complete the calculation, we must set

$$\frac{\partial J}{\partial \alpha} = \frac{\partial J}{\partial \beta} = 0, \tag{5.149}$$

and obtain the variationally optimum values of α and β. We find

$$\alpha_M = \frac{\Sigma_{a1} a}{\Sigma_{a1} a + \Sigma_{a2} L_2} \tag{5.150a}$$

and

$$\beta_M = \frac{q L_2}{\Sigma_{a1} a + \Sigma_{a2} L_2}\,. \tag{5.150b}$$

Thus the variationally optimum estimate of N_1 is

$$N_1 \approx J = 2\,\frac{\Sigma_{a1} a L_2 q}{\Sigma_{a1} a + \Sigma_{a2} L_2}\,. \tag{5.151}$$

PROBLEMS

5.1. Calculate the error in the critical mass of a bare, homogeneous reactor which results as a consequence of 1 percent of error in k. Assume that only the size would be adjusted to obtain criticality.

5.2. Determine the adjoint for the operator ∇^2. Remember $g(\mathbf{R}') = f(\mathbf{R}') = 0$.

5.3. Determine the adjoint of the operator K (multiplication by a complex constant).

5.4. If Ω_1 and Ω_2 are operators, prove $(\Omega_1 \Omega_2)^\dagger = \Omega_2{}^\dagger \Omega_1{}^\dagger$. Note that in general Ω_1 and Ω_2 do *not* comute, that is,

$$\Omega_1 \Omega_2 \neq \Omega_2 \Omega_1.$$

5.5. Using the supervariational principle, determine the eigenvectors and eigenvalues of the matrix

$$\mathbf{A} = \begin{pmatrix} -23 & 11 & 1 \\ 11 & -3 & -2 \\ 1 & -2 & 1 \end{pmatrix}.$$

Take the unperturbed matrix as

$$\mathbf{A} = \begin{pmatrix} -23 & 0 & 0 \\ 0 & -3 & 0 \\ 0 & 0 & 1 \end{pmatrix}.$$

5.6. Compute the buckling of a sphere by using the trial function shown at the top of page 134. (Optimize with respect to α and β).

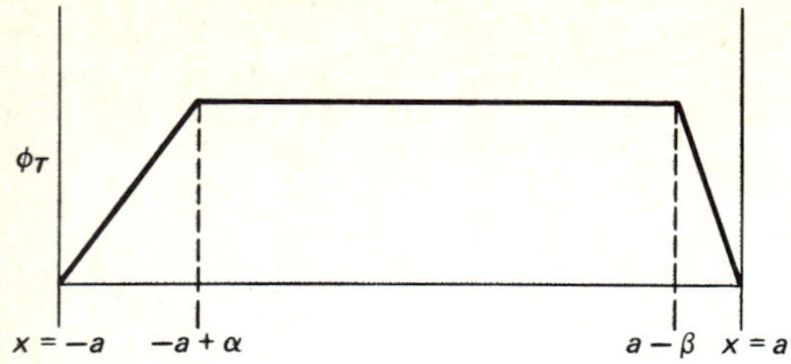

5.7. Compute the absorption rate in the slab shown in Fig. 5.3, using $\phi_T^\dagger$ as given in Eq. (5.144) and

$$\phi_T(x) = \begin{cases} \beta(1 - e^{-x/L_2}), & |x| > a, \\ \beta, & |x| < a. \end{cases}$$

5.8. The eigenvalues and eigenvectors of the matrix

$$\begin{pmatrix} 1 & 1 \\ 1 & -1 \end{pmatrix}$$

are

$$\lambda = \pm\sqrt{2}, \qquad \phi_\pm = \begin{pmatrix} & 1 \\ \pm\sqrt{2} & -1 \end{pmatrix}.$$

Using first-order perturbation theory, compute the eigenvalues of the matrix

$$\begin{pmatrix} \frac{1}{2} & 1 \\ 1 & -1 \end{pmatrix}.$$

5.9. Derive Eq. (5.27).

5.10. Prove that the operator $i\,d/dx$ with boundary conditions of Eq. (5.9) is self-adjoint.

5.11. Prove that the operator $i\,d/dx$ with boundary conditions $f(0) = f(d) = 0$ is *not* self-adjoint, but *is* hermitian (see footnote 3 on page 110).

6
Numerical Methods

6.1 INTRODUCTION

The analytical techniques described in Chap. 3 are useful in providing insight into the physical processes taking place in nuclear reactors. They also can be used for rough survey calculations. On the other hand, for detailed design information, numerical procedures are necessary. The major reason for this is that the construction of a typical reactor is much more complicated than the simple zone type of reactor treated in Chap. 3. In fact since the presence of fuel elements, coolant channels, control rods, etc., has to be taken into account, there is almost a continously varying distribution of material in the reactor.

Reactor calculations are not, in general, carried out in the one-speed approximation to which we still cling in the present chapter. However, the numerical techniques are well demonstrated by this approximation; extensions to the more practical multigroup calculations are then indicated in Chap. 8.

6.2 OUTER ITERATIONS

Consider the diffusion equation

$$-\nabla \cdot D(\mathbf{r})\nabla\phi(\mathbf{r}) + \Sigma_a(\mathbf{r})\phi(\mathbf{r}) = \nu_c \Sigma_f(\mathbf{r})\phi(\mathbf{r}), \tag{6.1}$$

which we propose to solve numerically. Further, let us rewrite Eq. (6.1) in operator form, a form which will permit simple generalization to the multi-group problem. (Note that we are using the eigenvalue ν_c as described in Sec. 3.1.)

$$A\phi = \nu_c F\phi, \tag{6.2}$$

where[1]

$$A = -\nabla \cdot D\nabla + \Sigma_a \tag{6.3a}$$

and

$$F = \Sigma_f . \tag{6.3b}$$

We now define a *source*

$$S(\mathbf{r}) = \nu_c F\phi(\mathbf{r}), \tag{6.4}$$

or

$$\phi(\mathbf{r}) = \frac{1}{\nu_c} F^{-1} S(\mathbf{r}). \tag{6.5}$$

Then $S(\mathbf{r})$ obeys the operator equation [with the use of (6.2) and (6.5)],

$$\frac{1}{\nu_c} S(\mathbf{r}) = PS(\mathbf{r}), \tag{6.6a}$$

where

$$P = FA^{-1}. \tag{6.6b}$$

Equation (6.6*a*) is used to establish convergence properties of the source, or outer, iteration, as will be discussed presently. However, it is not used for calculating, as the operator A^{-1} is not known explicitly. Further, it is known mathematically that (6.6*a*) has only one solution which is nowhere negative, and the corresponding eigenvalue is a maximum value (see Sec. 3.4).

Formally, however, Eq. (6.6*a*) can be solved as follows: A first guess at S, call it $S^{(0)}$, is made. Then one calculates

$$S^{(1)} = PS^{(0)}, \tag{6.7}$$

[1] The reason for this notation is that the multigroup equations will be almost identical in form.

and in general, continuing this procedure, we obtain

$$S^{(k)} = PS^{(k-1)} \tag{6.8a}$$

$$= (P)^k S^{(0)}. \tag{6.8b}$$

After a sufficient number of iterations, it is assumed that the difference between S^k and $S^{(k-1)}$ will be (except for a multiplicative factor) less than a predetermined value.[1] Then the iteration is said to have *converged*, and the ratio of two successive iterates yields the eigenvalue

$$\frac{S^{(k)}(\mathbf{r})}{S^{(k-1)}(\mathbf{r})} = \frac{1}{\nu_c}. \tag{6.9}$$

This same procedure can be used to solve any eigenvalue equation. As an example, let us calculate the eigenvalue and eigenfunction of the following matrix operator to within 1 percent:

$$P = \begin{pmatrix} 1 & 2 \\ 3 & 4 \end{pmatrix}. \tag{6.10}$$

We have

$$\begin{pmatrix} 1 & 2 \\ 3 & 4 \end{pmatrix}\begin{pmatrix} \phi_1 \\ \phi_2 \end{pmatrix} = \frac{1}{\nu_c}\begin{pmatrix} \phi_1 \\ \phi_2 \end{pmatrix}. \tag{6.11}$$

Take as the first guess

$$S^{(0)} = \begin{pmatrix} \phi_1{}^{(0)} \\ \phi_2{}^{(0)} \end{pmatrix} = \begin{pmatrix} 1 \\ 1 \end{pmatrix}. \tag{6.12}$$

Then,

$$S^{(1)} = \begin{pmatrix} 1 & 2 \\ 3 & 4 \end{pmatrix}\begin{pmatrix} 1 \\ 1 \end{pmatrix} \tag{6.13a}$$

$$= \begin{pmatrix} 3 \\ 7 \end{pmatrix} = \begin{pmatrix} \phi_1{}^{(1)} \\ \phi_2{}^{(1)} \end{pmatrix}. \tag{6.13b}$$

We note $\phi_1{}^{(1)}/\phi_1{}^{(0)} \neq \phi_2{}^{(1)}/\phi_2{}^{(0)}$ to within the predetermined 1 percent, so we continue.

$$S^{(2)} = \begin{pmatrix} 1 & 2 \\ 3 & 4 \end{pmatrix}\begin{pmatrix} 3 \\ 7 \end{pmatrix}, \tag{6.14a}$$

$$= \begin{pmatrix} 17 \\ 37 \end{pmatrix}. \tag{6.14b}$$

[1] Since $S^{(k)}(\mathbf{r})$ and $S^{(k-1)}(\mathbf{r})$ are functions of position, this statement means that their ratio will be a constant value $\pm\varepsilon$ at every point $\mathbf{r}$ (they will have the same spatial shape). Then the ratio of their *amplitudes* is the eigenvalue $1/\nu_c$.

Since $\frac{17}{3} = 5.67$, while $\frac{37}{7} = 5.30$, we are close to the required 1 percent accuracy. One more iteration is sufficient:

$$S^{(3)} = \begin{pmatrix} 1 & 2 \\ 3 & 4 \end{pmatrix} \begin{pmatrix} 17 \\ 37 \end{pmatrix} \tag{6.15a}$$

$$= \begin{pmatrix} 91 \\ 199 \end{pmatrix}, \tag{6.15b}$$

and $\frac{91}{17} = \frac{199}{37}$, well to within 1 percent. Thus, the eigenfunction is (proportional to)

$$S^{(3)} = \begin{pmatrix} 91 \\ 199 \end{pmatrix}, \tag{6.16a}$$

and the eigenvalue is

$$\frac{1}{\nu_c} = \frac{91}{17} = \frac{199}{37} = 5.36. \tag{6.16b}$$

One actually, generally, calculates $1/\nu_c$ from the ratio of the *norms* of $S^{(k)}$ to $S^{(k-1)}$, that is, the square root of the sums of the squares in this matrix example, or integration if P is a differential operator. Thus in our example,

$$\frac{1}{\nu_c} = \frac{\sqrt{(91)^2 + (199)^2}}{\sqrt{(17)^2 + (37)^2}}, \tag{6.17a}$$

while for a differential operator P,

$$\frac{1}{\nu_c} = \frac{\int d^3r |S^{(k)}(\mathbf{r})|^2}{\int d^3r |S^{(k-1)}(\mathbf{r})|^2}. \tag{6.17b}$$

If the iteration has converged sufficiently, the ratio of the norms should be equal to the ratios at any point $\mathbf{r}$ (or for any component ϕ_i in the case, illustrated above, that P is a matrix operator).

The method described above is referred to as the *power* method. In Sec. 6.5 we shall consider some of the mathematical proofs of convergence and some improvements on this method.[1]

Since a 2 × 2 matrix has, usually, two eigenvalues and two associated eigenfunctions, the question might well be asked, which of the two has been obtained in the above example? The answer will become clear in Sec. 6.5, where we show that the largest eigenvalue is obtained by this procedure.

[1] One difficulty with the method, which can be seen from our illustrative example, is the way successive iterates increase (or decrease) in magnitude. This can lead to serious problems in machine usage if the number of iterations required is very large.

Returning now to our reactor problem, let us discuss how the source iterations are carried out in practice. Equation (6-2) can be rewritten for the kth source iteration

$$A\phi^{(k)} = S^{(k)}, \tag{6.18}$$

and when $S^{(0)}$ is guessed, this equation is solved for $\phi^{(0)}$ by the methods described in the next section. When $\phi^{(0)}$ is known, $S^{(1)}$ is calculated from Eq. (6.8a) since v_c is not known,

$$S^{(k)} = PS^{(k-1)} = F\phi^{(k-1)}. \tag{6.19}$$

This procedure is continued until convergence takes place. (Comparison should be made between successive iterates of S, *not* of ϕ.)

The problem of finding ϕ when S is known, that of solving Eq. (6.18), is discussed in the next section.

6.3 FLUX CALCULATION: ONE-DIMENSIONAL CASE

In the previous section, we described the source, or outer, iterations which are used in practice to solve the neutron diffusion equation. In the process we passed over the problem of solving Eq. (6.18), that is, the calculation of $\phi^{(k)}$, where $S^{(k)}$ is given. (The calculation of $S^{(k+1)} = F\phi^{(k)}$ is trivial of course.) The equation for the flux is an inhomogeneous differential equation. For simple systems we have shown, in Chap. 2, how analytical solutions can be found. However, we are considering the problem here that the reactor is too complicated to permit analytical solutions, so that numerical methods must be used.

To this end, the equation

$$A\phi^{(k)} = S^{(k)}, \tag{6.20a}$$

or (by inserting the definition of A)

$$(-\nabla \cdot D\nabla + \Sigma_a)\phi^{(k)}(\mathbf{r}) = S^{(k)}(\mathbf{r}), \tag{6.20b}$$

is converted into a difference equation. In the present section we consider only one-dimensional geometry, so that Eq. (6.20b) becomes

$$\left[-\frac{d}{dx}\,D(x)\,\frac{d}{dx} + \Sigma_a(x)\right]\phi^{(k)}(x) = S^{(k)}(x). \tag{6.21}$$

The entire range of the x variable, which includes shield, reflector, blanket, core—whatever regions are to be considered in the diffusion equation—is divided into a number of intervals. The points between these intervals are labeled by x_i, and are called *mesh points*, while the distance between x_i and x_{i-1} is called h_{i-1} (see Fig. 6.1).

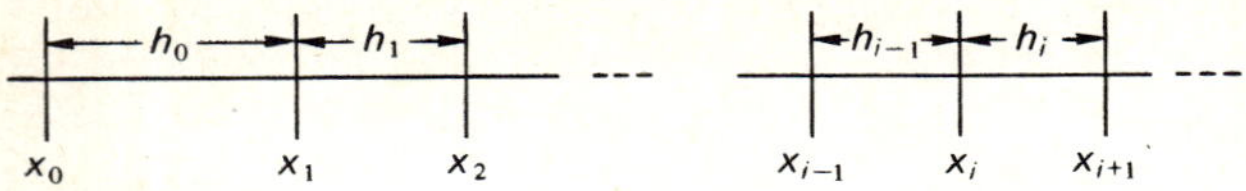

Fig. 6.1 Mesh for one-dimensional difference equation.

Note that the mesh points need not be equally spaced. In fact, in regions where the material properties of the reactor change rapidly, as near fuel and absorber elements, coolant channels, etc., one would naturally use a finer mesh spacing than, say, in the reflector.

Equation (6.21) is now integrated from $x_i - h_{i-1}/2$ to $x_i + h_i/2$ In integrating the terms $\Sigma_a(x)\phi(x)$ and $S(x)$, a number of assumptions can be made.[1] Let us merely assume that the mesh spacing is fine enough there that these terms can be considered constant over the range of integration, in which case

$$\int_{x_i - h_{i-1}/2}^{x_i + h_i/2} \Sigma_a(x)\phi(x)\,dx = \Sigma_{ai}\,\phi_i\,\frac{h_i + h_{i-1}}{2} \tag{6.22a}$$

and

$$\int_{x_i - h_{i-1}/2}^{x_i + h_i/2} S(x)\,dx = S_i\,\frac{h_i + h_{i-1}}{2}. \tag{6.22b}$$

Here we are using the notation

$$f(x_i) = f_i. \tag{6.23}$$

The integration of the derivative term gives

$$\int_{x_i - h_{i-1}/2}^{x_i + h_i/2} \frac{d}{dx}\,D(x)\,\frac{d}{dx}\,\phi(x)\,dx = D(x)\,\frac{d}{dx}\,\phi(x)\Big|_{x_i - h_{i-1}/2}^{x_i + h_i/2}. \tag{6.24}$$

In evaluating this expression we can use a two-point difference formula for the derivative

$$\frac{d}{dx}\,\phi(x) = \frac{\phi_{i+1} + \phi_i}{h_i}, \qquad x_i \le x \le x_{i+1}, \tag{6.25a}$$

and

$$\frac{d}{dx}\,\phi(x) = \frac{\phi_i - \phi_{i-1}}{h_{i-1}}, \qquad x_{i-1} \le x \le x_i. \tag{6.25b}$$

[1] Here we are suppressing the labels signifying the iteration step.

Also, we may approximate $D(x)$ at the point $x_i + h_i/2$ by its average value at x_i and x_{i+1},

$$D\left(x_i + \frac{h_i}{2}\right) = \frac{D_{i+1} + D_i}{2}, \tag{6.26a}$$

with an analogous expression for $D(x_i - h_{i-1}/2)$.

Then, the right-hand side of Eq. (6.24) becomes

$$\phi_{i+1}\frac{D_{i+1} + D_i}{2h_i} - \phi_i\left(\frac{D_{i+1} + D_i}{2h_i} + \frac{D_i + D_{i-1}}{2h_{i-1}}\right) + \phi_{i-1}\frac{D_i + D_{i-1}}{2h_{i-1}}. \tag{6.26b}$$

Combining all these results, we find that the difference equation approximation to Eq. (6.21) can be written in the form

$$a_{i,i-1}\phi_{i-1} + a_{i,i}\phi_i + a_{i,i+1}\phi_{i+1} = S_i, \tag{6.27a}$$

with

$$a_{i,i-1} = \frac{D_i + D_{i-1}}{h_{i-1}}\,\frac{1}{h_i + h_{i-1}}, \tag{6.27b}$$

$$a_{i,i} = \Sigma_{a_i} - \left(\frac{D_{i+1} + D_i}{h_i} + \frac{D_i + D_{i-1}}{h_{i-1}}\right)\frac{1}{h_i + h_{i-1}}, \tag{6.27c}$$

and

$$a_{i,i+1} = \frac{D_{i+1} + D_i}{h_i}\,\frac{1}{h_i + h_{i-1}}. \tag{6.27d}$$

We have outlined above only one of many possible ways by which this differential equation can be converted into a difference equation. Other schemes would lead to different coefficients. What is important here is that Eq. (6.27*a*) is a three-term equation, i.e., it includes values of the fluxes at three adjacent points.[1] The labels i on the coefficient $a_{i,j}$ indicate that the integration from which Eq. (6.27*a*) was obtained has been carried out over a cell centered at the mesh point i.

At the boundaries $i = 0$ and $i = N$, slightly different equations result. Consider, for example, $i = 0$. Then the integration can extend only from

[1] This follows from our having used a two-term difference formula, i.e., Eqs. (6.25), to represent $(d/dx)\,\phi(x)$.

$x = 0$ to $x = h_0/2$. We find Eqs. (6.22) carried over with no essential change (except, of course, $h_{i-1} = h_{-1} = 0$). However, Eq. (6.24) changes to

$$\int_0^{h_0/2} \frac{d}{dx} D(x) \frac{d}{dx} \phi(x)\, dx = \frac{D_1 + D_0}{2} \frac{\phi_1 - \phi_0}{h_0} - D_0 \frac{d\phi_0}{dx}. \tag{6.28}$$

Thus, the equation at the zeroth mesh point, i.e., the boundary point, is of the form

$$a_{00}\phi_0 + a_{01}\phi_1 + D_0 \frac{d\phi_0}{dx} = S_0. \tag{6.29}$$

Now the boundary conditions specify either $d\phi_0/dx$ (*current* boundary conditions) or the ratio $(1/\phi_0)(d\phi_0/dx)$. In either case $d\phi_0/dx$ can be eliminated from Eq. (6.29), and the first of the set of linear equations for the ϕ_i is a two-term system. (Very similar arguments apply to the last equation of the system, i.e., the one for the other boundary.)

A simple count shows that we have $N + 1$ inhomogeneous equations in the $N + 1$ unknown mesh-point fluxes $\phi_0, \phi_1, \ldots, \phi_N$. A solution can always be found.

The set of equations for the ϕ_i can be recast in matrix form

$$\begin{pmatrix} a_{00} & a_{01} & 0 & \cdots & \cdot & 0 \\ a_{10} & a_{11} & a_{12} & 0 & \cdot & \cdot \\ 0 & a_{21} & a_{22} & \cdots & \cdot & \cdot \\ \cdots & \cdots & \cdots & \cdots & a_{N-1,N-1} & a_{N-1,N} \\ 0 & \cdots & \cdots & \cdots & a_{N,N-1} & a_{N,N} \end{pmatrix} \begin{pmatrix} \phi_0 \\ \phi_1 \\ \vdots \\ \phi_N \end{pmatrix} = \begin{pmatrix} S_0 \\ S_1 \\ \vdots \\ S_N \end{pmatrix}, \tag{6.30}$$

or more compactly,

$$\mathbf{A\phi} = \mathbf{S}. \tag{6.31}$$

(Now we observe that in specifying $S(\mathbf{r})$ for the outer iterations described in Sec. 6.2, $N + 1$ values of the function $(S_0, \ldots, S_N)$ are specified, one at each mesh point. In comparing subsequent iterates, the ratios of $S_i^{(k)}$ to $S_i^{(k-1)}$ are found at each mesh point. When these ratios are all the same,[1] the source in subsequent iterations has the same shape, and the outer iteration has converged.)

To solve Eq. (6.31), a simple algorithm is available.[2] This algorithm, known as *gaussian elimination*, is available only to three-term equations, which means, in reactor physics, it can be applied to one-dimensional problems. (In the next section we discuss the basically different techniques which must be applied in two and three dimensions.)

[1] To within a prespecified error.

[2] E. Wachspress: "Iterative Solution of Elliptic Systems," Prentice-Hall, Inc., Englewood Cliffs, N.J., 1966.

The matrix **A** in Eq. (6.31) is said to be *tridiagonal*. If the first row of Eq. (6.30) is divided by a_{00}, one obtains

$$\begin{pmatrix} 1 & \dfrac{a_{01}}{a_{00}} & 0 & \cdots \\ a_{10} & a_{11} & a_{12} & \cdots \\ \cdots & \cdots & \cdots & \cdots \end{pmatrix}\begin{pmatrix} \phi_0 \\ \vdots \\ \phi_N \end{pmatrix} = \begin{pmatrix} \dfrac{S_0}{a_{00}} \\ S_1 \\ \vdots \\ S_N \end{pmatrix}. \tag{6.32}$$

Next we multiply the first row of (6.32) by $-a_{10}$, and add to the second row to obtain

$$\begin{pmatrix} 1 & \dfrac{a_{01}}{a_{00}} & 0 & \cdots \\ 0 & a_{11}-\dfrac{a_{01}a_{10}}{a_{00}} & a_{12} & \cdots \\ \\ \cdots & \cdots & \cdots & \cdots \end{pmatrix}\begin{pmatrix} \phi_0 \\ \vdots \\ \\ \phi_N \end{pmatrix} = \begin{pmatrix} \dfrac{S_0}{a_{00}} \\ S_1-\dfrac{a_{10}}{a_{00}}S_0 \\ S_2 \\ \vdots \\ S_N \end{pmatrix}. \tag{6.33}$$

Next one divides the second row by the diagonal element to obtain

$$\begin{pmatrix} 1 & \dfrac{a_{01}}{a_{00}} & 0 & \cdots & \cdots \\ 0 & 1 & \dfrac{a_{12}}{a_{11}-a_{10}a_{01}/a_{00}} & 0 & \cdots \\ \\ \cdots & \cdots & \cdots & \cdots & \cdots \end{pmatrix}\begin{pmatrix} \phi_0 \\ \vdots \\ \\ \phi_N \end{pmatrix} = \begin{pmatrix} \dfrac{S_0}{a_{00}} \\ \dfrac{S_1-(a_{10}/a_{00})S_0}{a_{11}-a_{10}a_{01}/a_{00}} \\ S_2 \\ \vdots \\ S_N \end{pmatrix}. \tag{6.34}$$

By continuing in this manner, the original matrix equation (6.30) may be reduced to the system:

$$\begin{pmatrix} 1 & h_0 & 0 & 0 & \cdots & 0 \\ 0 & 1 & h_1 & 0 & \cdots & 0 \\ 0 & 0 & 1 & h_2 & \cdots & 0 \\ \cdots & \cdots & \cdots & \cdots & \cdots & \cdots \\ \cdots & \cdots & \cdots & \cdots & 0 & 1 \end{pmatrix}\begin{pmatrix} \phi_0 \\ \vdots \\ \phi_N \end{pmatrix} = \begin{pmatrix} \alpha_1 \\ \vdots \\ \alpha_N \end{pmatrix}, \tag{6.35}$$

where the h_n and α_n are defined recursively as follows:

$$h_n = \frac{a_{n,n+1}}{a_{n,n}-a_{n,n-1}h_{n-1}}, \qquad h_0 = \frac{a_{01}}{a_{00}}, \tag{6.36}$$

$$\alpha_n = \frac{S_n - a_{n,n-1}S_{n-1}/a_{n-1,n-1}}{a_{n,n} - a_{n,n-1}a_{n-1,n}/a_{n,n}}, \quad \alpha_0 = \frac{S_0}{a_{10}}. \tag{6.37}$$

The set of Eqs. (6.35) is solved trivially, beginning with the Nth,

$$\phi_N = \alpha_N. \tag{6.38}$$

Then

$$\phi_{N-1} = -h_{N-1}\phi_N + \alpha_{N-1} \tag{6.39a}$$

$$= -h_{N-1}\alpha_N + \alpha_{N-1}. \tag{6.39b}$$

By continuing this procedure *upwards*, the flux at each mesh point can be obtained.

6.4 TWO- AND THREE-DIMENSIONAL PROBLEMS: INNER ITERATIONS

The Gauss elimination scheme described in the previous section amounted to a direct numerical inversion of the matrix **A**. This procedure was available because of the tridiagonal form of the **A** matrix, which, in turn, was obtained because a two-point difference formula, Eqs. (6.25), was used in one dimension. In two- or three-dimensional problems, even if a two-point difference formula is used, a tridiagonal **A** matrix is not obtained and a simple algorithm for the inversion of the **A** matrix is not available. A different procedure is used, namely, the method of inner iterations.

To understand this, first let us describe how the difference equations are obtained in two-dimensional geometry. For simplicity we restrict our attention to the xy plane. The $r - \theta$ (cylindrical) case has been described by Wachspress.[1] For the y two-dimensional problem, the x and y domains are divided into a grid, so that rectangular *unit cells* are obtained (see Fig. 6.2). Note that instead of using symbols such as h_i, h_{i-1}, etc., to express differences, it is customary to use the symbols R (for right; $R = x_{i+1} - x_i$), L (for left), T (for top), and B (for bottom). All these are explained in Fig. 6.2. We now multiply (6.18) by $dx\,dy$ and integrate over the area enclosed in the solid line in Fig. 6.2.

The procedure is to apply Gauss' theorem in order to express this surface integral as a line integral. Then, for example,

$$\int_{x_i-L/2}^{x_i+R/2} dx \int_{y_j-B/2}^{y_j+T/2} dy\, \nabla \cdot \mathbf{D}\, \nabla\phi$$

$$= \int_{pabc} D\frac{\partial\phi}{\partial n}\,dl + \int_{pcde} D\frac{\partial\phi}{\partial n}\,dl + \int_{plgf} D\frac{\partial\phi}{\partial n}\,dl + \int_{pgha} D\frac{\partial\phi}{\partial n}\,dl. \tag{6.40}$$

[1] E. Wachspress, *op. cit.*

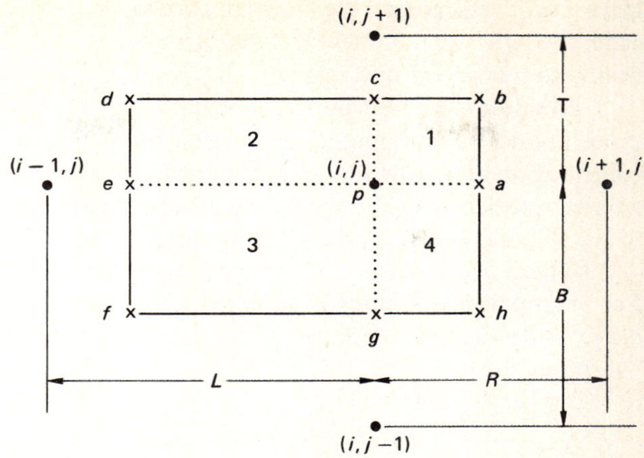

Fig. 6.2 Mesh for two-dimensional difference equation.

Table 6.1 Coordinates for Fig. 6.2

a	$\left(x_i + \frac{R}{2}, y_j\right)$
b	$\left(x_i + \frac{R}{2}, y_j + \frac{T}{2}\right)$
c	$\left(x_i, y_j + \frac{T}{2}\right)$
d	$\left(x_i - \frac{L}{2}, y_j + \frac{T}{2}\right)$
e	$\left(x_i - \frac{L}{2}, y_j\right)$
f	$\left(x_i - \frac{L}{2}, y_j - \frac{B}{2}\right)$
g	$\left(x_i, y_j - \frac{B}{2}\right)$
h	$\left(x_i + \frac{R}{2}, y_j - \frac{B}{2}\right)$

Here $\partial\phi/\partial n$ represents the normal derivative of ϕ, and the subscript, for example, *pabc* on the integral signs denotes region 1, over which the integration is to be carried out.

In adding the four integrals in Eq. (6.40), we note that there are certain cancelations. For example, in the first integral, there is a term including a contribution along the line segment from c to p. This is exactly canceled by a term in the second integral which is contributed by the line segment from p to c. All that remains finally are integrations over the exterior line segments.

These are most easily carried out by expanding in a Taylor's series. For example, in the integral along the line segment from a to b, derivative $\partial\phi/\partial x = \partial\phi/\partial n$, and we have

$$\left.\frac{\partial\phi}{\partial n}\right|_{x=R/2+x_i} = \phi_x(1) + \frac{R}{2}\,\phi_{xx}(1) + y\phi_{xy}(1) + \cdots, \tag{6.41a}$$

where subscripts denote differentiation. Integrating from a to b requires only an integration over dy, so that

$$\int_a^b D(x,y)\,\frac{\partial\phi}{\partial n}\,dl = D_1\left[\frac{T}{2}\,\phi_x(1) + \frac{TR}{4}\,\phi_{xx}(1) + \frac{T^2}{8}\,\phi_{xy}(1)\right]. \tag{6.41b}$$

Here we have introduced the notation D_1, $\phi_x(1)$, $\phi_{xx}(1)$, $\phi_{xy}(1)$ for the values of these quantities in quadrant 1 of Fig. 6.2.

A similar expression for the integral from d to e in quadrant 2 is

$$\int_d^e D(x,y)\,\frac{\partial\phi}{\partial n}\,dl = D_2\left[-\frac{T}{2}\,\phi_x(2) + \frac{TL}{4}\,\phi_{xx}(2) - \frac{T^2}{8}\,\phi_{xy}(2)\right]. \tag{6.42}$$

Noting from continuity of current that

$$D_1\phi_x(1) = D_2\phi_x(2) \tag{6.43a}$$

and

$$D_1\phi_{xy}(1) = D_2\phi_{xy}(2), \tag{6.43b}$$

we see that Eqs. (6.41) and (6.42) may be added to yield

$$\int_a^b D(x,y)\,\frac{\partial\phi}{\partial n}\,dl + \int_d^e D(x,y)\,\frac{\partial\phi}{\partial n}\,dl = \frac{D_1TR}{4}\,\phi_{xx}(1) + \frac{D_2TL}{4}\,\phi_{xx}(2). \tag{6.44}$$

We continue this procedure over the entire periphery *abcdefgh* and, in addition, use difference formulas for the derivatives, i.e.,

$$\phi_{xx}(1) = \frac{2(\phi_{i+1,j} - \phi_{ij})}{R^2} - \frac{2\phi_x(1)}{R} \tag{6.45a}$$

and

$$\phi_{xx}(2) = \frac{2(\phi_{i-1,j} - \phi_{ij})}{L^2} + \frac{2\phi_x(2)}{L}, \tag{6.45b}$$

and so forth. This leads finally to a five-point difference representation of the differential equation (6.18) in two dimensions [again using (6.43)],

$$\frac{TD_1 + BD_4}{2R}(\phi_{ij} - \phi_{i+1,j}) + \frac{TD_2 + BD_3}{2L}(\phi_{ij} - \phi_{i-1,j})$$
$$+ \frac{RD_1 + LD_2}{2T}(\phi_{ij} - \phi_{i,j+1}) + \frac{RD_4 + LD_3}{2B}(\phi_{ij} - \phi_{i,j-1}) = S_{ij}. \tag{6.46}$$

As in the one-dimensional problem, special equations must be devised for the fluxes at boundary points. The details will not be given here. Rather, the interested reader is referred to Wachspress' book[1] from which the foregoing treatment has been taken almost verbatim.

In order to solve the set of Eqs. (6.46) an iterative procedure is followed. First the equations must be recast in matrix form. The flux vector may be defined by running through an index, say i, from 0 to N, while the other is held fixed. Then the second index is run through, from 0 to M.[2] That is,

$$\mathbf{\Phi} = \begin{pmatrix} \phi_{00} \\ \phi_{10} \\ \vdots \\ \phi_{N0} \\ \phi_{01} \\ \vdots \\ \phi_{NM} \end{pmatrix}. \tag{6.47}$$

In the same way the matrix of coefficients, which we again call **A**, and the source vector **S** are constructed. So we are again led to a linear system

$$\mathbf{A\Phi} = \mathbf{S}. \tag{6.48}$$

The essential difference here from the one-dimensional problem is that the simple algorithm used there to solve this equation is no longer available because the matrix **A** is no longer tridiagonal. So the method of inner iterations is used.

Actually, Eq. (6.48) as it stands is not amenable to an iteration procedure because the dependent variable appears only once in the equation. However, we can break the matrix **A** into two parts,

$$\mathbf{A} = \mathbf{N} + \mathbf{D}, \tag{6.49}$$

[1] E. Wachspress, *op. cit.*, p. 35ff. The notation ϕ_{ij} means the value of the flux at the mesh point ij, that is, $\phi(x_i,y_j)$.

[2] We assume a total of $N+1$ mesh points along the x axis, and $M+1$ along the y axis. The flux vector has $(N+1)(M+1)$ components, and the **A** matrix has rank $(N+1)(M+1)$.

where $\mathbf{D}$ is the matrix of all the diagonal elements of $\mathbf{A}$. That is,

$$(\mathbf{D})_{jj'} = a_{jj}\,\delta_{jj'}. \tag{6.50}$$

Thus the inverse of $\mathbf{D}$ is trivial to calculate:

$$(\mathbf{D}^{-1})_{jj'} = \frac{1}{a_{jj}}\,\delta_{jj'}. \tag{6.51}$$

Now Eq. (6.48) can be rewritten in the form

$$\mathbf{\Phi} = \mathbf{D}^{-1}\mathbf{S} - \mathbf{D}^{-1}\mathbf{N}\mathbf{\Phi}, \tag{6.52}$$

and the iteration can proceed by guessing a $\mathbf{\Phi}^{(0)}$ at each of the $(N+1) \times (M+1)$ mesh points, inserting this guess into the right-hand side of Eq. (6.52), computing a $\mathbf{\Phi}^{(1)}$, reinserting this in the right-hand side, etc. This procedure is called *relaxation* because we begin with an estimated solution and relax it to the true solution.[1]

The method described above is called *simultaneous relaxation* because the flux value at every point of the mesh is relaxed simultaneously. Actually, a much more economical technique is that of *successive relaxation*. To understand this, rewrite Eq. (6.52) in the form

$$\begin{pmatrix} \phi_{00}^{(k)} \\ \vdots \\ \phi_{NM}^{(k)} \end{pmatrix} = \begin{pmatrix} \dfrac{1}{a_{00}} & 0 & 0 & \cdots & 0 \\ 0 & \dfrac{1}{a_{11}} & 0 & \cdots & \cdots \\ \cdots & \cdots & \cdots & \cdots & \cdots \\ \cdots & \cdots & \cdots & \cdots & \dfrac{1}{a_{NM,NM}} \end{pmatrix} \begin{pmatrix} S_{00} \\ \vdots \\ S_{NM} \end{pmatrix}$$

$$- \begin{pmatrix} 0 & \dfrac{a_{01}}{a_{00}} & \dfrac{a_{02}}{a_{00}} & 0 & \cdots & 0 \\ \dfrac{a_{10}}{a_{11}} & 0 & \dfrac{a_{12}}{a_{11}} & \cdots & \cdots & \cdots \\ \cdots & \cdots & \cdots & \cdots & \cdots & \cdots \\ \cdots & \cdots & \cdots & \cdots & \cdots & 0 \end{pmatrix} \begin{pmatrix} \phi_{00}^{(k-1)} \\ \vdots \\ \phi_{NM}^{(k-1)} \end{pmatrix}. \tag{6.53}$$

Here we explicitly have carried out the multiplication $\mathbf{D}^{-1}\mathbf{N}$.

[1] It must be admitted that no one who has tried to write a program of this type has ever found the process relaxing!

Equation (6.53) is really a set of $(N + 1) \cdot (M + 1)$ equations which give the ${\phi_{ij}}^{(k)}$ in terms of ${\phi_{ij}}^{(k-1)}$. Note that the equation for ${\phi_{00}}^{(k)}$ involves ${\phi_{j0}}^{(k-1)}$ with $1 \leq j \leq 3$.[1] The next equation is one in which ${\phi_{10}}^{(k)}$ is expressed in terms of ${\phi_{j0}}^{(k-1)}$ and $0 \leq j \leq 4$. But ${\phi_{00}}^{(k)}$ has been computed in the previous step. The idea of successive relaxation is to replace ${\phi_{00}}^{(k-1)}$ with the already computed ${\phi_{00}}^{(k)}$ and to do this wherever possible throughout the entire inner-iteration procedure. As might be expected, this method converges much faster than does the simple method of simultaneous relaxation described earlier. Mathematically, simultaneous relaxation can be expressed by rewriting Eq. (6.52) as

$$\mathbf{\Phi}^{(k)} = \mathbf{D}^{-1}\mathbf{S} - \mathbf{D}^{-1}\mathbf{N}\mathbf{\Phi}^{(k-1)}, \tag{6.54a}$$

while the equation for successive relaxation is

$$\mathbf{\Phi}^{(k)} = \mathbf{D}^{-1}\mathbf{S} - \mathbf{U}\mathbf{\Phi}^{(k-1)} - \mathbf{L}\mathbf{\Phi}^{(k)}. \tag{6.54b}$$

Here we see from Eq. (6.53) that the matrix **U** contains all the elements of $\mathbf{D}^{-1}\mathbf{N}$ *above* the diagonal, while **L** contains those below the diagonal.

The techniques for solving three-dimensional problems are very similar. Difference equations are obtained by integrating over a volume in the three-dimension cell. The simple scheme outlined above can easily be seen to lead to a nine-point formula in the three-dimensional problem. In any event, a matrix equation is set up, and relaxation techniques identical with those described above are employed.

In the next section we discuss some mathematical proofs that these iteration schemes actually converge to the correct solution.

6.5 PROOF OF CONVERGENCE

The schemes outlined in Secs. 6.2 and 6.4 for the outer and inner iterations, respectively, have been assumed to converge. Actually, a mathematical proof of convergence is required. Such a proof is beyond the scope of this book, but we shall indicate the sort of arguments which are used by the mathematicians when they are considering proofs of convergence.

Consider first the outer iterations. Here, we are considering the power method solution to Eq. (6.6*a*),

$$\mathbf{PS} = \lambda \mathbf{S}, \tag{6.55}$$

where we have written $\lambda = 1/\nu_c$. We assume that the operator **P** has associated with it a complete orthonormal set of eigenfunctions $\boldsymbol{\psi}_j$ with corresponding eigenvalues λ_j such that

$$\mathbf{P}\boldsymbol{\psi}_j = \lambda_j \boldsymbol{\psi}_j. \tag{6.56}$$

[1] We have explicitly used the fact that the **A** matrix represents a five-term linear system.

The initial source guess $\mathbf{S}^{(0)}$ is now expanded in terms of the $\boldsymbol{\psi}_j$,

$$\mathbf{S}^{(0)} = \sum_j c_j \boldsymbol{\psi}_j . \tag{6.57}$$

Since the kth iterate $\mathbf{S}^{(k)}$ is related to $\mathbf{S}^{(0)}$ by

$$\mathbf{S}^{(k)} = (\mathbf{P})^k \mathbf{S}^{(0)}, \tag{6.58}$$

we have, by virtue of Eq. (6.57),

$$\mathbf{S}^{(k)} = \sum_j c_j (\lambda_j)^k \boldsymbol{\psi}_j \tag{6.59a}$$

$$= (\lambda_0)^k \sum_j c_j \left(\frac{\lambda_j}{\lambda_0}\right)^k \boldsymbol{\psi}_j . \tag{6.59b}$$

Here λ_0 is the largest eigenvalue of all the λ_j. Thus, $(\lambda_j/\lambda_0)^k \to 0$ as $k \to \infty$ for $j \neq 0$, and Eq. (6.60b) approaches, in this limit,

$$\mathbf{S}^{(k)} = (\lambda_0)^k c_0 \boldsymbol{\psi}_0 . \tag{6.60}$$

This proves the convergence of $\mathbf{S}^{(k)}$ to $\boldsymbol{\psi}_0$, the eigenfunction of $\mathbf{P}$ associated with the *largest* eigenvalue λ_0 (and hence the *smallest* eigenvalue ν_c) on the assumption of the existence and completeness of the orthonormal set of eigenfunctions $\boldsymbol{\psi}_j$ and the existence of a largest eigenvalue λ_0 These problems have been studied by mathematicians, who find that these conditions are indeed fulfilled for the typical operators $\mathbf{P}$ that are encountered in reactor physics applications. In addition, they have shown that $\boldsymbol{\psi}_0$ is the only eigenfunction which is actually positive over the whole reactor volume, so that it is indeed the physically meaningful solution of Eq. (6.55).

It should be stressed that the Eq. (6.55) is only used for theoretical purposes. The actual solution of the reactor equations proceeds according to the steps outlined earlier. Although this point is not important in the one-group theory, it is of great moment in the multigroup solutions.

We now turn to the question of the convergence of inner iterations. This can be studied either for the problem of simultaneous relaxation [Equation (6.54a)] or successive relaxation (6.54b). Consider the former, and define an error vector $\boldsymbol{\varepsilon}_k$ on the kth iteration as the difference between $\boldsymbol{\Phi}^{(k)}$ and the exact solution, call it $\boldsymbol{\Phi}^{(\infty)}$:

$$\boldsymbol{\Phi}^{(\infty)} - \boldsymbol{\Phi}^{(k)} = \boldsymbol{\varepsilon}_k . \tag{6.61}$$

Clearly $\boldsymbol{\Phi}^{(\infty)}$ obeys Equation (6.52),

$$\boldsymbol{\Phi}^{(\infty)} = \mathbf{D}^{-1}\mathbf{S} - \mathbf{D}^{-1}\mathbf{N}\boldsymbol{\Phi}^{(\infty)}. \tag{6.62}$$

Thus since

$$\boldsymbol{\Phi}^{(k)} = \mathbf{D}^{-1}\mathbf{S} - \mathbf{D}^{-1}\mathbf{N}\boldsymbol{\Phi}^{(k-1)}, \tag{6.63}$$

$$\boldsymbol{\varepsilon}_k = -\mathbf{D}^{-1}\mathbf{N}\boldsymbol{\varepsilon}_{k-1}. \tag{6.64}$$

Define the eigenvalues and eigenfunctions of the matrix $\mathbf{D}^{-1}\mathbf{N}$ by

$$\mathbf{D}^{-1}\mathbf{N}\mathbf{X}_j = \lambda_j \mathbf{X}_j. \tag{6.65}$$

We now expand $\boldsymbol{\varepsilon}_k$ in terms of the $\mathbf{X}_j$.

$$\boldsymbol{\varepsilon}_k = \sum_j a_j^{(k)} \mathbf{X}_j,$$

so that

$$\boldsymbol{\varepsilon}_{k-1} = \sum_j a_j^{(k-1)} \mathbf{X}_j \tag{6.66}$$

or, since

$$\boldsymbol{\varepsilon}_k = (\mathbf{D}^{-1}\mathbf{N})^k \boldsymbol{\varepsilon}_0, \tag{6.67}$$

$$\boldsymbol{\varepsilon}_k = \sum_j a_j^{(0)} (\lambda_j)^{k-1} \mathbf{X}_j. \tag{6.68}$$

Thus $\boldsymbol{\varepsilon}_k \to 0$ as $k \to \infty$ if all the eigenvalues of the matrix $\mathbf{D}^{-1}\mathbf{N}$ are less than unity. The largest eigenvalue (in absolute value) of an operator is called the *spectral radius* of that operator. It has in fact been proved that the spectral radius of the operators which appear in the neutron diffusion equations are less than unity, so that convergence of the inner iteration scheme is guaranteed. In fact, it can be shown that the spectral radius for successive relaxation is *half* of that for the simultaneous relaxation method analyzed above. Hence, the rate of convergence in successive relaxation is twice as rapid.

This very sketchy analysis only provides an idea of the analysis that is necessary for a good understanding of the iteration schemes commonly used. A complete discussion may be found in various texts.[1]

6.6 EXTRAPOLATION METHODS

A number of techniques have been introduced to accelerate the convergence of both the outer and inner iterations. We shall discuss only one of them here, called the *extrapolation method.* Suppose, in an outer iteration, a first guess $\mathbf{S}^{(0)}$ is made, and $\mathbf{S}^{(1)}$ calculated. If it is assumed that $\mathbf{S}^{(0)}$ is a linear function of k, then the values of $\mathbf{S}^{(0)}$ and $\mathbf{S}^{(1)}$ could be extrapolated to $\mathbf{S}^{(2)}$ according to

$$\mathbf{S}^{(2)} = \mathbf{S}^{(1)} + (\mathbf{S}^{(1)} - \mathbf{S}^{(0)}). \tag{6.69}$$

Then $\mathbf{S}^{(2)}$ could be called $\mathbf{S}^{(1)}$. Generally, such a procedure is followed except that an extrapolation parameter α is introduced so that one uses

$$\mathbf{S}^{(2)} = \mathbf{S}^{(1)} + \alpha(\mathbf{S}^{(1)} - \mathbf{S}^{(0)}), \tag{6.70}$$

[1] E. Wachspress, *op. cit.*

or better, since $\mathbf{S}^{(2)}$ using (6.8),

$$\mathbf{S}^{(2)} = \mathbf{P}\mathbf{S}^{(0)} + \alpha(\mathbf{P} - 1)\mathbf{S}^{(0)} \tag{6.71a}$$

$$= (\alpha + 1)\mathbf{P}\mathbf{S}^{(0)} - \alpha\mathbf{S}^{(0)}. \tag{6.71b}$$

Here α is chosen between 0 and 1. Of course, $\alpha = 0$ corresponds to the unextrapolated power method. The optimum choice of α is a complicated procedure, and is discussed at length by Wachspress.[1] The same procedure can be applied to inner iterations, in which case the procedure is called *overrelaxation.* Thus, one speaks of *successive overrelaxation* (SOR).

An extension of the simple extrapolation would be to use *all* the previous iterates to generate the next guess. This procedure is sometimes used, and it turns out that the optimum choice of coefficients of the previous k iterates is related to the Tchebychef polynomial of order k. For this reason, the method is referred to as the *Tchebychef polynomial method.* For more information on these, as well as other methods for dealing with inner iterations (e.g., direct matrix inversion and the so-called ADI procedure) the reader is again referred to Wachspress.[2]

6.7 BUCKLING ITERATION, FLUX SYNTHESIS

Sometimes it is desirable to carry out a two-dimensional calculation, say, when only a one-dimensional code is available, or perhaps a three-dimensional calculation when only one- or two-dimensional codes are available. This may be accomplished by a method called *flux synthesis*, or *buckling iteration.*

We note from the critical equation (3.17), which may be written in the form

$$1 = \nu\Sigma_f(\Sigma_a + DB^2)^{-1}, \tag{6.72}$$

that DB^2 may be interpreted as a leakage cross section. Also, we recall from Sec. (3.5) that for two- and three-dimensional systems, the buckling B^2 can be written as the sum of buckling components along the various coordinate axes. Thus, for example, in a cylindrical system we find

$$B^2 = B_\rho{}^2 + B_z{}^2, \tag{6.73}$$

where $B_\rho{}^2 = (2.405/R')^2$ is called the *radial buckling*, and $B_z{}^2 = (\pi/2H')^2$ is called the *axial buckling.*

[1] E. Wachspress, *op. cit.*
[2] E. Wachspress, *ibid.*

In addition, ∇^2 can be written as a sum of components by using Eq. (3.51), which gives

$$\nabla^2 = \nabla_\rho^2 + \nabla_z^2, \tag{6.74}$$

where

$$\nabla_\rho^2 = \frac{\partial^2}{\partial \rho^2} + \frac{1}{\rho}\frac{\partial}{\partial \rho} \tag{6.75a}$$

and

$$\nabla_z^2 = \frac{\partial^2}{\partial z^2}. \tag{6.75b}$$

We now assume flux separability, writing

$$\phi(\rho,z) = Z(z)R(\rho). \tag{6.76}$$

We shall first perform an axial calculation. That is, we take, as a first approximation,

$$R(\rho) \approx J_0(B_\rho \rho) \tag{6.77}$$

and guess at B_ρ, for example, by taking the geometric buckling of a bare cylinder (see Sec. 3.5*b*). Now

$$-D(\nabla_\rho^2 + \nabla_z^2)\phi(\rho,z) + (\Sigma_a - \nu\Sigma_f)\phi(\rho,z) = 0; \tag{6.78}$$

since

$$\nabla^2 J_0(B_\rho \rho) = -B_\rho^2 J_0(B_\rho \rho), \tag{6.79}$$

Eq. (6.78) reduces to

$$(-D\nabla_z^2 + DB_\rho^2 + \Sigma_a - \nu\Sigma_f)Z(z) = 0. \tag{6.80}$$

This equation can be solved (numerically) as a one-dimensional equation, since we have assumed an initial B_ρ^2. Once $Z(z)$ is known, the axial leakage can be determined.

$$\text{Axial leakage} = \int_{\text{ends}} J_z \, dS, \tag{6.81a}$$

where

$$J_z = -D\frac{\partial Z}{\partial z} R(\rho). \tag{6.81b}$$

Thus, Eq. (6.81*a*) reduces to

$$\text{Axial leakage} = \int_0^R \rho \, d\rho \, R(\rho)[J_z(h) + J_z(-h)]. \tag{6.81c}$$

This leakage can be equated with the integral, over the volume, of $DB_z^2\phi(\rho,z)$,

$$\text{Axial leakage} = B_z^2 \int_{-h}^{h} dz \int_0^R \rho \, d\rho \, D(\rho,z)Z(z)R(\rho). \tag{6.82}$$

Since all the quantities in Eq. (6.81c) and (6.82) are known (or in the case of $R(\rho)$, assumed known), B_z^2 can be determined.

This value of B_z^2 is now used in an equation analogous to (6.80), for the radial function $R(\rho)$. Thus, a radial buckling might be computed in the manner described above for the calculation of B_z^2 and compared with the original guess. If it does not agree, this procedure can be iterated, although it does not necessarily converge for every case.

If we consider a two-dimensional problem with 100 mesh points in each direction, the total number of mesh points is 10^4, and the machine time is at least proportional to this number. Thus, 100 one-dimensional calculations might be made at the same expense in machine time as 1 two-dimensional calculation. As considerably fewer than 100 iterations should be required to make this flux synthesis converge, one can see its immediate advantage. However, it should be noted that this method is not so accurate as a full two-dimensional calculation; so for detailed design calculation the latter must be preferred.

Implicit in this method is the assumption that the flux is separable, i.e., that the radial flux distribution is the same at every height z and the axial distribution is the same at every radius. Of course, in an actual reactor the flux is never really separable, so that the synthesis procedure is inherently less accurate than a full multidimensional calculation. However, if the reactor is constructed in a fairly symmetric manner, flux separability may be a reasonable approximation, and answers provided by synthesis calculations would be suitable for many purposes. For detailed design studies, however, the synthesis method is generally not adequate, and multidimensional calculations are required. More details have been given in a number of papers.[1]

PROBLEMS

6.1. Derive the nine-point difference equations for a three-dimensional x, y, z coordinate system.

6.2. Derive the five-point difference equations for an infinite cylinder ($\beta\theta$). (Do not assume azimuthal symmetry.)

6.3. Derive the five-point difference equations for a finite cylinder with azimuthal symmetry (ρ, z).

[1] See, e.g., E. Wachspress, *op. cit.*, for further references.

7 Reactor Control

7.1 INTRODUCTION

The reasons for the necessity of varying the reactivity for an operating reactor have already been discussed in a previous chapter. These, we recall, included the necessity of making a reactor slightly supercritical or subcritical in order to increase or decrease the power level, the necessity of compensating for fuel depletion and fission-product poisoning, as well as the need to compensate for small fluctuations in reactivity caused by temperature and other effects. In the present chapter we discuss some methods for calculating the effect on ρ of inserting a poison element (i.e., a control rod) into or withdrawing one from an operating reactor as well as some means of estimating the reactivity changes caused by fuel depletion and fission-product poisoning.

The most straightforward method for calculating the reactivity worth of a control rod is to carry out a numerical solution of the transport equation(s) by techniques similar to those described in Chap. 6.[1] Thus a number of

[1] Diffusion theory is not adequate because control rods generally have high-absorption cross sections.

mesh points are located on the boundaries and throughout the interior of the control rod (because we expect the flux to be rapidly varying there). We shall not consider this approach further in the present chapter. We shall, however, discuss one or two analytical approaches which can sometimes prove useful if, for one reason or another, the numerical method is not used.

7.2 DIRECT CALCULATION OF ROD WORTH

One approach to the calculation of control-rod worth which immediately suggests itself is the use of the variational or perturbation theoretical methods described in Chap. 5. Such a procedure is perhaps feasible for a fast reactor, where because of the general decrease of absorption cross sections with increasing neutron energy, the cross section of the rod may be small. However, in thermal reactors the nuclear cross sections of the rods are so large that these methods are not very accurate.

Another approach is the direct analytical solution of the diffusion equation for a reactor core containing one or more rods. While such calculations can indeed be carried out for sufficiently simple geometrical cases, their practical application is limited. However, just for the sake of demonstrating how such calculations might be carried out, we here describe the methods which might be used to compute the worth of control *slabs* in a slab reactor.[1]

As a first example, consider a slab reactor of width $2a$ with a control slab of width $2d$ in the center. If we assume that the slab is *black*, i.e., it absorbs all neutrons which strike it, then the two halves of the reactor are effectively decoupled into two reactors of extrapolated width $a - d + 2z_0$, where z_0 is the extrapolation length for the reactor core.

In any event, the reactivity of the entire system is simply equal to the reactivity of each half separately; so calling

$$a_1 = a - d + 2z_0 \tag{7.1}$$

and a' the extrapolated half-width of the bare reactor, then with the control element removed[2]

$$\nu_c = \frac{D(\pi/2a')^2 + \Sigma_a}{\Sigma_f}, \tag{7.2}$$

[1] In the literature some calculations are described for control rods in cylindrical reactors, but we shall not go into these here, because the mathematical complexities scarcely justify the practical results available from such calculations. See, however, Sec. 7.5.

[2] Here we assume that when the control element is removed, somehow core material replaces it.

while when it is inserted,

$$\nu_{c1} = \frac{D(\pi/2a_1)^2 + \Sigma_a}{\Sigma_f}. \tag{7.3}$$

Then the δk associated with the rod is

$$\delta k = \frac{\nu_c - \nu_{c1}}{\nu} = \frac{D}{\nu\Sigma_f}\frac{\pi^2}{4}\left(\frac{1}{a'^2} - \frac{1}{a_1{}^2}\right). \tag{7.4}$$

The introduction of a black control element may *not* completely decouple two parts of a reactor, for example, if a rod is inserted into a cylindrical reactor. Then the flux must be found by the solution of the diffusion equation, subject to the appropriate boundary conditions at the surface of the rod and the surface of the reactor. Then, ν_c can be found simply from

$$\nu_c = \frac{\int -\nabla \cdot D\nabla\phi \, d^3r + \int \Sigma_a\phi \, d^3r}{\int \Sigma_f\phi \, d^3r}, \tag{7.5}$$

where ϕ is the exact solution.

7.3 THE SELF–SHIELDING FACTOR

Clearly, if poison were inserted uniformly into a reactor and not in the form of a rod, or discrete element, it would be easier to calculate the reactivity worth of the poison. The difficulty that arises in the calculation for lumped poisons is that there is a large flux depression within the poison lump and this factor is very difficult to compute.

Suppose, for example, our reactor were composed of two regions labeled *A* and *B*. Region *A* is assumed to be the poison lump. Figure 7.1 shows the special case of two slabs separated by a poison slab.

The regions *A* and *B* will have different absorption cross sections (say, $\Sigma_a{}^A$ and $\Sigma_a{}^B$). We can denote the flux for this composition as ϕ_1. Clearly we must specify considerably more than the absorption cross sections to determine ϕ_1. We must know the fission cross sections, sources, diffusion lengths, and the various other parameters. Let us assume that all these factors are specified and that in fact ϕ_1 is known.

Now suppose that we replace the material in region *A* with material identical with that in region *B* (including sources). The flux in this case can be denoted by ϕ_2. A representative sketch of these two fluxes in slab geometry is given in Fig. 7.1.

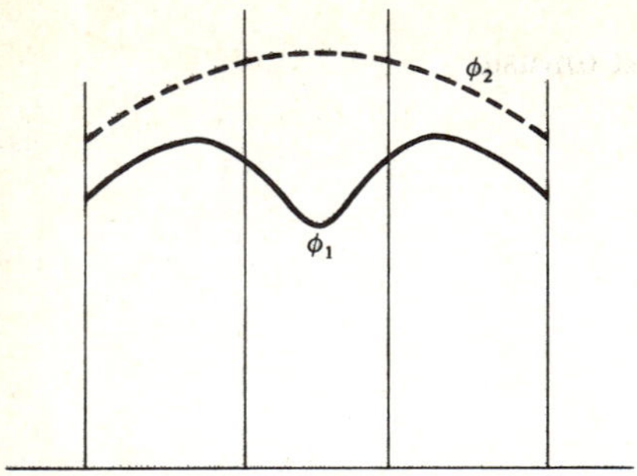

Fig. 7.1 The solid line is the flux when the absorption cross section in A is $\Sigma_a{}^A$ and B is $\Sigma_a{}^B$; the broken line is the flux when the two regions have been "homogenized."

The average fluxes in region A in these two cases are given by

$$\bar{\phi}_{1A} = \frac{1}{V_A}\int_{V_A} d^3r\, \phi_1(\mathbf{r}) \tag{7.6a}$$

$$\bar{\phi}_{2A} = \frac{1}{V_A}\int_{V_A} d^3r\, \phi_2(\mathbf{r}). \tag{7.6b}$$

The self-shielding factor of the poison in region A is defined to be

$$F = \frac{\bar{\phi}_{1A}}{\bar{\phi}_{2A}}. \tag{7.6c}$$

Then the absorption rate in A for case 1 is

$$R_A = \bar{\phi}_{1A} V_A \Sigma_a{}^A \tag{7.7a}$$

$$= \bar{\phi}_{2A} V_A \Sigma_a{}^A F. \tag{7.7b}$$

What we have accomplished here is to relate the absorption rate in a poison lump to the flux which would be present if the poison lump were replaced by material identical with the surrounding material. Eventually we shall use this result to obtain a method for "smearing" the poison lump over the entire region $A + B$.

The reason for introducing the self-shielding factor is that F does not depend strongly on the detailed features of region B. One can compute F once for a specific composition of region B. Then, in subsequent calculations, for moderate changes in the composition of B, one can use this value of F and compute the absorption rate by the (simpler) computation of $\bar{\phi}_{2A}$ instead of the more complicated computation of $\bar{\phi}_{1A}$. (The calculation of $\bar{\phi}_{2A}$ is simpler than that of $\bar{\phi}_{1A}$ because $\bar{\phi}_{2A}$ refers to a *homogeneous* medium.) If F can be accurately calculated in diffusion theory, the extra complication of introducing it is not worth the savings in computer time. Poison elements are usually composed of materials with very large absorption cross sections. We have already noted, in Chap. 2, that diffusion theory is not accurate for materials with large absorption cross sections. Simply stated, the main reason for introducing F is that the neutron distribution must be computed from the transport theory. The foregoing arguments about the relative insensitivity of F to slight changes in the composition of region B must still hold; however, the reason for introducing F is, as stated, the importance of transport effects in poison elements.

To show how F can be computed, we must make a definition and prove a theorem.

DEFINITION 7.1

Consider two homogeneous bodies A and B which may or may not be simply connected[1] and may or may not incorporate all space. P_{AB} *is defined as the probability that neutrons, emitted uniformly and isotropically in body A, will be absorbed in body B. Clearly, we also define quantities* P_{BA}, P_{AA}, *and* P_{BB}. *If bodies A and B incorporate all space, then by definition,*

$$P_{AB} + P_{AA} + P_{BA} + P_{BB} = 1. \tag{7.8}$$

In this case, P_{AB} is called the *escape probability* for neutrons born in body A, and it is denoted by P_A. The term P_{BA} is simply the escape probability for body B.

Theorem 7.1

$$P_{AB} = P_{BA} \frac{V_B \Sigma_a^B}{V_A \Sigma_a^A}. \tag{7.9}$$

Here V_B/V_A *is the volume ratio of body B to body A, while of course* Σ_a^A *and* Σ_a^B *represent their absorption cross sections. Note that* P_{BA} *refers to a problem with the source in region B.*

To prove this theorem we need to prove the following.

Lemma The Green's function of the one-speed diffusion equation is a symmetric function of its argument, i.e.,

$$G(\mathbf{r}_1 \to \mathbf{r}_2) = G(\mathbf{r}_2 \to \mathbf{r}_1). \tag{7.10}$$

Actually, this lemma is trivially true for the infinite medium because, as we have seen, in that case $G = G(|\mathbf{r}_1 - \mathbf{r}_2|)$. However, the lemma is actually true for any system. We know it is true if we recall that the Green's function of

[1] That is "body" A may be a *number* of bodies of arbitrary shape but identical composition scattered throughout the medium, and body B, the same.

an operator and its adjoint are related by the interchange of source and field variables (that is, $\mathbf{r}_1$ and $\mathbf{r}_2$) and that the one-speed diffusion operator is self-adjoint. We can also prove the lemma directly in the following way: Write down the equations obeyed by Green's functions[1] $G(\mathbf{r}_1 \to \mathbf{r})$ and $G(\mathbf{r}_2 \to \mathbf{r})$:

$$-\nabla \cdot D\,\nabla G(\mathbf{r}_1 \to \mathbf{r}) + \Sigma_a(\mathbf{r})G(\mathbf{r}_1 \to \mathbf{r}) = \delta(\mathbf{r} \to \mathbf{r}_1), \tag{7.11a}$$

$$-\nabla \cdot D\,\nabla G(\mathbf{r}_2 \to \mathbf{r}) + \Sigma_a(\mathbf{r})G(\mathbf{r}_2 \to \mathbf{r}) = \delta(\mathbf{r} \to \mathbf{r}_2). \tag{7.11b}$$

Now multiply Eq. (7.11*a*) by $G(\mathbf{r}_2 \to \mathbf{r})$, Eq. (7.11*b*) by $G(\mathbf{r}_1 \to \mathbf{r})$, integrate over $\mathbf{r}$, and subtract. The first terms, i.e., those with the ∇ operator, may be integrated by parts, in which case they are seen to be identical (the surface contribution may be assumed to vanish by virtue of the usual boundary condition). Similarly the absorption terms subtract, and carrying out the integration over the delta distributions, we find (7.10).

Now let us return to the theorem.

Proof If there is a uniform source q in region A, the flux in region B is

$$\phi^B(\mathbf{r}) = q \int_A d^3r'\, G(\mathbf{r}' \to \mathbf{r}). \tag{7.12}$$

Then P_{AB} is just the total number of absorptions in B divided by the total number of neutrons produced in A:

$$P_{AB} = \Sigma_a{}^B \int_B \frac{\phi^B(\mathbf{r})\, d^3r}{qV_A} \tag{7.13a}$$

$$= \frac{\Sigma_a{}^B}{V_A} \int_B d^3r \int_A d^3r'\, G(\mathbf{r}' \to \mathbf{r}). \tag{7.13b}$$

In the same way we find that P_{BA} can be written

$$P_{BA} = \frac{\Sigma_a{}^A}{V_B} \int_A d^3r \int_B d^3r'\, G(\mathbf{r}' \to \mathbf{r}). \tag{7.13c}$$

Comparing (7.13*b*) and (7.13*c*) and recalling the symmetry of $G(\mathbf{r}' \to \mathbf{r})$, we see that (7.9) is indeed true and the theorem is proved.

Let us now assume that region B contains a uniform source q. Then the absorption rate in A is just the number of neutrons produced in B times P_{BA},

$$R_A = qV_B P_{BA}. \tag{7.14a}$$

[1] We are using diffusion theory to prove the lemma. It can also be shown in transport theory: K. M. Case and P. F. Zweifel: "Linear Transport Theory," Addison-Wesley Publishing Company, Inc., Reading, Mass., 1967.

This quantity is related to the average flux in region A by (7.7a) so that

$$\bar{\phi}_{1A} = \frac{qV_B P_{BA}}{V_A \Sigma_a{}^A}. \tag{7.14b}$$

If regions A and B combined are very large, then ϕ_2 (the flux when the material of A is replaced with material identical with that in B) is spatially constant [see Eq. (2.107b)],

$$\phi_2 = \frac{q}{\Sigma_a{}^B} = \bar{\phi}_{2A}. \tag{7.14c}$$

Since the self-shielding factor is defined to be the ratio of $\bar{\phi}_{1A}$ and $\bar{\phi}_{2A}$, we have

$$F = \frac{\Sigma_a{}^B V_B}{\Sigma_a{}^A V_A} P_{BA}, \tag{7.15a}$$

or using (7.9),

$$F = P_{AB}. \tag{7.15b}$$

Here we have determined F when region B contains a constant isotropic source. We shall see in Chap. 9 that the distribution of thermal neutrons being produced by slowing-down of fission spectrum neutrons is approximated by a constant isotropic source. Thus we can expect that the self-shielding factor we have derived here will apply for thermal neutrons in reactors.

We may wish to smear the poison out over regions A and B. Considering the self-shielding effects we have just discussed, we need to determine the effective absorption cross section to add to the combined regions. As we show in Fig. 7.2, regions A and B need not occupy the whole reactor.

Specifically, we need a flux ϕ_3 when the absorption cross section in regions A and B is $\Sigma_a{}^{\text{eff}} + (V_B/V)\Sigma_a{}^B$, and $\Sigma_a{}^{\text{eff}}$ is chosen to give the absorption rate (7.14a) and $V = V_A + V_B$. We still assume that V is large enough to use the infinite medium solution,

$$\phi_3 = \frac{q}{V_B \Sigma_a{}^B/V + \Sigma_a{}^{\text{eff}}}. \tag{7.16}$$

The absorption rate from the added poison $\Sigma_a{}^{\text{eff}}$ is

$$R = \phi_3 V \Sigma_a{}^{\text{eff}}. \tag{7.17a}$$

This is the quantity we want equal to the absorption rate in the original poison element (7.14a), that is,

$$\phi_3 V \Sigma_a{}^{\text{eff}} = R_A = qV_B P_{BA} \tag{7.17b}$$

$$= q \frac{\Sigma_a{}^A V_A}{\Sigma_a{}^B} F. \tag{7.17c}$$

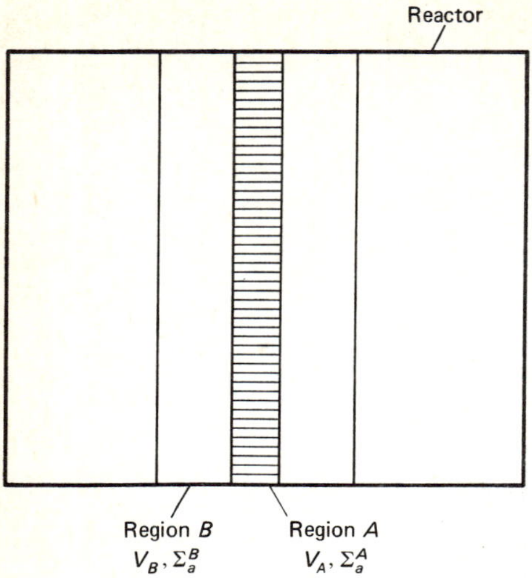

Fig. 7.2 When the material in A and B is smeared uniformly throughout $A+B$, $\Sigma_a{}^A$ must be replaced by $\Sigma_a{}^{\text{eff}}$ as given by Eq. (7.18). In that equation F refers to the self-shielding factor of region A.

Solving this equation for $\Sigma_a{}^{\text{eff}}$, we find

$$\Sigma_a{}^{\text{eff}} = \frac{V_B V_A}{V^2} F \frac{\Sigma_a{}^A}{1 - (\Sigma_a{}^A V_A / \Sigma_a{}^B V) F}. \tag{7.18}$$

The calculation of escape probabilities in diffusion theory is quite trivial. In fact, escape probabilities have been computed by exact, one-speed transport theory, and numerical values have been widely tabulated.[1] However, it must be admitted that most tabulations are of *first-flight* escape probabilities (see Sec. 7.4.).

For the moment let us ignore the inaccuracy of diffusion theory for such calculations, and compute the escape probability for a slab of width $2a$ with infinite *reflectors* (Fig. 7.3).

We have in region A

$$\left(-D^A \frac{d^2}{dx^2} + \Sigma_a{}^A\right) \phi_A(x) = q, \tag{17.19a}$$

and in B (the reflector region)

$$\left(-D^B \frac{d^2}{dx^2} + \Sigma_a{}^B\right) \phi_B(x) = 0. \tag{7.19b}$$

[1] K. M. Case, F. de Hoffman, and G. Placzek: "Introduction to the Theory of Neutron Diffusion," Los Alamos Scientific Laboratory, Los Alamos, N. Mex., 1953.

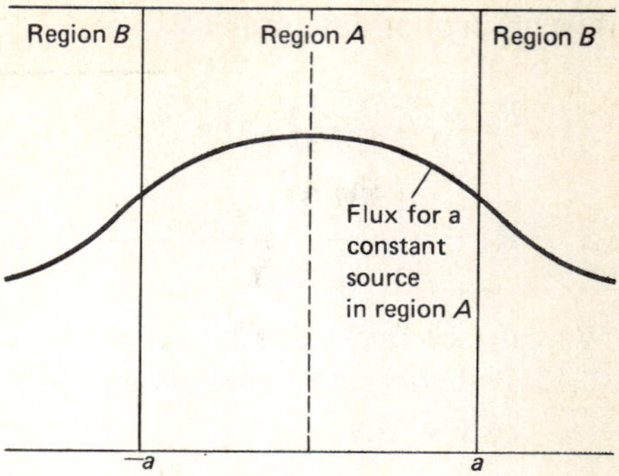

Fig. 7.3 Reflected slab for calculation of escape probability.

Then

$$\phi_A(x) = \frac{q}{\Sigma_a^A} + C_A \cosh \kappa_A x, \tag{7.20a}$$

and

$$\phi_B(x) = C_B e^{-\kappa_B x}, \tag{7.20b}$$

where we have used the fact that $\phi(x) = \phi(-x)$.[1] We can find the coefficient C_A by equating the fluxes and currents of ϕ_A and ϕ_B at $x = a$:

$$C_A = -\frac{q}{\Sigma_a^A}\left(\frac{\kappa_A D^A}{\kappa_B D^B} \sinh \kappa_A a + \cosh \kappa_A a\right)^{-1}. \tag{7.21}$$

Now we can compute

$$P_{AA} = \frac{\Sigma_a^A \int_A \phi_A(x)\, dx}{2aq} \tag{7.22a}$$

and of course

$$P_{AB} = 1 - P_{AA}. \tag{7.22b}$$

Using (7.20*a*) for the flux and (7.21) for the coefficient C_A, we can carry out the integration in (7.22) to obtain

$$P_{AB} = \frac{\sinh \kappa_A a}{\kappa_A a}\left(\cosh \kappa_A a + \frac{D^A \kappa_A}{D^B \kappa_B} \sinh \kappa_A a\right)^{-1}. \tag{7.23a}$$

[1] Here $\kappa = \sqrt{\Sigma_a/D}$.

The absorption rate in the slab is now given by (7.7) and (7.15*b*):

$$R_A = \bar{\phi}_B V \frac{V_A}{V} P_{AB} \Sigma_a{}^A \tag{7.23b}$$

$$R_A = \frac{q}{\Sigma_a{}^B} \frac{V_A \sinh \kappa_A a}{\kappa_A a} \frac{1}{(D^A \kappa_A / D^B \kappa_B) \sinh \kappa_A a + \cosh \kappa_A a}. \tag{7.23c}$$

We can see that this rate is in agreement with the variational calculation (5.151) when $\kappa_A a$ is small enough to permit the expansions

$$\cosh \kappa_A a \approx 1, \tag{7.24a}$$

$$\sinh \kappa_A a \approx \kappa_A a. \tag{7.24b}$$

(*Note*: $D\kappa = L$.)

As a practical matter, most self-shielding calculations are applied to problems of heavy absorbers, in which we have already noted that diffusion theory is inaccurate. For this reason, it is better to use the tabulated and plotted escape probabilities, referred to earlier, which are obtained by more accurate theories. The calculations above were included mostly for their academic interest, and to give an idea of how self-shielding factors are used in reactor calculations.

7.4 FIRST–FLIGHT ESCAPE PROBABILITY

We have seen in the preceding section that the self-shielding factor is closely related to the escape probability P_{AB}. Frequently the escape probability is closely approximated by the first-flight escape probability P_0^*, which is considerably easier to calculate. The *first-flight escape probability* P_0^* (sometimes called *collision probability*) is defined as the probability that a neutron's flight ends outside of the region in which the neutron was produced. The diffusion theory calculation outlined above calculated true escape probabilities. If both regions are purely absorbing (i.e., not scattering), then these two quantities are identical.

The difficulty here is that true escape probabilities are easy to calculate in the diffusion approximation, but the diffusion theory is not accurate for heavy absorbers. Using more sophisticated transport calculations, however, makes it very difficult to take scattering into account.

Consider the special case in which region A is an isolated, convex object as in Fig. 7.4. Then once a neutron has escaped through the surface of this object, along $\mathbf{R}(\mathbf{\Omega}, v)$ say, it cannot return without undergoing a collision in B. Thus, we need only calculate the probability that the neutron reaches the surface in this case. Of course the first-flight escape probability from an isolated, convex object (which we denote by P_0) does not depend upon the material surrounding the object.

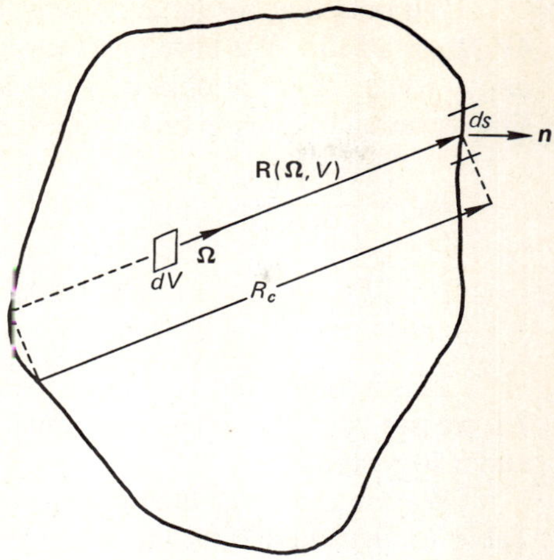

Fig. 7.4 **n** is the normal to the surface dS; V_0 is the total volume and S_0 the total surface area.

Examining Fig. 7.4, we can see that P_0 is given by

$$P_0 = \frac{1}{4\pi V_0}\int_{V_0} dV \int d\Omega \exp[-\Sigma_T{}^A R(\mathbf{\Omega}, v)]. \tag{7.25}$$

It is convenient to choose the volume element dV as

$$dV = dR(\mathbf{\Omega},v)\, dS\, \mathbf{\Omega}\cdot\mathbf{n}, \qquad \mathbf{\Omega}\cdot\mathbf{n} > 0. \tag{7.26}$$

Then P_0 becomes

$$P_0 = \frac{1}{4\pi V_0}\int_{\mathbf{\Omega}\cdot\mathbf{n}>0} d\mathbf{\Omega} \int_{S_0} dS\, \mathbf{\Omega}\cdot\mathbf{n} \int_0^{R_c} dR \exp(-\Sigma_T{}^A R) \tag{7.27a}$$

$$= \frac{1}{4\pi V_0}\int_{\mathbf{\Omega}\cdot\mathbf{n}>0} d\mathbf{\Omega} \int_{S_0} dS\, \mathbf{\Omega}\cdot\mathbf{n}\, \frac{1-\exp(-\Sigma_T{}^A R_c)}{\Sigma_T^A}. \tag{7.27b}$$

For very small objects, $\Sigma_T{}^A R_c \ll 1$, and every neutron in the object must escape so that

$$P_0{}^{\text{small}} = 1. \tag{7.28a}$$

Also for very large objects, $\Sigma_T{}^A R_c \gg 1$ and

$$P_0{}^{\text{large}} \approx \frac{1}{4\pi V_0}\int_{\mathbf{\Omega}\cdot\mathbf{n}>0} d\mathbf{\Omega} \int_{S_0} dS\, \mathbf{\Omega}\cdot\mathbf{n}\, \frac{1}{\Sigma_T^A} \tag{7.28b}$$

$$= \frac{1}{4\pi V_0 \Sigma_T{}^A}\int_{S_0} dS \int_0^{2\pi} d\phi \int_0^1 d\mu\, \mu \tag{7.28c}$$

$$= \frac{S_0}{4 V_0 \Sigma_T{}^A}. \tag{7.28d}$$

We note that for large objects, $4V_0/S_0$ is large, and for small objects, $4V_0/S_0$ is small. We have just shown that for small objects, P_0 is equal to unity, while for large objects, P_0 is inversely proportional to $4V_0/S_0$. A simple, rational function of $4V_0/S_0$ with these limits is

$$P_0{}^R = \frac{1}{1 + \bar{r}\Sigma_T{}^A}, \tag{7.29a}$$

$$\bar{r} = \frac{4V_0}{S_0}. \tag{7.29b}$$

This is called the *Wigner rational approximation* to P_0. From Table 7.1 we see that the Wigner rational approximation is accurate to within about 20 percent.

We can do better than (7.29) in approximating P_0. We can see from Table 7.1 that there is, at most, a 9 percent difference in P_0 among the three geometries. Thus we can expect to find an approximation to P_0 which is accurate to within about 5 percent. In fact

$$P_0{}^r = \frac{10 + 5\bar{r}\Sigma_T{}^A + 5(\bar{r}\Sigma_T{}^A)^2}{10 + 14r\Sigma_T{}^A + 5(\bar{r}\Sigma_T{}^A)^2 + 5(\bar{r}\Sigma_T{}^A)^3} \tag{7.29c}$$

gives the P_0 listed in Table 7.1 to within 6 percent.

As we have stated, the preceding comments apply only to the first-flight escapes from isolated, convex absorbers. For closely spaced absorber lumps or concave surfaces on the absorber, the computations are considerably more complicated. In these cases, it is possible for a neutron to leave and reenter the surfaces of absorber lumps without undergoing a collision.

Table 7.1 The escape probability P_0 as a function of mean chord length over mean free path for various geometries and in the rational approximation, where $\bar{r} = 4V_0/S_0$

$\bar{r}\Sigma_T{}^A$	*Sphere*	*Cylinder*	*Slab*	*Rational*
0.04	0.978	0.974	0.952	0.962
0.1	0.947	0.939	0.902	0.909
0.2	0.896	0.885	0.837	0.823
0.3	0.850	0.819	0.785	0.769
0.5	0.767	0.753	0.701	0.667
1	0.607	0.596	0.557	0.500
2	0.411	0.407	0.390	0.333
3	0.302	0.302	0.295	0.250
5	0.193	0.193	0.193	0.167
10	0.099	0.099	0.100	0.091

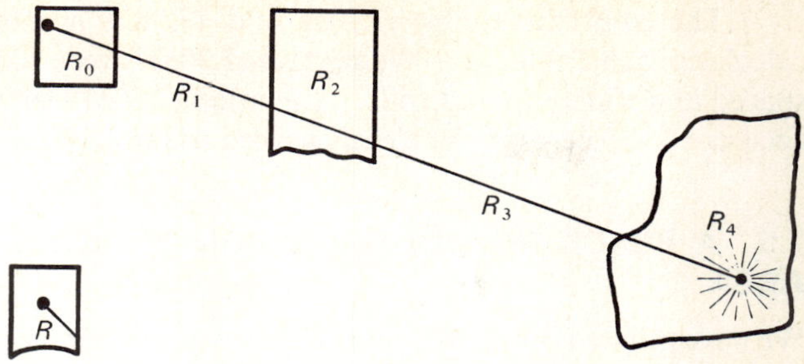

Fig. 7.5 Typical flight path of neutron in homogeneous assembly.

Other absorbers may be quite near the first absorber. Then it is possible that a neutron, leaving the surface of the first absorber and traversing one or more of the regions surrounding the absorber, will undergo its next collision in another absorber (see Fig. 7.5). Then we must compute P_0^* including this possibility. We shall not derive this result. However, to an excellent approximation (see Table 7.2), we can write

$$P_0^* = P_0 \frac{1 - c}{1 - (1 - \Sigma_T{}^A \bar{r} P_0)c}, \tag{7.30a}$$

where c is what is called the *Dancoff-Ginsberg correction*. It is tabulated in a number of references (for example, ANL 5800). An alternative and somewhat less accurate approximation for P_0^* is

$$P_0^*(\bar{r}) = P_0 \left(\frac{\bar{r}}{1 - c}\right). \tag{7.30b}$$

That is, P_0^* for closely spaced absorbers is approximately equal to the escape probability for an isolated absorber for which the volume-to-surface ratio increased by a factor of $1/(1 - c)$.

Table 7.2 Various approximations for P_0^* for rods in a square lattice in an equal volume of water

Rod diameter	$\bar{r}\Sigma_T{}^A$	c	P_0^*	$\frac{P_0(1 - c)}{1 - (1 - \bar{r}\Sigma_T{}^A P_0)c}$	$P_0 \left(\frac{\bar{r}}{1 - c}\right)$
0.25 in	0.5	0.464	0.572	0.571	0.611
0.25 in	1	0.464	0.397	0.393	0.425
0.25 in	6	0.464	0.088	0.088	0.089
0.60 in	0.5	0.208	0.691	0.689	0.710
0.60 in	1	0.208	0.522	0.520	0.535
0.60 in	6	0.208	0.133	0.130	0.131

The approximations (7.30a) and (7.30b) are compared with exact calculations in Table 7.2. We see that (7.30a) is very accurate for both closely spaced and widely spaced rods. Equation (7.30b) seems to be somewhat less accurate ($\sim$10 percent) for closely spaced rods.

7.5 A BLACK ROD IN DIFFUSION THEORY: ABSORPTION AREA

We have pointed out, in previous sections of this chapter, that diffusion theory cannot predict accurately the absorption in highly absorbing (black) media. However, it is possible, by using the self-shielding factor, to reduce the *effective* cross section to a point that diffusion theory becomes valid, which is actually the whole point of the self-shielding factor anyway. In this section we apply diffusion theory to the calculation of the reactivity of a black control rod in a reactor, recognizing that if the rod is really black, the results have little validity. However, the ideas can be used to understand what happens in the "gray" case, i.e., the situation in which the absorber effective cross section has been reduced by the introduction of the self-shielding factor.

Our black rod, region A, is immersed in a finite region B containing a uniform source (Fig. 7.6).

Earlier we described how we might perform this computation for a black control slab inserted in a slab reactor. We performed this calculation by assuming that the extrapolated flux vanished inside the control slab according to

$$\left.\frac{\phi(x)}{d\phi/dx}\right|_{\text{boundary}} = 0.71/\Sigma_{\text{tr}}{}^{B} \tag{7.31a}$$

$$= d. \tag{7.31b}$$

When we try to do the same calculation in a more complicated geometry, we cannot use an equation as simple as (7.31). In particular, we cannot use (7.31) for the case of the black cylindrical control rod.

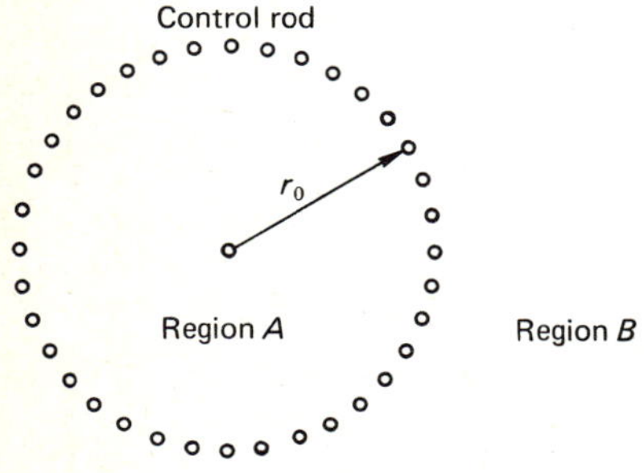

Fig. 7.6 Black control rod.

Several authors[1] have studied the appropriate boundary conditions for black cylindrical control rods. We can summarize their results as

$$\left.\frac{\phi}{d\phi/d\rho}\right|_{r0} = d(r_0) \tag{7.32a}$$

$$d(r_0) = \frac{0.71 + 0.31/(r_0 \Sigma_{tr}{}^B + 0.50)}{\Sigma_{tr}{}^B}. \tag{7.32b}$$

This result is sketched in Fig. 7.7. We see that in the limit of large r_0, the result goes over into the well-known result (7.31). Of course (7.32) is an approximate expression in the spirit we used to obtain Eq. (2.60).

We begin by solving the diffusion equation in region B,

$$(-D^B\nabla^2 + \Sigma_a{}^B)\phi(\rho) = q, \tag{7.33a}$$

or recalling ∇^2 in cylindrical coordinates (2.116),

$$\left(-D^B \frac{1}{\rho}\frac{d}{d\rho}\rho\frac{d}{d\rho} + \Sigma_a{}^B\right)\phi(\rho) = q. \tag{7.33b}$$

The particular solution of (7.33) is

$$\phi_p(\rho) = \frac{q}{\Sigma_a{}^B}. \tag{7.34a}$$

[1] For example, Alvin M. Weinberg and Eugene P. Wigner: "The Physical Theory of Neutron Chain Reactors," The University of Chicago Press, Chicago, 1958.

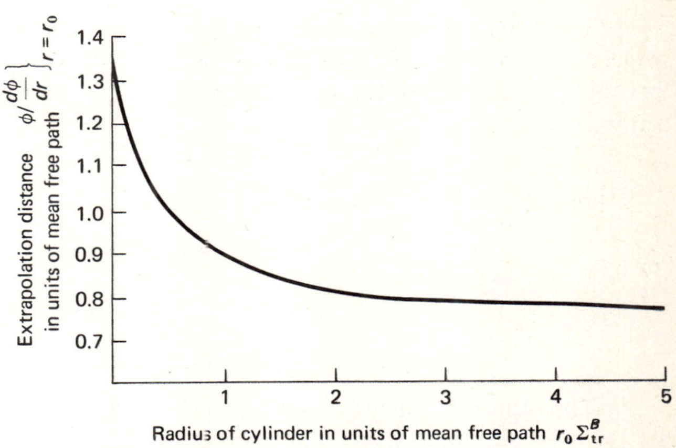

Fig. 7.7 Extrapolated distance for cylinder.

The general solution of the homogeneous equation is given again by (2.125*a*). Since we assume region B is infinite, we only need the term which vanishes at $\rho = \infty$. This gives

$$\phi_H(\rho) = CK_0(\kappa_B \rho), \tag{7.34b}$$

where

$$\kappa_B = \sqrt{\frac{\Sigma_a{}^B}{D^B}}\,. \tag{7.34c}$$

The complete solution is

$$\phi(\rho) = \phi_p + \phi_H(\rho), \tag{7.35a}$$

$$= \frac{q}{\Sigma_a{}^B} + CK_0(\kappa_B \rho). \tag{7.35b}$$

The constant C is determined by applying the boundary condition (7.32*b*),

$$\frac{q/\Sigma_a{}^B + CK_0(\kappa_B r_0)}{-C\kappa_B K_1(\kappa_B r_0)} = d(r_0), \tag{7.36a}$$

or

$$C = -\frac{q/\Sigma_a{}^B}{K_0(\kappa_B r_0) + \kappa_B\, d(r_0)\, K_1(\kappa_B r_0)}. \tag{7.36b}$$

Using this result, we can write the total flux in region B as

$$\phi(\rho) = \frac{q}{\Sigma_a{}^B}\left[1 - \frac{K_0(\kappa_B \rho)}{K_0(\kappa_B r_0) + \kappa_B\, d(r_0)\, K_1(\kappa_B r_0)}\right]. \tag{7.37}$$

The rod is black; thus the current $-D^B(d/d\rho)\phi(\rho)$ at the boundary r_0 will be entirely inward. Then the number of neutrons absorbed per second per unit surface area of the rod is equal to the magnitude of the current at the surface,

$$|J(r_0)| = \left| D^B \frac{d}{d\rho}\phi(\rho) \right|_{r_0}, \tag{7.38a}$$

or by using (7.37),

$$|J(r_0)| = \frac{q}{\kappa_B}\frac{K_1(\kappa_B r_0)}{K_0(\kappa_B r_0) + \kappa_B\, d(r_0)\, K_1(\kappa_B r_0)}. \tag{7.38b}$$

The number of neutrons absorbed by the rod per second per unit length of the rod will be $|J(r_0)|$ times the circumference of the rod $2\pi r_0$. The number of neutrons per unit length absorbed by the rod, divided by the number of neutrons per unit volume produced outside the rod, q, is called the *absorption area* of the rod, A_a.

$$A_a = \frac{2\pi r_0}{\kappa_B} \frac{K_1(\kappa_B r_0)}{K_0(\kappa_B r_0) + \kappa_B\, d(r_0)\, K_1(\kappa_B r_0)}. \tag{7.39a}$$

Several curves of A_a versus rod radius r_0 are shown in Fig. 7.8.

This concept of absorption area was introduced by Hurwitz and Roe.[1] They introduce some equivalence relations for other, more complicated geometrical shapes. That is, they write

$$A_a = \frac{2\pi r_{\text{eff}}}{\kappa_B} \frac{K_1(\kappa_B r_{\text{eff}})}{K_0(\kappa_B r_{\text{eff}}) + \kappa_B\, d(r_{\text{eff}})\, K_1(\kappa_B r_{\text{eff}})}, \tag{7.39b}$$

[1] H. Hurwitz and G. Roe: *KAPL* 1136, Mar. 24, 1955.

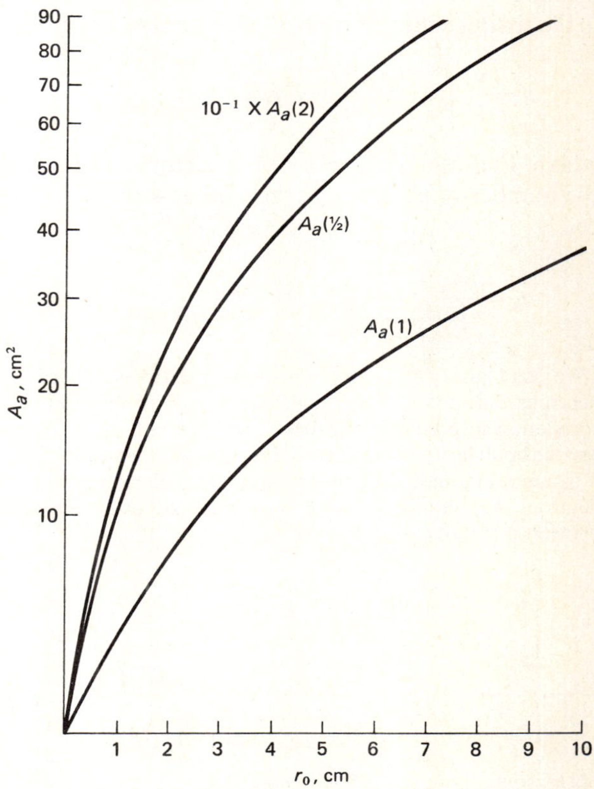

Fig. 7.8 Absorption areas $Aa(\kappa^B)$ for cylindrical control rods as functions of radius r_0.

where r_{eff} is an effective radius of an *equivalent cylinder*. For example, the absorption area for the three shapes shown in Fig. 7.9 are given by (7.39*b*) with

$$\text{Black cross} \qquad r_{\text{eff}} = \frac{d}{\sqrt{2}}. \tag{7.40a}$$

$$\text{Black riband} \qquad r_{\text{eff}} = \frac{d}{2}. \tag{7.40b}$$

$$\text{Black ellipse} \qquad r_{\text{eff}} = \frac{d+b}{2}. \tag{7.40c}$$

We can use the absorption area to obtain an estimate of the effective reactivity decrease in a critical reactor resulting from the insertion of a length of black control rod, l. First we assume that the source of thermal neutrons is the fission source and that it is uniformly distributed over the moderator.[1]

$$q = \frac{\nu_c}{V}\int_{\text{reactor}} \Sigma_f \phi(\mathbf{r})\, d^3r, \tag{7.41}$$

where V is the volume of the reactor. Then from the definition of A_a, the number of neutrons per second absorbed in a length l of a black control rod is

$$n_a = \frac{lA_a}{V}\int_{\text{reactor}} \nu_c \Sigma_f \phi(\mathbf{r})\, d^3r. \tag{7.42a}$$

[1] We shall justify this approximation in Chap. 9. It says simply that the fast fission neutrons diffuse widely through the reactor before thermalizing, so that they have no correlation in position with their "birth place," i.e., a particular fuel element. Also, it ignores boundary effects, and so it breaks down for small, highly enriched reactors. Finally, it assumes that mean free paths for epithermal neutrons are longer than those for thermal neutrons, so that spatial inhomogeneities do not have a drastic effect on the epithermal neutron distribution.

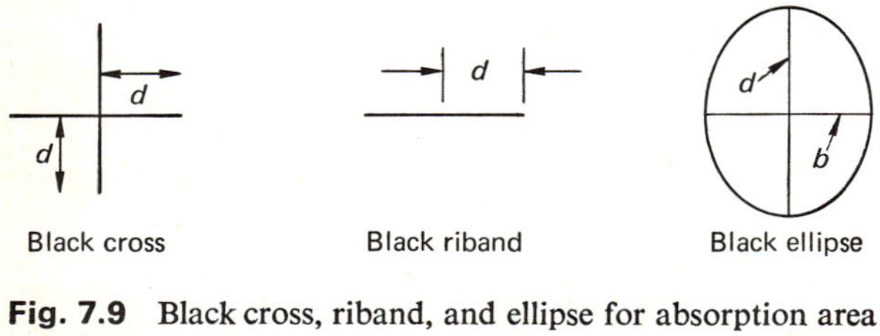

Fig. 7.9 Black cross, riband, and ellipse for absorption area calculations.

Recall that we choose ν_c to be the increased value of the number of secondaries per fission required to ensure criticality. Then the increase in the number of fission neutrons is

$$\delta n_f = (\nu_c - \nu) \int_{\text{reactor}} \Sigma_f \phi(\mathbf{r})\, d^3r. \tag{7.42b}$$

Then we can ensure criticality by setting n_a equal to δn_f, which gives

$$\nu_c = \frac{\nu}{1 - lA_a/V} \tag{7.42c}$$

or

$$\delta k = \frac{\nu - \nu_c}{\nu} = \frac{1}{1 - V/lA_a}, \tag{7.42d}$$

which for small rods ($lA_a \ll V$) gives

$$\delta k = -\frac{lA_a}{V}. \tag{7.42e}$$

These Eqs. (7.42) are, of course, only crude approximations for the worth of a rod.

7.6 FUEL DEPLETION EFFECTS

If a reactor operates at an appreciable power level for any length of time, a number of effects occur which change the critical conditions of the reactor. These have already been described qualitatively in Sec. 3.3, but in the next two sections we shall try to give some quantitative understanding of these effects. Actually, detailed calculations are very difficult, and surely beyond the scope of this book, so that we can give only the most basic ideas.

Consider, first, fuel burnup, and assume, for convenience, the fuel is U^{235} and that no U^{238} is present. Then the fuel depletion is proportional to the power density and, in fact,

$$\frac{dN_{25}(\mathbf{r},t)}{dt} = -\Sigma_a{}^{25}\phi(\mathbf{r},t) \tag{7.43a}$$

$$= -\sigma_a{}^{25}N_{25}(\mathbf{r},t)\phi(\mathbf{r},t), \tag{7.43b}$$

ignoring, for the moment, self-shielding.[1] This equation can be integrated to give

$$N_{25}(\mathbf{r},t) = N_{25}(\mathbf{r},0) \exp\left[-\sigma_a{}^{25}\int_0^t \phi(\mathbf{r},t')\, dt'\right]. \tag{7.44}$$

[1] We are using the standard abbreviations ${}_{92}U^{235} = 25$, ${}_{92}U^{235} = 28$ and ${}_{94}Pu^{239} = 49$. These isotopes are labeled by two-digit numbers, the first of which is the second digit of Z, the atomic number, and the second is the last digit of A, the mass number.

Of course a similar equation would apply for the burnup of any other material in the reactor which absorbs neutrons. Note the exponential dependence of N_{25} on $\int \phi(\mathbf{r},t')\,dt'$, which is a measure of the effective length of time the reactor has been operating. If the reactor has been operating at constant *flux*, ϕ is independent of time. However, reactors generally operate at constant power; since N_{25} is changing with t, then ϕ will also change with time. That is, ϕ is really a functional of N_{25}, because as the uranium burns out the resultant decreased cross section will tend to make the flux rise to keep the product constant.[1]

The long-term time-dependent reactor equations are integrated in small time steps, in each of which $\phi(\mathbf{r})$ is assumed independent of t. Then at each step a new N_{25} is calculated, the corresponding $\phi(\mathbf{r})$ calculated, and the procedure repeated. (This direct difference approach to integration of the initial value problem is in some ways analogous to the numerical integration of the steady-state reactor equations described in Chap. 6.) Another practical difficulty is that as the fuel is depleted, the control rods will be withdrawn to maintain a critical condition. This also tends to change the flux distribution in the reactor. Again, however, the short-time-step approximation is generally adequate.

We now may consider correcting Eq. (7.43*b*) to include the effect of self-shielding. In this case, the Wigner rational approximation, Eq. (7.29*a*), may be used, which yields

$$\frac{dN_{25}(\mathbf{r},t)}{dt} = -\sigma_a^{25}\phi(\mathbf{r},t)\frac{N_{25}(\mathbf{r},t)}{1 + V'N_{25}(\mathbf{r},t)}, \tag{7.45}$$

where we have abbreviated

$$V' = \frac{4V_0}{S_0}\sigma_a^{25}. \tag{7.46}$$

Equation (7.45) may also be integrated rather simply to give

$$N_{25}(t)\exp\{-V'[N_{25}(t) - N_{25}(0)]\} = N_{25}(0)\exp\left[-\sigma_a^{25}\int_0^t \phi(t')\,dt'\right]. \tag{7.47}$$

The remarks made on the applicability of Eq. (7.44) apply to (7.34) as well.

The procedure sketched above gives a reasonably practical scheme for calculating the amount of burnup of various types of nuclei as a reactor is operated. However, the calculation of plutonium buildup is a more difficult

[1] For a reactor containing much U^{238}, Pu^{239} (and higher isotopes) will be built up, so that the flux strongly depends on their concentration as well; see the following text.

problem. One reason is that the plutonium, which is formed by the radiative capture of neutrons in U^{238}, also burns up through fission. In addition, higher isotopes of Pu^{239} are formed by successive capture of neutrons. Some of these higher isotopes are fissionable, others are merely poisons, but the cross sections of these higher isotopes are *not* extremely well known.

Another problem is that absorption in U^{238} occurs primarily in the resonance region, although there is a certain amount of capture of thermal neutrons as well. If we recall from Chap. 1 our definition of p, the resonance escape probability, and P_F, the fast nonleakage probability, then it is clear that the number of neutrons absorbed per second per unit reactor volume in resonances of U^{238} will be given by[1]

$$n = \nu\Sigma_f \phi P_F(1 - p) \tag{7.48a}$$

$$= \nu\sigma_f^{25} N_{25} P_F(1 - p)\phi. \tag{7.48b}$$

If we now include thermal absorption and fast fission in U^{238} as well, we have (ignoring the time delay between the capture of a neutron by a U^{238} nucleus and the subsequent production of a Pu^{239} nucleus)

$$\frac{dN_{28}(t)}{dt} = -\phi[(\sigma_a^{28} + \sigma_f^{28})N_{28}(t) + \nu^{25}\sigma_f^{25}N_{25}(t)P_F(1 - p) + \nu^{49}\sigma_f^{49}N_{49}(t)P_F(1 - p)], \tag{7.49a}$$

$$\frac{dN_{49}(t)}{dt} = \phi[\sigma_a^{28}N_{28}(t) + \nu^{25}\sigma_f^{25}N_{25}(t)P_F(1 - p) + \nu^{49}\sigma_f^{49}N_{49}(t)P_F(1 - p) - \sigma_a^{49}N_{49}(t)]. \tag{7.49b}$$

The terms including σ_f^{49} take into account fission in the plutonium; we ignore the slight difference in P_F of fission neutrons for plutonium and for uranium.

The integration of these equations is facilitated if we ignore the fractional change in U^{238} concentration, so that N_{28} can be assumed constant (thus p is constant as well). Equation (7.49*b*) can then be integrated by introducing the *integrating factor*, i.e., we use the identity

$$\begin{aligned} &\frac{d}{dt}N_{49}(t) + [\sigma_a^{49} - \nu^{49}\sigma_f^{49}P_F(1 - p)]\phi(t)N_{49}(t) \\ &= \exp\left\{-[\sigma_a^{49} - \nu^{49}\sigma_f^{49}P_F(1 - p)]\int_0^t \phi(t')\,dt'\right\} \\ &\times \frac{d}{dt}\left\{N_{49}(t)\exp\{+[\sigma_a^{49} - \nu^{49}\sigma_f^{49}P_F(1 - p)]\int_0^t \phi(t)\,dt'\}\right\}. \end{aligned} \tag{7.50}$$

[1] We assume that leakage occurs first, followed by resonance capture. This is physically reasonable. Equation (7.48*a*) then follows directly from the definitions P_F, p, and $\nu\Sigma_f\phi$.

Thus, by writing

$$A^{49} = \sigma_a^{49} - \nu^{49}\sigma_f^{49}P_F(1-p), \tag{7.51a}$$

Eq. (7.49*b*) becomes

$$N_{49}(t) = N_{49}(t)\exp\left[-A^{49}\int_0^t \phi(t')\,dt'\right]$$
$$+\int_0^t dt'[\nu^{25}\sigma_f^{25}N_{25}(t')P_F(1-p) + \sigma_a^{28}N_{28}\,\phi(t')]$$
$$\times \exp\left[-A^{49}\int_{t'}^t \phi(t'')\,dt''\right]. \tag{7.51b}$$

In this equation, $N_{25}(t)$ is to be read from Eq. (7.47). The assumption that N_{28} is constant may be relaxed to the extent that only p is assumed constant. Then Eq. (7.49) may be integrated simultaneously, most conveniently by the method of Laplace transforms (see Appendix C). The details are omitted.

After $N_{49}(t)$ has been computed in a given time step, the residual plutonium concentration must be considered fuel. Then a new criticality and flux calculation is carried out, the control adjusted for criticality, and the process continued to the end of life. More elaborate calculations than those sketched above might include spatial calculations, for example, of the plutonium distribution throughout the uranium rod.

In thermal reactors, the plutonium production takes place largely in the reactor core, since there is in general a low enrichment of fuel. Thus there is a great deal of U^{238} in the core. The production in fast reactors is somewhat different, however. There is still a great deal of U^{238} in the core (although less than for thermal reactors), but a large fraction of the *conversions* take place in a blanket, exterior to the core. In any case, the flux distribution used in the above calculation is found from (group) diffusion calculations, so that it does not make a great deal of difference for computational purposes where the conversion takes place.

For reasons which were described in Chap 1, only fast reactors can *breed*, i.e., produce more plutonium than U^{235} consumed (although thermal reactors which breed U^{233} from Th^{232} are in principle possible).

Calculations of the type described in this section are exceedingly important for practical purposes, since economic considerations require obtaining the highest possible burnup from fuel elements in reactors. (This is true because of the high cost of fuel element fabrication and, to a lesser extent, the cost of reprocessing, i.e., recovery of the Pu^{249} and unburnt U^{235}).

The endurance of a fuel rod is generally expressed in the units of megawatt days per metric ton of initial fuel, which is merely an expression of the amount of energy received per unit mass of original fuel mixture. Clearly the energy output is given by

$$E = C \int d^3r \int_0^t dt' \, [\sigma_f^{25} N_{25}(t') + \sigma_f^{49} N_{49}(t')] \phi(t'), \tag{7.52}$$

where the time integration is carried out over the life history of the reactor. The factor C represents the energy per fission; the factors $N_{25}(t)$, $N_{49}(t)$, and $\phi(t)$ are obtained from the formulas given earlier in this section by using discrete time steps as described.

7.7 FISSION-PRODUCT POISONING

As a concomitant to fuel depletion, fission products are built up, and some of these (or all) absorb neutrons. This effect is known as *fission-product poisoning*. Fission products can be divided roughly into two classes, saturating and nonsaturating. The saturating fission products are those with cross sections so large that after a short time their production rate is balanced by their own burnout, and a constant density (equilibrium concentration) obtains. The most important of the saturating fission products is certainly Xe^{135}; it is discussed in Sec. 7.8. Second in importance is Sm^{149}. Other saturating fission products include Sm^{150}, Gd, two Eu isotopes, and an isotope of Rb. They can all be treated in the same manner as is Xe^{135}.

The nonsaturating fission products should again be divided into three groups, two of which actually saturate. There is a group with an average cross section of 80 b/nucleus,[1] which saturates in 50 to 100 days of operation (for a typical pressurized water reactor). A second group with an average cross section of 50 b/nucleus saturates in about 6 months, while a third group with a 30 b/nucleus cross section is essentially nonsaturating.[2]

Each of these three groups of "non-saturating" poisons should be treated separately. Here, for simplicity, we shall ignore the saturation properties of the 80- and 50-b groups, and assume simply that 160 b/nucleus of nonsaturating fission-product poison is released into the system at the rate of two poison nuclei per fission.

A rather interesting problem is the question of treating the *partially saturating* fission products, those discussed above which saturate in 50 to

[1] Hence 160 b/fission since each fission yields two fission products. Typical curves are plotted in Fig. 7.10.

[2] This average cross section is supposed to include the absorption effect, if any, of the nuclei which result from neutron absorption in the primary nuclei. Actually, the effect of these secondary absorbers is not large. Also, the effective cross sections of the different groups vary somewhat, because of differences in fission yield, from one fissile nuclide to another. Let us ignore this obnoxious complication.

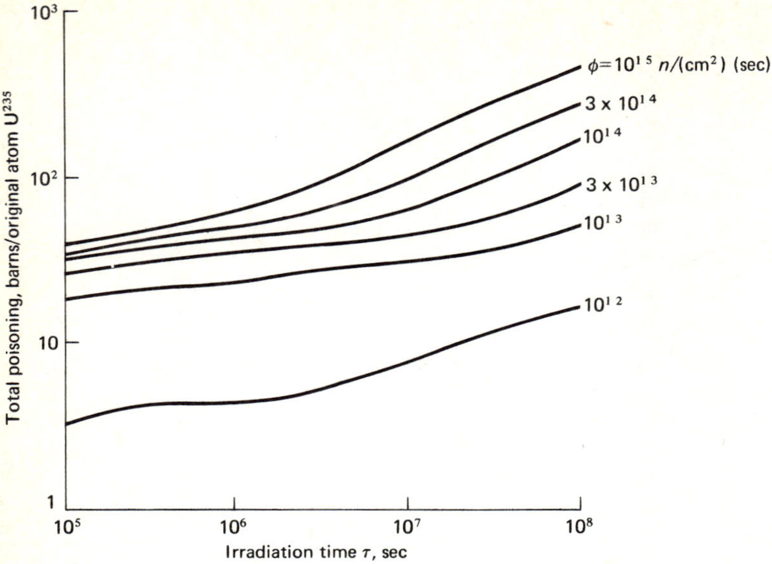

Fig. 7.10 Buildup of total fission product poisoning with irradiation time and thermal neutron flux.

100 days. We see from Eq. (7.54*a*) below that at equilibrium $dN_p/dt = 0$, the partially saturating fission-product density becomes

$$N_{p0}{}^{\text{PS}} = \frac{2\gamma_p(\sigma_f{}^{25}N_{25} + \sigma_f{}^{49}N_{49})}{\bar{\sigma}_a}, \tag{7.53}$$

where, of course, $\bar{\sigma}_a = 80$ b. The interesting point is that on a time scale as large as 50 to 100 days, one can expect N_{25} and N_{49} to change, so that that $N_{p0}{}^{\text{PS}}$ is not *really* a constant. [Similarly, the equilibrium concentration of xenon changes for the same reason, see Eq. (7.54*b*).] The use of short (20-day) time steps to integrate the burnup equation, as discussed in Sec. 7.6, circumvents this difficulty for both the saturating and partially saturating fission products.

To estimate the effect of this fission-product poisoning, it is necessary to know the density N_p of fission products at any time during the operating history of the reactor. This is calculated in much the same way as the Pu^{249} concentration is calculated. If we write

$$\frac{dN_p}{dt} = 2\gamma_p(\sigma_f{}^{25}N_{25} + \sigma_f{}^{49}N_{49})\phi - \bar{\sigma}_a N_p \phi, \tag{7.54a}$$

then

$$N_p(t) = 2\gamma_p \sigma_f^{25} \int_0^t \left[N_{25}(t') + \frac{\sigma_f^{49}}{\sigma_f^{25}} N_{49}(t') \right] \phi(t') \times \exp\left[-\sigma_a \int_{t'}^{t} dt'' \, \phi(t'') \right] dt', \qquad (7.54b)$$

where we have of course set $N_p(0) = 0$. Since $N_{25}(t')$, $N_{49}(t')$, and $\phi(t')$ are known, Eq. (7.54*b*) can be evaluated. The factor γ_p represents the fraction of those fissions which lead to fission products of the fission-product group under consideration (i.e., it is the *yield*). In the spirit of footnote 2 on page 177 we ignore the difference in yields between uranium and plutonium fission.

7.8 XENON POISONING

The isotope Xe^{135} is of extreme importance in thermal reactors because it has a resonance in the thermal range ($\approx$0.08 eV) whose average thermal cross section is about 8×10^6 b. This is the largest known cross section of any isotope and, in fact, is almost equal to the maximum possible cross section obtainable, a quantity which can be estimated from nuclear theory.

Some Xe^{135} is found directly as a fission product, but more is produced as the result of a radioactive-decay chain which begins with Te^{135} (half-life, 2 min) and proceeds through I^{135} (half-life, 6.7 h) to Xe^{135}. The xenon itself is radioactive with a half-life of 9.2 h.

As a reactor begins operation, there is initially no iodine or xenon present. After a certain length of time, *equilibrium* concentrations of these quantities are approached, in which the production of xenon by direct formation plus decay of I^{135} is just balanced by its removal by radioactive decay plus burnup. At the same time, an equilibrium concentration of I^{135} is obtained.

An interesting phenomenon occurs after reactor shutdown. In this case the xenon is no longer destroyed by burnout, but it is still produced by the decay of the iodine, which is, of course, still present in its equilibrium concentration. Thus, for a time the xenon concentration actually increases. Eventually the radioactive decay of the xenon dominates, and the xenon concentration approaches zero. The peak xenon condition, which typically occurs at about 12 h after shutdown for a flux of the order of 2×10^{12} thermal neutrons/cm^{-2} s^{-1}, may increase the xenon reactivity by as much as an order of magnitude. The effect is, of course, more pronounced the higher the flux of the reactor, since then the burnout term is relatively more important and its loss after shutdown has a correspondingly more important effect.

To study this effect, we first write down the equation describing the iodine concentration as a function of time (because of its very short half-life, we shall ignore the Te^{135} and assume I^{135} is produced directly by fission).

Denoting the iodine density by I(t), we have

$$\frac{d}{dt}\mathrm{I}(t) = \gamma_1\Sigma_f\phi - \sigma_1\mathrm{I}(t)\phi - \lambda_1\mathrm{I}(t), \tag{7.55}$$

where γ_1 is the fraction of fissions which lead to formation of iodine,[1] $\sigma_1 = 8$ b is the iodine-absorption cross section, and $\lambda_1 = 2.89 \times 10^{-5}$ s^{-1} is the iodine-decay constant. From these numbers, we see that even for $\phi = 10^{15}$ cm^{-2} s^{-1}, which is a very high flux, $\sigma_1\phi \ll \lambda_1$, so that Eq. (7.55) reduces to

$$\frac{d}{dt}\mathrm{I}(t) + \lambda_1\mathrm{I}(t) = \gamma_1\Sigma_f\phi \tag{7.56a}$$

or

$$(\mathrm{I}t) = e^{-\lambda_1 t}\left[\mathrm{I}(0) + \gamma_1\dot{\Sigma}_f \int_0^t \phi(t')e^{\lambda_1 t'}\,dt'\right]. \tag{7.56b}$$

If the reactor starts up from the "clean" condition, i.e., no iodine present, I(0) = 0. Also, let us consider that $\phi(t)$ goes from zero to a constant value ϕ_0 in a short time.[2] In this case,

$$\mathrm{I}(t) = \frac{\gamma_1\Sigma_f\phi_0}{\lambda_1}(1 - e^{-\lambda_1 t}). \tag{7.57}$$

In any event, the equilibrium iodine concentration I_0 is seen to be

$$\mathrm{I}_0 = \frac{\gamma_1\Sigma_f\phi_0}{\lambda_1}. \tag{7.58}$$

If the reactor is shut down at time $t = 0$ with equilibrium iodine concentration present, then *after* shutdown

$$\mathrm{I}(t) = \mathrm{I}_0 e^{-\lambda_1 t} \tag{7.59}$$

where we have assumed a rapid shutdown [if the shutdown is carried out over a long period of time, Eq. (7.56b) applies, with I(0) = I_0 and, of course, $\phi(t)$ determined by the shut-down procedure].

We now consider the xenon concentration X(t), which obeys the equation

$$\frac{d}{dt}\mathrm{X}(t) = \lambda_1\mathrm{I}(t) + \gamma_2\Sigma_f\phi - \lambda_2\mathrm{X}(t) - \sigma_2\phi\mathrm{X}(t). \tag{7.60}$$

[1] This number differs for different fissionable isotopes. For U^{235}, $\gamma_1 = 0.001$; for Pu^{239}, $\gamma_1 = 0.055$; while for U^{233}, $\gamma_1 = 0.051$ (see ANL 5800, p. 394).

[2] We may consider $\phi(t)$ a constant for the purposes of this calculation because the time changes here are short term compared with the long-term effects (fuel depletion, buildup, etc.) which cause the flux to change.

Here γ_2 represents the *direct* production of xenon as a fission product (rather than as a result of the Te to I chain), λ_2 is the xenon-decay constant, and σ_2 is the xenon cross section shown in Fig. 7.11. Numerically[1]

$$\gamma_2 = 0.0003, \tag{7.61a}$$

$$\lambda_2 = 2.09 \times 10^{-5}\ \mathrm{s}^{-1}, \tag{7.61b}$$

$$\sigma_2 = 3 \times 10^6\ \mathrm{b}. \tag{7.61c}$$

(The value σ_2 is an average over the thermal neutron spectrum in the reactor; see Chap. 9.)

Equation (7.60) is readily solved:

$$\mathrm{X}(t) = \exp\left\{-\int_0^t [\lambda_2 + \sigma_2 \phi(t')\, dt']\right\}\left(\int_0^t \exp\left\{\int_0^{t'} [\lambda_2 + \sigma_2 \phi(t'')]\, dt''\right\} \times [\lambda_1 \mathrm{I}(t') + \gamma_2 \Sigma_f \phi(t')]\, dt' + \mathrm{X}(0)\right), \tag{7.62}$$

where, of course, $\mathrm{I}(t')$ is obtained from Eq. (7.56*b*) or (7.57). Under the same assumptions that led to Eq. (7.56) from (7.56*b*), Eq. (7.62) reduces to

$$\mathrm{X}(t) = \mathrm{X}_0\left[1 + \frac{1}{\gamma_1 + \gamma_2}\left(\frac{\gamma_1 \lambda_1}{\lambda_2 - \lambda_1 + \sigma_2 \phi_0} - \gamma_2\right) e^{-(\lambda_2 + \sigma_2 \phi_0)t} - \frac{\gamma_1}{\gamma_1 + \gamma_2}\, \frac{\lambda_2 + \sigma_2 \phi_0}{\lambda_2 - \lambda_1 + \sigma_2 \phi_0}\, e^{-\lambda_1 t}\right] \tag{7.63}$$

where we have taken $\mathrm{X}(0) = 0$.

[1] For I^{233} and Pu^{239} fission, $\gamma_2 = 0$.

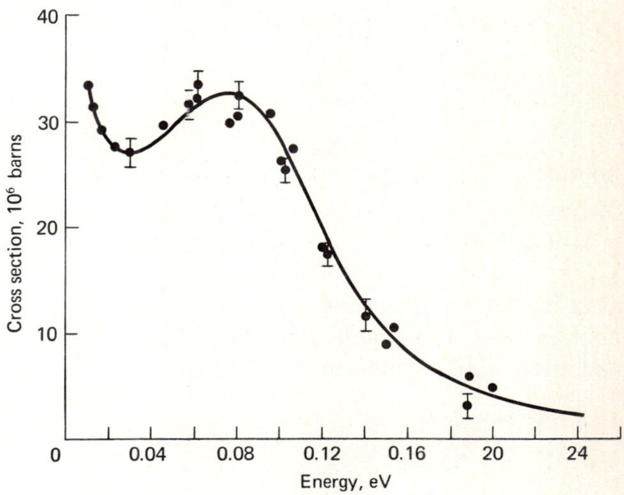

Fig. 7.11 Total cross section of Xe^{135}.

Here the equilibrium xenon concentration X_0 is obtained from Eq. (7.60) by setting $\dot{X} = 0$:

$$X_0 = \frac{\lambda_1 I_0 + \gamma_2 \Sigma_f \phi_0}{\lambda_2 + \sigma_2 \phi_0} \tag{7.64a}$$

$$= \frac{(\gamma_1 + \gamma_2)\Sigma_f \phi_0}{\lambda_2 + \sigma_2 \phi_0}, \tag{7.64b}$$

where Eq. (7.58) has been used.

Suppose we consider the effect of xenon on a reactor operating in the steady-state condition. We see that the diffusion equation becomes[1]

$$-D\nabla^2\phi_0 + \left[\Sigma_a' + \frac{\sigma_2(\gamma_1 + \gamma_2)\Sigma_f \phi_0}{\lambda_2 + \sigma_2 \phi_0}\right]\phi_0 = \nu\Sigma_f \phi_0, \tag{7.65}$$

i.e., the effect of the xenon is to introduce a nonlinear term into the diffusion equation! Consider the high flux limit, however, that is, $\sigma_2 \phi_0 \gg \lambda_2$. Then

$$-D\nabla^2\phi_0 + \Sigma_a' \phi_0 = \nu\Sigma_f \phi_0 - (\gamma_1 + \gamma_2)\Sigma_f \phi_0. \tag{7.66}$$

This limit is easy to treat, since we see the effect is to replace ν with $(\nu - \gamma_1 - \gamma_2)$. Thus $\nu_c = \nu + \gamma_1 + \gamma_2$, and

$$\rho = \frac{k-1}{k} \tag{7.67a}$$

$$= \frac{\nu/\nu_c - 1}{\nu/\nu_c} \tag{7.67b}$$

$$= \frac{\nu - \nu_c}{\nu} \tag{7.67c}$$

becomes

$$\rho_{xe} = -\frac{\gamma_1 + \gamma_2}{\nu}. \tag{7.67d}$$

That is, the reactor initially must have *an excess* reactivity of this amount built in (and presumably held down by control rods) in order to permit operation with equilibrium xenon present. Since $\gamma_1 + \gamma_2 \approx 0.06$, we see

$$\rho_{xe} = -0.024 \tag{7.68}$$

$$\approx -\$3.30. \tag{7.69}$$

[1] Σ_a' represents the absorption cross section of all materials other than xenon.

This is clearly the maximum amount of (equilibrium) xenon reactivity.

For smaller values of the flux, we may attempt to linearize Eq. (7.65). First, we multiply by the factor $\lambda_2 + \sigma_2 \phi_0$. Hence

$$-D\lambda_2 \nabla^2 \phi_0 - D\sigma_2 \phi_0 \nabla^2 \phi_0 + \Sigma_a' \lambda_2 \phi_0 + \sigma_2 \Sigma_a' {\phi_0}^2 + \sigma_2(\gamma_1 + \gamma_2 - \nu)\Sigma_f {\phi_0}^2 = \nu \Sigma_f \lambda_2 \phi_0. \tag{7.70}$$

If we assume that the flux is only slightly different from a value ψ which obeys the linear equation (with no xenon)

$$-D\nabla^2 \psi + \Sigma_a' \psi = \nu \Sigma_f \psi, \tag{7.71}$$

we can write

$$\phi_0 = \psi + \varepsilon \tag{7.72}$$

and ignore quantities proportional to ε^2. Inserting (7.72) into (7.70) gives

$$-D\lambda_2 \nabla^2(\psi + \varepsilon) - D\sigma_2(\psi \nabla^2 \psi + \varepsilon \nabla^2 \psi + \psi \nabla^2 \varepsilon) + \Sigma_a' \lambda_2(\psi + \varepsilon) + \sigma_2[\Sigma_a' + (\gamma_1 + \gamma_2 - \nu)\Sigma_f](\psi^2 + 2\varepsilon\psi) = \nu \Sigma_f \lambda_2(\psi + \varepsilon). \tag{7.73}$$

This is a linear equation (in ε) because ψ is already known [as the solution to Eq. (7.71)]. From this equation, ν_c can be found, and hence ρ. This is not the only possible method of linearizing a nonlinear equation, but it is one standard technique. Another approach would be to solve Eq. (7.70) on a computer.

In Fig. 7.12 we show the relative absorption in xenon compared with that in U^{235} for a highly enriched reactor; note that here this value approaches an asymptotic limit.

We now turn to the question, discussed qualitatively in the first part of the present section, of the xenon concentration following shutdown. For simplicity, we assume that the reactor is operating at equilibrium and that the flux drops rapidly to zero at time $t = 0$, in which case Eq. (7.59), for the iodine concentration, applies.

$$I(t) = I_0 e^{-\lambda_1 t} \tag{7.74a}$$

$$= \frac{\gamma_1 \Sigma_f}{\lambda_1} \phi_0 e^{-\lambda_1 t}. \tag{7.74b}$$

For $X(t)$, we have Eq. (7.62), with $\varphi(t) = 0$ and $X(0) = X_0$ [Eq. (7.64b)].

$$X(t) = e^{-\lambda_2 t}\left[\int_0^t dt'\, e^{(\lambda_2 - \lambda_1)t'} \gamma_1 \Sigma_f \phi_0 + \frac{(\gamma_1 + \gamma_2)\Sigma_f \phi_0}{\lambda_2 + \sigma_2 \phi_0}\right]. \tag{7.75}$$

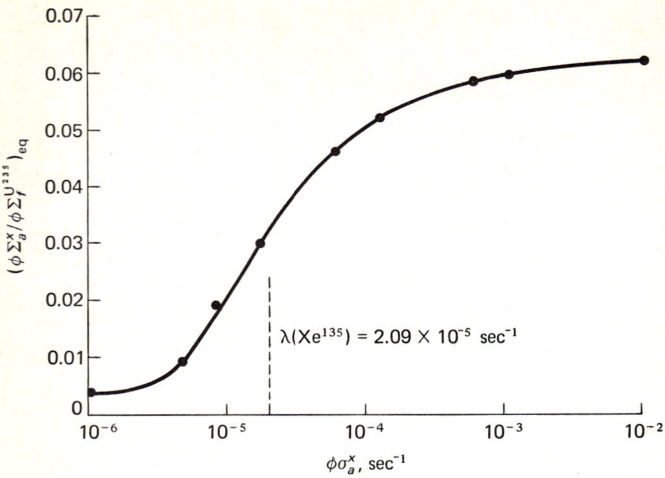

Fig. 7.12 Relative neutron absorption by equilibrium xenon in a highly enriched (U^{235}) thermal reactor.

Carrying out the integration, we have

$$X(t) = e^{-\lambda_2 t}\left[\frac{\gamma_1\Sigma_f\phi_0}{\lambda_2 - \lambda_1}(e^{(\lambda_2-\lambda_1)t} - 1) + \frac{(\gamma_1+\gamma_2)\Sigma_f\phi_0}{\lambda_2 + \sigma_2\phi_0}\right]. \tag{7.76}$$

We note $X(0) = X_0$. Also,

$$\dot{X}(0) = \gamma_1\Sigma_f\phi_0\left(1 - \frac{\lambda_2}{\lambda_2+\sigma_2\phi_0}\frac{\gamma_1+\gamma_2}{\gamma_1}\right) \tag{7.77}$$

is positive if

$$\frac{\lambda_2}{\lambda_2+\sigma_2\phi_0} < \frac{\gamma_1}{\gamma_1+\gamma_2}, \tag{7.78a}$$

i.e., if

$$\frac{\sigma_2\phi_0}{\lambda_2} > \frac{\gamma_2}{\gamma_1} \approx 0.05 \tag{7.78b}$$

or

$$\phi_0 > 5 \times 10^{11} \text{ neutrons/cm}^2\ \text{s}^{-1}. \tag{7.79}$$

Thus for values of ϕ_0 in excess of this value, the xenon concentration will originally increase after shutdown, and [see Eq. (7.75)] approach zero with a lifetime of $1/\lambda_2 \approx 14$ h. The behavior of this concentration is shown in

Fig. 7.13. The time of peak xenon concentration can be found by setting $\dot{X}(t) = 0$ and solving for t.

$$\dot{X}(t) = -\lambda_2 e^{-\lambda_2 t}\left[\frac{\gamma_1 \Sigma_f \phi_0}{\lambda_2 - \lambda_1}(e^{(\lambda_2 - \lambda_1)t} - 1) + \frac{(\gamma_1 + \gamma_2)\Sigma_f \phi_0}{\lambda_2 + \sigma_2 \phi_0}\right] + e^{-\lambda_1 t}\gamma_1 \Sigma_f \phi_0$$
$$= 0. \quad (7.80)$$

The reactivity effect of the peak xenon can be found by the methods described earlier for equilibrium.

Another method which may be used to calculate the reactivity worth of any given xenon concentration (although it is grossly approximate) is to use the formula

$$k = \frac{\eta f \varepsilon p}{1 + L^2 B^2} \quad (7.81a)$$

and note, since xenon does not affect ε or p,

$$\frac{\delta k}{k} = \frac{\delta f}{f} - \frac{B^2}{1 + L^2 B^2}\,\delta L^2. \quad (7.81b)$$

Since f, the thermal utilization, is given by

$$f = \frac{\Sigma_a^u}{\Sigma_a}, \quad (7.82)$$

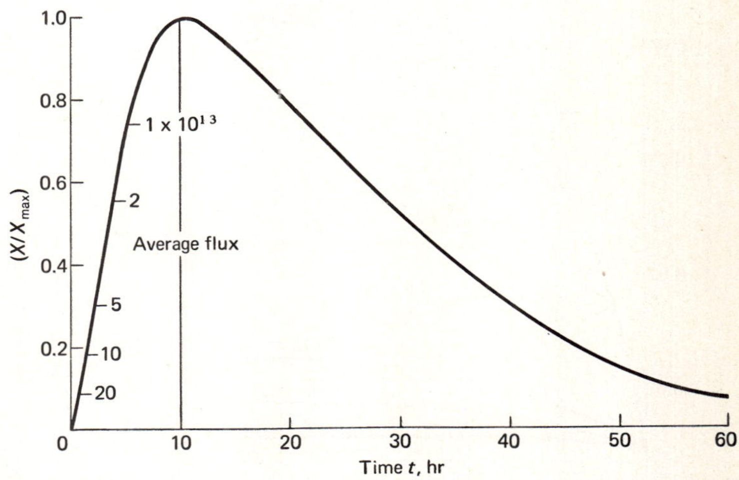

Fig. 7.13 Flux-independent curve of xenon concentration as a function of time after complete shutdown. [Equation (7.79) applies.]

where Σ_a is the total core absorption cross section, we have

$$\delta f = \frac{\Sigma_a{}^u}{\Sigma_a' + \Sigma_a{}^{xe}} - \frac{\Sigma_0{}^u}{\Sigma_a'} = -\frac{f}{1 + \Sigma_a'/\Sigma_a{}^{xe}}, \tag{7.83}$$

where we are using the notation that f is the thermal utilization of the clean core. Similarly,

$$\delta L^2 = \frac{\delta D}{\Sigma_a} = L_0{}^2 \frac{1}{1 + \Sigma_a'/\Sigma_a{}^{xe}}. \tag{7.84}$$

Thus we see the important quantity $1/(1 + \Sigma_a'/\Sigma_a{}^{xe})$ enters both terms. This method of computing ρ is not precise because it assumes that the xenon concentration does not change the spatial flux shape. Otherwise, Eq. (7.81) would not be valid.

7.9 OTHER SATURATING FISSION PRODUCTS

As we discussed in Sec. 7.5, Sm^{149} is, next to Xe^{135}, the most important saturating fission product. It differs from Xe^{135} only in being nonradioactive. The precursor of Sm^{149} is Pm^{149}, with $\lambda_1 = 4.1 \times 10^{-6}$ s^{-1} as its decay constant.[1] The absorption cross section σ_2 of Sm^{149} is 5.3×10^4 b. As

[1] We are using notation identical with that in the previous section, except that the subscript "1" here refers to Pm^{149}, and "2" to Sm^{149}.

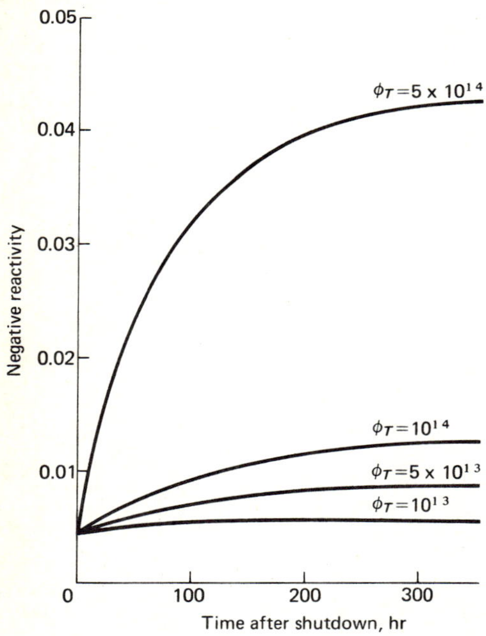

Fig. 7.14 Sm^{149} buildup after shutdown for various values of the operating flux.

with iodine, σ_1, that is, absorption of neutrons by the Pm^{149}, can be ignored. Also, $\gamma_2 = 0$ (there is no direct production of Sm^{149} in fission) and $\lambda_2 = 0$ (Sm^{149} is not radioactive). With these values, the results of the previous section can be taken over directly. By using the high flux limit already described in the previous section, the curves plotted in Fig. 7.14 may be derived. Note that samarium builds up after shutdown to a value determined by the precursor concentration; because samarium is not radioactive, there is no falloff, as with xenon.

PROBLEMS

7.1. Using first-order perturbation theory, calculate the δk for a control slab in a bare slab reactor. Compare your answer with Eq. (7.4).

7.2. What is the shutdown margin of a bare slab reactor of width $2a$ with a black control slab of width $2d$ whose center is a distance b to the right of the reactor centerline?

7.3. Calculate P_{AB} for a sphere, using diffusion theory.

7.4. Calculate P_{AB} for an infinite cylinder of radius R, using diffusion theory.

7.5. Derive Eq. (7.47).

7.6. Derive the exact expression for P_0 for a slab.

7.7. Show that A_a [Eq. (7.39a)] increases linearly with r_0 for large r_0.

7.8. Calculate the reactivity effect of Sm^{149} after shutdown as a function of time. Numerically evaluate this expression for long times.

7.9. For a reactor with a peak specific power of 1 MW/liter, compute the peak xenon reactivity.

8 Energy–Dependent Problems: The Multigroup Method

8.1 INTRODUCTION

In the first seven chapters of this book we have studied reactor physics under the assumption that all neutrons had a single energy, i.e., we have used the one-speed or one-group approximation. Although this assumption is very different from reality, there is nonetheless a sound reason for having approached reactor physics in this way. The first reason is pedagogical. Most of the mathematical techniques introduced in the one-group approximation can be taken over with only minor modification to the multigroup (i.e., energy-dependent) method. In fact, the necessary modifications are described in the present chapter, insofar as criticality and flux calculations are concerned. The second reason is that there are special problems for which the one-group method may be a satisfactory approximation, particularly for rough *survey* calculations. Examples are the xenon poisoning, fuel depletion,[1] and plutonium burnup calculations described in Chap. 7.

[1] The depletion calculations are really *pseudo one group* since resonance capture must be accounted for, see Sec. 7.6.

8.2 THE MULTIGROUP METHOD

In the multigroup method we assume that instead of a single energy, the neutron distribution in a reactor can be represented by a number of discrete energy groups. The number of groups used depends upon the accuracy desired and the machine capabilities available, but generally ranges between 2 and 18. (Clearly the larger the number of groups, the higher the accuracy since one is attempting to represent the actual continuous energy distribution in the reactor by these discrete groups.) The groups are numbered from 1 for the highest energy to N for the lowest (in the N-group scheme). For a thermal reactor the Nth group typically is the thermal group.[1]

The flux in the Kth group will be represented by ϕ_K. Similarly, there is a capture cross section Σ_{aK}, a diffusion coefficient D_K, and a fission cross section Σ_{fK} associated with the Kth group. In addition, χ_K will be defined as the fraction of fission neutrons emitted into the Kth group, while ν_K is the mean number of neutrons emitted when a fission is caused by a neutron of the Kth group. Since ν is only a very slowly varying function of energy, however, we shall drop the subscript K on ν_K. (This approximation is not valid for fast reactors. However, the group dependence of ν can be incorporated with only trivial differences in formulas.) Finally there is a cross section Σ_{ij}, which represents the scattering (elastic or inelastic) from group i to group j.[2]

With these definitions we note that the fission source for the ith group can be written as

$$\chi_i \nu \sum_k \phi_k \Sigma_{fk} \equiv \chi_i \nu S(\mathbf{r}). \tag{8.1}$$

In other words, the fission source in each group has the same spatial behavior; it differs from group to group only by the amplitude factor χ_i. This fact will turn out to be of great importance in setting up the numerical scheme for solving the multigroup equations.

In addition to the fission source, there is a slowing-down source in every group except the first.[3] This clearly has the form, for the ith group,

$$\sum_{k<i} \Sigma_{ki} \phi_k . \tag{8.2}$$

Neutrons are removed from the ith group both by absorption Σ_{ai} and by scattering to lower-energy groups. The latter contribution is given by

$$\phi_i \sum_{k<i} \Sigma_{ik} \equiv \phi_i \Sigma_{si}, \tag{8.3}$$

where we have defined the slowing-down cross section Σ_{si} for the ith group.

[1] Although in some cases the thermal region consists of two or even more groups.

[2] For the time being we shall assume all these quantities (or *group constants*, as they are called) to be known. In the next section we derive explicit expressions for them and in Chap. 9 we discuss how they may be calculated.

[3] We ignore upscattering; this is valid unless more than one thermal group is used.

Collecting all these terms, we see that the diffusion equation for neutrons in the ith group can be written

$$-\nabla \cdot D_i \nabla \phi_i(\mathbf{r}) + (\Sigma_{ai} + \Sigma_{si})\phi_i(\mathbf{r}) = \nu x_i S(\mathbf{r}), \tag{8.4a}$$

$$S(\mathbf{r}) = \sum_k \phi_k(\mathbf{r})\Sigma_{fk}. \tag{8.4b}$$

This set of equations can conveniently be written in matrix form

$$\mathbf{M\Phi} = \nu\mathbf{F\Phi}, \tag{8.5}$$

where $\mathbf{\Phi}$ is a column vector whose elements are the group fluxes

$$\mathbf{\Phi} = \begin{pmatrix} \phi_1 \\ \phi_2 \\ \vdots \\ \phi_N \end{pmatrix}. \tag{8.6}$$

$\mathbf{M}$ is the multigroup matrix operator

$$\tag{8.7a}$$

This matrix, we note, is *triangular* (that is, it has only zero elements above the main diagonal) a fact which greatly facilitates the solution of Eq. (8.5). Also, for convenience, we have defined the *removal* cross section Σ_{Rk} for the group as the sum of absorption and slowing-down cross sections.

The matrix $\mathbf{F}$ is simply

$$\mathbf{F} = \begin{pmatrix} \chi_1\Sigma_{f1} & \chi_1\Sigma_{f2} & \chi_1\Sigma_{f3} & \cdots & \chi_1\Sigma_{fN} \\ \chi_2\Sigma_{f1} & \chi_2\Sigma_{f2} & \cdots & \cdots & \chi_2\Sigma_{fN} \\ \cdots & \cdots & \cdots & \cdots & \cdots \\ \chi_N\Sigma_{f1} & \chi_N\Sigma_{f2} & \cdots & \cdots & \chi_N\Sigma_{fN} \end{pmatrix}. \tag{8.7b}$$

Alternatively, we note that

$$\mathbf{F\Phi} = \boldsymbol{\chi} S(\mathbf{r}), \tag{8.8}$$

where $\boldsymbol{\chi}$ is the column vector

$$\boldsymbol{\chi} = \begin{pmatrix} \chi_1 \\ \vdots \\ \chi_N \end{pmatrix}, \tag{8.9}$$

and $S(\mathbf{r})$ is the fission source defined already in Eq. (8.1). Then Eq. (8.5) can be written as

$$\mathbf{M}\boldsymbol{\Phi} = v\chi S(\mathbf{r}). \tag{8.10}$$

These two equations, namely, (8.5) and (8.10), are the two basic forms for the multigroup equation whose solution we shall consider in later sections of this chapter.

8.3 DERIVATION OF THE MULTIGROUP EQUATIONS

We have written down, using our intuition and the results of the first seven chapters, the multigroup diffusion equations. In Chap. 2, we derived the energy-dependent transport equation (2.27), which should exactly describe the behavior of the neutrons, in a reactor. We might ask how these two sets of equations are connected. In answering this question, we shall also find a way to determine the multigroup constants. We first derive the energy-dependent P_1 equations; the analogous one-speed equations were obtained in Chap. 2. The procedure here is quite similar.

Consider the angular flux $\Phi(\mathbf{r},\mathbf{u},t)$ defined in Eq. (2.26). It is more convenient to work in terms of the variables E and $\boldsymbol{\Omega}$ (the neutron energy and the direction of $\mathbf{u}$). We have defined $\Phi(\mathbf{r},E,\boldsymbol{\Omega})$ such that $(1/u)\Phi(\mathbf{r},E,\boldsymbol{\Omega})\, d^3r\, dE\, d\boldsymbol{\Omega}$ is the expected number of neutrons in the phase volume $d^3r\, dE\, d\boldsymbol{\Omega}$ at $\mathbf{r}$, E, $\boldsymbol{\Omega}$ at time t. Since both $\Phi(\mathbf{r},\mathbf{u})$ and $\Phi(\mathbf{r},E,\boldsymbol{\Omega})$ must predict the same number of neutrons in a particular element of phase space, we must have

$$\Phi(\mathbf{r},E,\boldsymbol{\Omega})\, d\boldsymbol{\Omega}\, dE = \Phi(\mathbf{r},\mathbf{u})\, d^3u. \tag{8.11a}$$

Then using $E = \frac{1}{2} mu^2$ and $d^3u = u^2\, du\, d\boldsymbol{\Omega}$, we have

$$\Phi(\mathbf{r},\mathbf{u}) = \sqrt{\frac{m^3}{2E}}\, \Phi(\mathbf{r},E,\boldsymbol{\Omega}). \tag{8.11b}$$

We can determine the time-independent transport equation for $\Phi(\mathbf{r},E,\boldsymbol{\Omega})$ from (2.27c),

$$[\boldsymbol{\Omega}\cdot\nabla + \Sigma_T(\mathbf{r},E)]\Phi(\mathbf{r},E,\boldsymbol{\Omega})$$
$$= S(\mathbf{r},E,\boldsymbol{\Omega}) + \int d\boldsymbol{\Omega}' \int dE'\Sigma_s(E'\to E, \boldsymbol{\Omega}'\cdot\boldsymbol{\Omega})\Phi(\mathbf{r},E',\boldsymbol{\Omega}'), \tag{8.12}$$

where the source and $\Sigma_s(E'\to E, \boldsymbol{\Omega}'\cdot\boldsymbol{\Omega})$ are given by

$$S(\mathbf{r},E,\boldsymbol{\Omega}) = \sqrt{\frac{2E}{m^3}}\, S(\mathbf{r},\mathbf{u}) \tag{8.13a}$$

and

$$\Sigma_s(E'\to E, \boldsymbol{\Omega}'\cdot\boldsymbol{\Omega}) = \sqrt{\frac{2E}{m^3}}\, \Sigma_s(u'\to u, \boldsymbol{\Omega}'\cdot\boldsymbol{\Omega}). \tag{8.13b}$$

Integrating (8.12) over $\mathbf{\Omega}$, we obtain

$$\nabla \cdot \mathbf{J}(\mathbf{r},E) + \Sigma_T(\mathbf{r},E)\phi(\mathbf{r},E) = S(\mathbf{r},E) + \int dE'\,\Sigma_s{}^0(E' \to E)\phi(\mathbf{r},E'), \tag{8.14}$$

where, of course, we have

$$\phi(\mathbf{r},E) = \int d\mathbf{\Omega}\,\Phi(\mathbf{r},E,\mathbf{\Omega}) \tag{8.15a}$$

and

$$\mathbf{J}(\mathbf{r},E) = \int d\mathbf{\Omega}\,\mathbf{\Omega}\Phi(\mathbf{r},E,\mathbf{\Omega}). \tag{8.15b}$$

Also

$$S(\mathbf{r},E) = \int d\mathbf{\Omega}\,S(\mathbf{r},E,\mathbf{\Omega}), \tag{8.15c}$$

and[1]

$$\Sigma_s{}^0(E' \to E) = \int d\mathbf{\Omega}\,\Sigma_s(E' \to E,\, \mathbf{\Omega}' \cdot \mathbf{\Omega}). \tag{8.15d}$$

Now, as we did in one-speed theory [Eq. (2.41)], we assume that we can expand Φ in a two-term power series in $\mathbf{\Omega}$:

$$\Phi(\mathbf{r},E,\mathbf{\Omega}) = \frac{1}{4\pi}\,[\phi(\mathbf{r},E) + 3\mathbf{\Omega} \cdot \mathbf{J}(\mathbf{r},E)]. \tag{8.16}$$

If we insert this expansion in (8.12), we obtain

$$\begin{aligned}\frac{1}{4\pi}\,[\mathbf{\Omega} \cdot \nabla + \Sigma_T(\mathbf{r},E)][\phi(\mathbf{r},E) + 3\mathbf{\Omega} \cdot \mathbf{J}(\mathbf{r},E)] &= S(\mathbf{r},E,\mathbf{\Omega}) \\ &+ \frac{1}{4\pi}\int dE' \left[\Sigma_s{}^0(E' \to E)\phi(\mathbf{r},E') + 3\mathbf{J}(\mathbf{r},E)\int d\mathbf{\Omega}'\,\mathbf{\Omega}'\Sigma_s(E' \to E,\, \mathbf{\Omega}' \cdot \mathbf{\Omega})\right].\end{aligned} \tag{8.17}$$

where we have used (8.15d). From Eq. (2.45) we can see that

$$\int d\mathbf{\Omega}\,\mathbf{\Omega}'\Sigma_s(E' \to E,\, \mathbf{\Omega}' \cdot \mathbf{\Omega}) = \mathbf{\Omega}\Sigma_s{}^1(E' \to E), \tag{8.18a}$$

[1] In Eq. (8.15d), the quantity $\Sigma_s{}^0(E' \to E)$ is equal to $\Sigma_s(E')P(E' \to E)$ of Sec. 9.2. The angularly dependent quantity $\Sigma_s(E' \to E,\, \mathbf{\Omega}' \cdot \mathbf{\Omega})$ is not used *in toto* in this text, but only its zeroth and first moments [see Eq. (8.22) for the first moment].

where

$$\Sigma_s^{\ 1}(E' \rightarrow E) = 2\pi \int_{-1}^{1} d\mu_0 \ \mu_0 \, \Sigma_s(E' \rightarrow E, \mu_0). \tag{8.18b}$$

Thus we can rewrite (8.17) as

$$\frac{1}{4\pi} [\mathbf{\Omega} \cdot \nabla + \Sigma_T(\mathbf{r},E)][\phi(\mathbf{r},E) + 3\mathbf{\Omega} \cdot \mathbf{J}(\mathbf{r},E)] = S(\mathbf{r},E,\mathbf{\Omega})$$

$$+ \frac{1}{4\pi} \int dE' \, [\Sigma_s^{\ 0}(E' \rightarrow E)\phi(\mathbf{r},E) + 3\mathbf{\Omega}\mathbf{J}(\mathbf{r},E')\Sigma_s^{\ 1}(E' \rightarrow E)]. \tag{8.19}$$

We shall assume that the source $S(\mathbf{r}, E, \mathbf{\Omega})$ is independent of $\mathbf{\Omega}$. Then multiplying (8.19) by $\mathbf{\Omega}$ and integrating over $\mathbf{\Omega}$ [again using (2.39)], we obtain

$$\tfrac{1}{3}\nabla\phi(\mathbf{r},E) + \Sigma_T(\mathbf{r},E)\mathbf{J}(\mathbf{r},E) = \int dE' \, \mathbf{J}(\mathbf{r},E')\Sigma_s^{\ 1}(E' \rightarrow E). \tag{8.20}$$

Equations (8.14) and (8.20) are referred to as the P_1 *equations*.[1] We see that (8.20) provides us with a connection between $\mathbf{J}$ and ϕ. It is not so simple, however, as Fick's rule, which we obtained for the one-speed method in Chap. 2.

As we explained in the first two sections of this chapter, our aim here is to convert the P_1 equations into equations for the multigroup fluxes defined in terms of the energy intervals shown in Fig. 8.1.

$$\phi_k(\mathbf{r}) = \int_{E_{k+1}}^{E_k} dE \, \phi(\mathbf{r},E). \tag{8.21}$$

Under usual circumstances, we should assume that we were applying this discrete method because $\phi(\mathbf{r},E)$ and the cross sections were almost constant over the energy intervals. A glance at Fig. 1.5 and 1.6 are sufficient to convince us that it is certainly not possible to include the many thousands of energy intervals needed to ensure that the cross sections are almost constant over each interval. Thus, we must carefully consider the justification of our use of the multigroup equations with only a few groups.

In deriving the multigroup diffusion equations, we must first consider the equation connecting the current and the flux, Eq. (8.20). The presence

[1] The term P_1 refers to expansions in the angular variable up to linear terms, corresponding to an expansion in Legendre polynomials P_l up to $l = 1$.

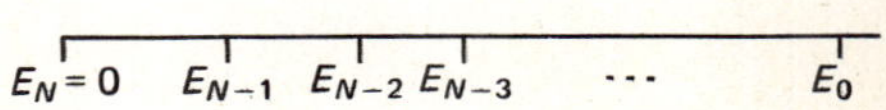

Fig. 8.1 Energy intervals. (We use the largest index $N+1$ to represent $E = 0$ to give agreement with the ordering of lethargy intervals used later.)

of the integral term on the right-hand side of (8.20) causes us some difficulty, in that we do not have a simple energy-dependent Fick's rule as we did in Eq. (2.49). We can recall that $\Sigma_s^{\,1}(E' \to E)$ is related to the average scattering angle $\bar{\mu}$ in the form of Eq. (2.45):

$$\Sigma_s^{\,1}(E' \to E) = \bar{\mu}\Sigma_s^{\,0}(E' \to E). \tag{8.22}$$

Then for an isotropic scattering cross section, $\Sigma_s^{\,1}(E' \to E)$ is zero and Eq. (8.20) defines an energy-dependent diffusion coefficent

$$\mathbf{J}(\mathbf{r},E) = -D(E)\,\nabla\phi(\mathbf{r},E), \tag{8.23a}$$

where

$$D(E) = \frac{1}{3\Sigma_T(\mathbf{r},E)}. \tag{8.23b}$$

In most cases, the inelastic scattering is not isotropic and an energy-dependent diffusion constant cannot be defined. As we shall see, this does *not* preclude obtaining multigroup diffusion equations.

We have defined the multigroup fluxes in Eq. (8.21). We can define the multigroup currents in a similar fashion,

$$J_k(\mathbf{r}) = \int_{E_{k+1}}^{E_k} dE\ \mathbf{J}(\mathbf{r},E). \tag{8.24}$$

Let us now integrate Eq. (8.20) over the kth energy interval,

$$\int_{E_{k+1}}^{E_k} dE\ [\tfrac{1}{3}\nabla\phi(\mathbf{r},E) + \Sigma_T(\mathbf{r},E)\mathbf{J}(\mathbf{r},E)]$$
$$= \sum_{k'=1}^{N} \int_{E_{k'+1}}^{E_{k'}} dE' \int_{E_{k+1}}^{E_k} dE\ \mathbf{J}(\mathbf{r},E')\Sigma_s^{\,1}(E' \to E). \tag{8.25a}$$

In (8.25a) we have broken the E' integration up into a sum of integrals over the N energy intervals. Equation (8.25a) gives the following for the z component of $\mathbf{J}$:

$$\int_{E_{k+1}}^{E_k} \left[dE\,\frac{1}{3}\frac{\partial}{\partial z}\,\phi(\mathbf{r}, E) + \Sigma_T(\mathbf{r},E)J_z(\mathbf{r},E)\right]$$
$$= \sum_{k'=1}^{N} \int_{E_{k'+1}}^{E_{k'}} dE' \int_{E_{k+1}}^{E_k} dE\ J_z(\mathbf{r},E')\Sigma_s^{\,1}(E' \to E). \tag{8.25b}$$

Now we shall approximate the right-hand side of (8.25b) by including only the term $k' = k$ in the sum. This approximation gives

$$\frac{1}{3}\frac{\partial}{\partial z}\,\phi_k(\mathbf{r}) + \int_{E_{k+1}}^{E_k} dE\ J_z(\mathbf{r},E)\left[\Sigma_T(\mathbf{r},E) - \int_{E_{k+1}}^{E_k} dE'\ \Sigma_s^{\,1}(E \to E')\right] = 0, \tag{8.25c}$$

where we interchanged the integration variables E and E' in the term on the right-hand side of (8.25*b*) and moved this term over to the left-hand side to obtain (8.25*c*); and, of course, we have used (8.21). Now we need to justify not including the terms $k' \neq k$ in (8.25). Basically this justification is valid because of the general rule that the cross section $\Sigma_s(E' \to E, \mathbf{\Omega}' \cdot \mathbf{\Omega})$ vanishes if $|E' - E|$ is large enough. Then we see that, for E and E' each in a different energy interval, $\Sigma_s{}^1(E' \to E)$ should be much smaller than it is for E and E' in the same interval.[1]

We can use (8.25*c*) to determine the diffusion coefficient for the kth interval. That is, we write

$$3 \int_{E_{k+1}}^{E_k} dE\, J_z(\mathbf{r},E)\left[\Sigma_T(\mathbf{r},E) - \int_{E_{k+1}}^{E_k} dE'\, \Sigma_s{}^1(E' \to E)\right] = \frac{J_{zk}(\mathbf{r})}{D_k(\mathbf{r})}, \tag{8.26}$$

where $J_{zk}(\mathbf{r})$ is given in Eq. (8.24). Equation (8.26) defines $D_k(\mathbf{r})$. If this technique is to be useful, the diffusion constant defined in (8.26) can depend strongly only upon the material composition of the reactor. Furthermore, we should expect to obtain the same value for the diffusion constant by using J_x and J_y.

This leads us to a multigroup version of Fick's rule,

$$\mathbf{J}_k(\mathbf{r}) = -D_k \nabla \phi_k(\mathbf{r}). \tag{8.27}$$

Now we can multiply (8.14) by dE and integrate over the kth energy interval:

$$\int_{E_{k+1}}^{E_k} dE\, [\nabla \cdot \mathbf{J}(\mathbf{r},E) + \Sigma_T(\mathbf{r},E)\phi(\mathbf{r},E)] = \int_{E_{k+1}}^{E_k} dE\, S(\mathbf{r},E)$$

$$+ \sum_{k'=1}^{N} \int_{E_{k'+1}}^{E_{k'}} dE' \int_{E_{k+1}}^{E_k} dE\, \Sigma_s{}^0(E' \to E)\phi(\mathbf{r},E'). \tag{8.28}$$

Now we can use the definitions of the group constants,

$$\Sigma_{Tk}(\mathbf{r}) = \int_{E_{k+1}}^{E_k} dE \frac{\Sigma_T(\mathbf{r},E)\phi(\mathbf{r},E)}{\phi_k(\mathbf{r})}, \tag{8.29a}$$

$$\Sigma_{k'k}(\mathbf{r}) = \int_{E_{k'+1}}^{E_{k'}} dE' \int_{E_{k+1}}^{E_k} dE \frac{\Sigma_S{}^0(E' \to E)\phi(\mathbf{r},E')}{\phi_{k'}(\mathbf{r})}, \tag{8.29b}$$

$$S_k(\mathbf{r}) = \int_{E_{k+1}}^{E_k} dE\, S(\mathbf{r},E). \tag{8.29c}$$

[1] These results clearly depend upon the choice of the energy intervals and the nature of the moderator. We discuss the slowing-down equations and various approximations to them in Chap. 9.

In fissionable materials $S(\mathbf{r},E)$ is the fission source discussed in Chap. 1 (e.g., Fig. 1.7). The fission source can be written

$$S_f(\mathbf{r},E) = N(E)\int dE'\Sigma_f(\mathbf{r},E')\phi(\mathbf{r},E'), \tag{8.30a}$$

where $N(E)$ is the fission spectrum discussed in Chap. 1. As we have mentioned, frequently one finds $N(E)$ written as

$$N(E) = \nu\chi(E), \tag{8.30b}$$

where ν is the mean number of neutrons produced per fission, and the integral of $\chi(E)$ over all energies is unity.

For the fission source, $S_k(\mathbf{r})$ can be written

$$S_{fk}(\mathbf{r}) = \nu\chi_k \sum_{k'} \phi_{k'}(\mathbf{r})\Sigma_{fk'}(\mathbf{r}), \tag{8.31a}$$

where $\Sigma_{fk'}$ and χ_k are defined as

$$\Sigma_{fk'}(\mathbf{r}) = \int_{E_{k'+1}}^{E_{k'}} dE\,\frac{\phi(\mathbf{r},E)\Sigma_f(E)}{\phi_{k'}(\mathbf{r})}, \tag{8.31b}$$

and

$$\chi_k = \int_{E_{k+1}}^{E_k} dE\,\chi(E). \tag{8.31c}$$

We see that (8.31*a*) is identical with Eq. (8.1), and (8.31*b*) and (8.31*c*) give precise definitions to the group constants.

Using these definitions in Eq. (8.28), we obtain

$$\nabla\cdot\mathbf{J}_k(\mathbf{r}) + \Sigma_{Tk}(\mathbf{r})\phi_k(\mathbf{r}) = \nu\chi_k\sum_{k'}\Sigma_{fk'}(\mathbf{r})\phi_{k'}(\mathbf{r}) + \sum_{k'}\Sigma_{k'k}(\mathbf{r})\phi_{k'}(\mathbf{r}). \tag{8.32}$$

Since we are ignoring upscattering, $\Sigma_{k'k} = 0$ for $k' > k$. Also, it is convenient to include the $k' = k$ term with the total cross section by defining the removal cross section,[1]

$$\Sigma_{Rk}(\mathbf{r}) = \Sigma_{Tk}(\mathbf{r}) - \Sigma_{kk}(\mathbf{r}). \tag{8.33}$$

Now, using Fick's rule (8.27), we can rewrite (8.32) as

$$-\nabla\cdot D_k(\mathbf{r})\nabla\phi_k(\mathbf{r}) + \Sigma_{Rk}(\mathbf{r})\phi_k(\mathbf{r}) = \nu\chi_k\sum_{k'}\Sigma_{fk'}(\mathbf{r})\phi_{k'}(\mathbf{r}) + \sum_{k'<k}\Sigma_{k'k}(\mathbf{r})\phi_{k'}(\mathbf{r}), \tag{8.34}$$

which we can see is identical with Eq. (8.10).

[1] This definition agrees with the one used in (8.7*a*).

Consider the definitions of the group constants in Eqs. (8.29) and (8.31). Clearly these group constants depend upon the energy-dependent fluxes $\phi(\mathbf{r},E)$. Thus, the solution of the equations must be known in order to determine the group constants! The game that is played (and this is discussed in detail in Chap. 9) is to use a spectrum, i.e., an energy-dependent flux, for averaging the group constants, this spectrum being obtained from a simplified (infinite-medium) calculation.

In any event, if groups are chosen in a sufficiently narrow range of energy, then the "averaging" scheme does not make much difference. However, as we mentioned earlier, our energy groups are not nearly small enough to expect $\phi(\mathbf{r},E)$ or the cross sections to be constant over an energy group. Since the cross sections and fluxes vary significantly over an energy interval, the group constants will depend strongly upon the material composition of the reactor (more than just through the material densities appearing in the macroscopic cross sections). This dependence will result because energy dependence of $\phi(\mathbf{r},E)$ in an energy group will be determined by the material composition, and the energy dependence can be expected to change significantly as the composition changes. This effect results in an extra dependence upon composition as can be seen in Eqs. (8.29). Thus the spectral procedure referred to above is needed. This approximation is valid because among the different compositions arising in the operation of a reactor or among different reactors of a specific type, we do not expect to see strong variations.[1] Then we expect that the group constants will be reasonably constant for moderate changes in composition.

We might point out here that a serious analysis of the operating characteristics of a reactor over its lifetime must include recalculations of the group constants as the composition changes because of such effects as burnup of U^{235} and production of Pu^{239} In fact, BWRs (boiling-water reactors) are quite difficult to analyze because the group constants must be recomputed every time the steam fraction increases or decreases.

We shall discuss the problems involved in computing group constants in more detail in Chaps. 9 and 10.

8.4 ANALYTICAL SOLUTIONS OF THE MULTIGROUP EQUATIONS

For very simple problems, i.e., with few groups or few regions (over each of which the group cross sections are separately constant), analytical solutions of the multigroup equations are feasible. Actually, these analytical solutions are of little practical value, so that we shall consider only a particularly simple case which will help us to gain insight into the calcuation of group constants.

[1] Except, of course, reactors with boiling coolants such as BWRs.

Consider then the two-group equations for a bare, homogeneous reactor. First, let us rewrite the definitions of some of the group constants for this special case. The fluxes are (recall that $E_3 = 0$ in this case)

$$\phi_2(\mathbf{r}) = \int_0^{E_2} dE\,\phi(\mathbf{r},E), \tag{8.35a}$$

$$\phi_1(\mathbf{r}) = \int_{E_2}^{E_1} dE\,\phi(\mathbf{r},E). \tag{8.35b}$$

Usually, in the two-group equations, E_2 is taken to be about 1 eV. Then the group labeled 2 is referred to as the *thermal group*, and the group labeled 1 as the *fast group*. We must bear this in mind when we compute the fission source

$$\chi_2 = \int_0^{E_1} dE\,\chi(E), \tag{8.36a}$$

$$\chi_1 = \int_{E_2}^{E_3} dE\,\chi(E). \tag{8.36b}$$

Since $E_2 \approx 1$ eV and $E_3 \approx 10^7$ eV, we see $\chi_1 = 1$ and $\chi_2 = 0$; the fission source appears only in group 1 (although it contains contributions *from* both group 1 and group 2).

$$S_{f1}(\mathbf{r}) = \nu\phi_1(\mathbf{r})\Sigma_{f1}(\mathbf{r}) + \nu\phi_2(\mathbf{r})\Sigma_{f2}(\mathbf{r}). \tag{8.36c}$$

Another simplification is in the removal cross section in group 2. Since there is no slowing down out of group 2, the integral of the scattering cross section over the final energies in group 2 equals the total scattering cross section:

$$\int_0^{E_2} dE\,\Sigma_s(E' \to E) = \Sigma_s(E'), \qquad 0 \le E' \le E_2\,. \tag{8.37a}$$

Comparing this result with Eq. (8.29*b*), we see that Σ_{22} is just the average scattering cross section in group 2,

$$\Sigma_{22} = \Sigma_{S2} = \int_0^{E_2} dE\,\frac{\Sigma_S(E)\phi(\mathbf{r},E)}{\phi_2(\mathbf{r})}. \tag{8.37b}$$

Then the removal cross section for group 2 is just equal to the average absorption cross section for group 2,

$$\Sigma_{R2}(\mathbf{r}) = \Sigma_{T2}(\mathbf{r}) - \Sigma_{S2}(\mathbf{r}) = \Sigma_{a2}(\mathbf{r}). \tag{8.37c}$$

All the other group constants have the definitions we noted in (8.29), (8.31), and (8.33), and the two-group diffusion equations are

$$-D_1\nabla^2\phi_1(\mathbf{r}) + \Sigma_{R1}\phi_1(\mathbf{r}) = \nu\Sigma_{f1}\phi_1(\mathbf{r}) + \nu\Sigma_{f2}\,\phi_2(\mathbf{r}), \tag{8.38a}$$

$$-D_2\nabla^2\phi_2(\mathbf{r}) + \Sigma_{a2}\,\phi_2(\mathbf{r}) = \Sigma_{12}\,\phi_1(\mathbf{r}). \tag{8.38b}$$

We are assuming that the reactor is bare and homogeneous. Then the boundary conditions are [1]

$$\phi_1(\mathbf{R}') = \phi_2(\mathbf{R}') = 0. \tag{8.38c}$$

Now let us look for solutions of (8.38) in terms of geometrical buckling introduced in Chap. 3,

$$(\nabla^2 + B_g^2)\phi_1(\mathbf{r}) = 0, \tag{8.39a}$$

$$(\nabla^2 + B_g^2)\phi_2(\mathbf{r}) = 0, \tag{8.39b}$$

where ϕ_1 and ϕ_2 satisfy boundary condition (8.38*c*). (This, of course, determines B_g.) Using (8.39*a*) and (8.39*b*), we can write (8.38*a*) and (8.38*b*) as

$$(D_1 B_g^2 + \Sigma_{R1} - \nu\Sigma_{f1})\phi_1 - \nu\Sigma_{f2}\,\phi_2 = 0, \tag{8.40a}$$

$$-\Sigma_{12}\,\phi_1 + (D_2 B_g^2 + \Sigma_{a2})\phi_2 = 0. \tag{8.40b}$$

The criticality condition is obtained by noting that the homogeneous equations (8.40) for ϕ_1 and ϕ_2 will have a nonzero solution only if the determinant of the coefficient matrix vanishes.

$$\begin{vmatrix} D_1 B_g^2 + \Sigma_{R1} - \nu\Sigma_{f1} & -\nu\Sigma_{f2} \\ -\Sigma_{12} & D_2 B_g^2 + \Sigma_{a2} \end{vmatrix} = 0. \tag{8.41}$$

By expanding the determinant and using the definitions of the fast and thermal diffusion lengths,

$$L_2^2 = \frac{D_2}{\Sigma_{a2}}, \tag{8.42a}$$

$$L_1^2 = \frac{D_1}{\Sigma_{R1} - \nu\Sigma_{f1}}, \tag{8.42b}$$

the criticality condition reads

$$(1 + B_g^2 L_1^2)(1 + B_g^2 L_2^2) = \frac{\nu\Sigma_{f2}\,\Sigma_{12}}{\Sigma_{a2}(\Sigma_{R1} - \nu\Sigma_{f1})}. \tag{8.43}$$

Several of the quantities appearing in (8.43) have been discussed earlier in the context of one-group theory. In Eq. (3.80), we pointed out that $\nu\Sigma_{f2}/\Sigma_{a2}$ equals the thermal utilization f times the fuel efficiency η for thermal neutrons,

$$\frac{\nu\Sigma_{f2}}{\Sigma_{a2}} = \eta f. \tag{8.44}$$

[1] We assume that the extrapolated boundaries are the same for the fast and the thermal groups. We justify this assumption by the fact that z_0 is much smaller than the dimensions in both the fast and thermal groups.

The quantities $(1 + B_g^2 L_1^2)^{-1}$ and $(1 + B_g^2 L_2^2)^{-1}$ are the fast and thermal nonleakage probabilities as shown in (3.87) for the one-group approximation.

$$P_T = \frac{1}{1 + B_g^2 L_2^2}, \tag{8.45a}$$

$$P_F = \frac{1}{1 + B_g^2 L_1^2}. \tag{8.45b}$$

If we do not include leakage, the number of neutrons leaving group 1 per second is $\phi_1 \Sigma_{R1}$, while the number slowing down into group 2 per second is $\phi_1 \Sigma_{12}$. The ratio of these numbers is equal to the probability that a neutron is not absorbed in group 1 (ignoring leakage). Since most of the absorption in group 1 occurs in resonances, as we pointed out in Chap. 1, this quantity is referred to as the *resonance escape probability p*,

$$p = \frac{\Sigma_{12}}{\Sigma_{R1}}. \tag{8.46}$$

Recall that the fast fission factor ε is defined to be the ratio of the total number of neutrons produced by fission, $\nu\Sigma_{f1}\phi_1 + \nu\Sigma_{f2}\phi_2$, to the number produced by thermal neutrons, $\nu\Sigma_{f2}\phi_2$, in the absence of leakage

$$\varepsilon = \frac{\nu\Sigma_{f1}\phi_1 + \nu\Sigma_{f2}\phi_2}{\nu\Sigma_{f2}\phi_2}. \tag{8.47a}$$

We can use (8.40*a*) (with $B_g = 0$, since ε is defined for zero leakage) to express $\nu\Sigma_{f2}\phi_2$ in terms of ϕ_1. Substituting this in (8.47*a*), we obtain

$$\varepsilon = \frac{\Sigma_{R1}}{\Sigma_{R1} - \nu\Sigma_{f1}}. \tag{8.47b}$$

Now we shall combine these results to show that (8.43) is identical with the six-factor formula introduced in Chap. 1 [Eq. (1.12)]. Using the definitions of P_F and P_T, (8.45*a*) and (8.45*b*), in (8.43), we obtain

$$1 = P_F P_T \frac{\nu\Sigma_{f2}}{\Sigma_{a2}} \frac{\Sigma_{12}}{\Sigma_{R1}} \frac{\Sigma_{R1}}{\Sigma_{R1} - \nu\Sigma_{f1}}. \tag{8.48a}$$

We have factored (8.48*a*) to indicate the possible application of (8.47*b*), (8.46), and (8.44) for the fast fission factor, the resonance escape probability, and ηf, respectively. These equations in (8.48*a*) result in the six-factor criticality condition,

$$1 = P_F P_T \eta f p \varepsilon. \tag{8.48b}$$

Thus far we have related the two-group bare-reactor formulas to earlier, more general, results. Let us recall (3.89), the expression for the reactor power,

$$P = 3.2 \times 10^{-11} (\bar{\phi}_1 \Sigma_{f1} + \bar{\phi}_2 \Sigma_{f2}) \quad \text{W}, \tag{8.49}$$

where

$$\bar{\phi}_1 = \int_{\text{reactor}} d^3r\, \phi_1(\mathbf{r}), \tag{8.50a}$$

and

$$\bar{\phi}_2 = \int_{\text{reactor}} d^3r\, \phi_2(\mathbf{r}). \tag{8.50b}$$

If we integrate (8.40*b*) over the reactor volume and use (8.49) we can find $\bar{\phi}_1$ and $\bar{\phi}_2$ in terms of the power P in watts.

$$\bar{\phi}_1 = \frac{(D_2 B_g^2 + \Sigma_{a2})\Sigma_{f2}}{(D_2 B_g^2 + \Sigma_{a2})\Sigma_{f1} + \Sigma_{f2}\Sigma_{12}} \frac{P}{3.2 \times 10^{-11}}, \tag{8.51a}$$

$$\bar{\phi}_2 = \frac{\Sigma_{12}\Sigma_{f2}}{(D_2 B_g^2 + \Sigma_{a2})\Sigma_{f1} + \Sigma_{f2}\Sigma_{12}} \frac{P}{3.2 \times 10^{-11}}. \tag{8.51b}$$

It is of some interest to see how the two-group results can be "collapsed" to one group. (Such a procedure is frequently used in practical calculation, although the collapsing might be from 50 groups to 4 groups, for example.) If we add Eqs. (8.40*a*) and (8.40*b*), and integrate over the reactor volume, we obtain

$$(D_1\bar{\phi}_1 + D_2\bar{\phi}_2)B_g^2 + (\Sigma_{11}\bar{\phi}_1 + \Sigma_{a2}\bar{\phi}_2 - \Sigma_{12}\bar{\phi}_2) - \nu\Sigma_{f1}\bar{\phi}_1 - \nu\Sigma_{f2}\bar{\phi}_2 = 0. \tag{8.52}$$

The one-group constants are obtained by writing (8.52) in terms of the total flux $\bar{\phi}_1 + \bar{\phi}_2$.

$$\bar{D}B_g^2(\bar{\phi}_1 + \bar{\phi}_2) + \bar{\Sigma}_a(\bar{\phi}_1 + \bar{\phi}_2) - \nu\bar{\Sigma}_f(\bar{\phi}_1 + \bar{\phi}_2) = 0, \tag{8.53}$$

where

$$\bar{D} = \frac{D_1\bar{\phi}_1 + D_2\bar{\phi}_2}{\bar{\phi}_1 + \bar{\phi}_2} \tag{8.54a}$$

$$= \frac{(D_2 B_g^2 + \Sigma_{a2})D_1 + \Sigma_{12} D_2}{D_2 B_g^2 + \Sigma_{a2} + \Sigma_{12}}. \tag{8.54b}$$

$$\bar{\Sigma}_a = \frac{\Sigma_{R1}\bar{\phi}_1 + \Sigma_{a2}\bar{\phi}_2 - \Sigma_{12}\bar{\phi}_1}{\bar{\phi}_1 + \bar{\phi}_2} \tag{8.55a}$$

$$= \frac{(\Sigma_{R1} - \Sigma_{12})(D_2 B_g^2 + \Sigma_{a2}) + \Sigma_{a2}\Sigma_{12}}{D_2 B_g^2 + \Sigma_{a2} + \Sigma_{12}}. \tag{8.55b}$$

$$\nu\bar{\Sigma}_f = \nu\frac{\Sigma_{f1}\bar{\phi}_1 + \Sigma_{f2}\bar{\phi}_2}{\bar{\phi}_1 + \bar{\phi}_2} \tag{8.56a}$$

$$= \nu\frac{(D_2 B_g^2 + \Sigma_{a2})\Sigma_{f1} + \Sigma_{12}\Sigma_{f2}}{D_2 B_g^2 + \Sigma_{a2} + \Sigma_{12}}. \tag{8.56b}$$

The criticality condition (8.41) provides a further relation among the quantities appearing in these equations. A reactor is usually maintained critical by inserting and removing control rods which absorb neutrons in the thermal group. Thus, it is most convenient to express the criticality as a condition on Σ_{a2}.

$$\Sigma_{a2} = \frac{\nu\Sigma_{12}\Sigma_{f2}}{D_1 B_g^2 + \Sigma_{R1} - \nu\Sigma_{f1}} - D_2 B_g^2. \tag{8.57}$$

If we ignore the fission in all but the thermal group, we can easily expand the procedure, described above for two groups, to N groups, in which case we find a formula similar to (8.48b),

$$\frac{\eta f \prod_{i=1}^{N-1} p_i}{\prod_{i=1}^{N} (1 + L_i^2 B_g^2)} = 1, \tag{8.58}$$

where

$$p_i = \sum_{k>i} \frac{\Sigma_{ik}}{\Sigma_{Ri}}, \tag{8.59a}$$

$$p = \prod_{i=1}^{N-1} p_i; \tag{8.59b}$$

p_i is interpreted as the resonance escape probability for the ith group. Similarly,

$$L_i^2 = \frac{D_i}{\Sigma_{R1}} \tag{8.60}$$

is the ith group diffusion length. If B_g^2 is small (as is generally the case), then

$$\prod_{i=1}^{N} (1 + L_i^2 B_g^2) \approx 1 + B_g^2 \sum_{i=1}^{N} L_i^2 \tag{8.61a}$$

and the sum of all the L_i^2 is generally denoted by M^2 and called the migration area

$$M^2 = \sum_i L_i^2. \tag{8.61b}$$

Although the equations treated here, and the critical equations, are valid only for thermal reactors, they give some knowledge of the group constants and will be of material assistance to us in calculating the constants which are applicable to any type of reactor.

8.5 NUMERICAL SOLUTION OF THE MULTIGROUP EQUATIONS

The numerical solution of the multigroup equations in the form of (8.8) is so similar to the numerical solution of the one-speed diffusion equations discussed in Chap. 6 that very little further comment is necessary. In the first outer iteration we guess a source function $S^{(0)}(\mathbf{r})$. Note that it is *not* necessary to guess a flux in each group; only a single function of position $S^{(0)}(\mathbf{r})$ is necessary.

Consider, then, Eq. (8.10), which, being a matrix equation, is really a shorthand method of writing N equations (N = number of groups). If we refer to the definition of **M**, Eq. (8.7), we see that the first of these equations is [we are replacing $S(\mathbf{r})$ with $S^{(0)}(\mathbf{r})$, where $S^{(0)}(\mathbf{r})$ is our first guess for $S(\mathbf{r})$]

$$(-\nabla \cdot D_1 \nabla + \Sigma_{R1})\phi_1 = \nu\chi_1 S^{(0)}(\mathbf{r}). \tag{8.62a}$$

This is precisely the one-speed diffusion equation whose (numerical) solution is discussed extensively in Chap. 6.

The second of Eqs. (8.10) is

$$(-\nabla \cdot D_2 \nabla + \Sigma_{R2})\phi_2 = \nu\chi_2 S^{(0)}(\mathbf{r}) + \Sigma_{12}\phi_1. \tag{8.62b}$$

Again, this has the form of the one-speed diffusion equation. On the assumption that (8.62a) has been solved previously, the flux ϕ_1 on the right-hand side is a known function, and is merely added to the source term. In this way, the N equations (8.10) can be solved successively. The simplicity here is due to the fact that the matrix **M** is triangular, which, in turn, arises from the assumption that in a collision, neutrons never gain energy (i.e., there is no upscattering).

When the N fluxes $\phi_1, \phi_2, \ldots, \phi_N$ have been obtained, a new source function $S^{(1)}(\mathbf{r})$ is calculated from (8.4b). Then new fluxes are computed, and a second source iterate obtained, and so forth. The procedure is identical with the outer-iteration procedure described in Chap. 6.

If upscattering is permitted in the multigroup matrix, then the equations do not decouple in this simple fashion. In that case some number of equations must be solved simultaneously, the number depending upon the number of groups to which upscattering is permitted. Procedures for these calculations are discussed by Wachspress.[1]

8.6 INTEGRAL EQUATION FORMULATION

In Sec. 3.8 we treated the one-speed critical problem by an integral equation formulation, and pointed out, at the time, that the treatment could be applied to more general problems.

[1] E. Wachspress, "Iterative Solution of Elliptic Systems," Prentice-Hall, Inc., Englewood Cliffs, N.J., 1966.

In fact the results derived there, if interpreted properly, can be applied to the energy-dependent problem immediately.

The critical equation derived in Sec. 3.8 was

$$1 = \nu\Sigma_f \bar{G}_\infty(B^2), \tag{8.63}$$

where $\bar{G}_\infty$ was the Fourier transform of the infinite-medium Green's function $G_\infty(\mathbf{r}-\mathbf{r}')$. For the one-speed diffusion treatment of the reactor, we had Eq. (3.104).

$$\bar{G}_\infty = \frac{1}{\Sigma_a(1+L^2B^2)}, \quad \text{one speed.} \tag{8.64}$$

Suppose, in our present problem, we write $\bar{G}_\infty$ as the convolution of a slowing-down term $g_\infty(|\mathbf{r}'-\mathbf{r}|)$ with the diffusion kernel (2.104). Then writing

$$G_\infty(\mathbf{r}) = \int d^3r'\, g_\infty(|\mathbf{r}'-\mathbf{r}|)\frac{e^{-r'L}}{4\pi Dr'}, \tag{8.65a}$$

we have

$$\bar{G}_\infty(B^2) = \bar{g}_\infty(B^2)\frac{1}{\Sigma_a(1+L^2B^2)}. \tag{8.65b}$$

To obtain (8.65*b*), we have used the convolution theorem discussed in Appendix C. Also, we shall use L with no subscript to denote the thermal diffusion length. Thus, the critical equation becomes

$$\frac{\eta f \bar{g}_\infty(B^2)}{1+L^2B^2} = 1, \tag{8.66}$$

and in analogy with our previous interpretation of the criticality equation, we can interpret $\bar{g}_\infty(B^2)$ as the product of the probabilities that a neutron will not leak out of the system and will not be absorbed while it is slowing down.[1] In the two-group model (i.e., if the slowing-down process is described by *one* fast group),

$$\bar{g}_\infty(B^2) = \frac{p}{1+L_F{}^2B^2}, \tag{8.67}$$

where, of course, L_F is the diffusion length in the fast group. In the case of two fast groups

$$\bar{g}_\infty(B^2) = \frac{p_1p_2}{(1+L_{1F}{}^2B^2)(1+L_{2F}{}^2B^2)}, \tag{8.68}$$

[1] We are ignoring fast fission. Also, note that the nonleakage and noncapture probabilities are not really independent, although it is not a bad approximation to assume that they are.

and so forth. In the next chapter we shall present another method, the Fermi age method, which gives a better estimate of $\bar{g}_\infty(B^2)$ than do the multigroup methods.[1] However, it can only be applied directly to criticality calculations for simple systems. Practical calculations still require numerical multigroup methods, but the age theory analysis will assist us in determining the group constants.

Some further information may be gleaned from the integral equation formulation. Suppose we write, for the thermal flux $\phi_T(\mathbf{r})$, the integral equation (3.98),

$$\phi_T(\mathbf{r}) = \nu\Sigma_f \int \phi_T(\mathbf{r}')G_\infty(|\mathbf{r}' - \mathbf{r}|)\, d^3r'. \tag{8.69}$$

Now we know that ϕ_T obeys the thermal diffusion equation with a *slowing-down source*. We know, therefore, from (8.69) that it is proportional to the lowest eigenmode $\chi_0(\mathbf{r})$ of the Helmholtz equation with a buckling $B_m{}^2$ which is the positive solution of Eq. (8.66) (all this was shown in general in Sec. 3.8). However, the flux at *any* energy E can be written in the form of Eq. (8.69), that is,

$$\phi(E,\mathbf{r}) = \int d^3r'\, \nu\Sigma_f\, \phi_T(\mathbf{r}')G_\infty(E,|\mathbf{r}' - \mathbf{r}|). \tag{8.70}$$

Thus, the arguments we have just used in (8.69) can be applied to (8.70) to show that $\phi(E,\mathbf{r})$ is proportional to $\chi_0(\mathbf{r})$, and hence that $\phi(E,\mathbf{r})$ is separable into a product $F(E)\chi_0(\mathbf{r})$.

This analysis is valid only for a bare, homogeneous reactor in which the extrapolated end-point connection can be assumed to be energy independent.[2]

8.7 MEASUREMENT OF k_∞

a THE FRENCH INTERMEDIATE EXPERIMENT

Very soon after the discovery of fission in 1938, it was found that additional neutrons were emitted in the fission process. The exact number of these neutrons was not known, although preliminary measurements had indicated this number to be somewhere between two and three.

Immediately, the possibility of a chain reaction occurred to a large number of people. Whether or not *criticality* could be obtained depended

[1] Note Fermi age theory is only applicable to heavy moderators like graphite. For H_2O and D_2O other methods must be used.

[2] For a rigorous analysis, see Joel Ferziger and P. F. Zweifel: "The Theory of Neutron Slowing Down in Nuclear Reactors," Pergamon Press, New York, 1966.

upon a number of unknown factors—in today's terminology, ε, p, and f (rough measurements of η already existed). Again translating the early ideas into today's language, the scientists working in the field then realized the requirement that k_∞ must be greater than unity if criticality were to be achieved. It was also of interest to know by how much k_∞ exceeded unity in order to know if a chain reaction were *practical*, but the earliest measurements did not give this information.

It is important to understand the significance of the concept of k_∞. If this concept did not exist, and if there were no way to determine the magnitude of this quantity, there would be no way to determine in advance whether or not a particular reactor configuration could achieve criticality without actually building the complete reactor. Even if the reactor were built and turned out to be subcritical, it would not be known, to any degree of certainty, whether the reactor was subcritical merely because it was too small ($k_\infty > 1$, $k < 1$).

The French solved this problem in the following ingenious way. A small, spherical aluminum container was filled with the core material. (They considered homogeneous mixtures of H_2O and natural uranium oxide and of D_2O and natural uranium oxide, and heterogeneous arrangements of paraffin and natural uranium oxide. Enriched uranium did not exist at that time.) Surrounding this sphere, which we shall call the *core*, was a large region of H_2O, which we shall call the *reflector*. This reflector was so large that it would, for all practical purposes, be considered infinite. Small foils were positioned in the core and reflector so that a radial thermal-flux plot could be obtained by activation analysis. A neutron source was positioned in the center of the core (see Fig. 8.2).

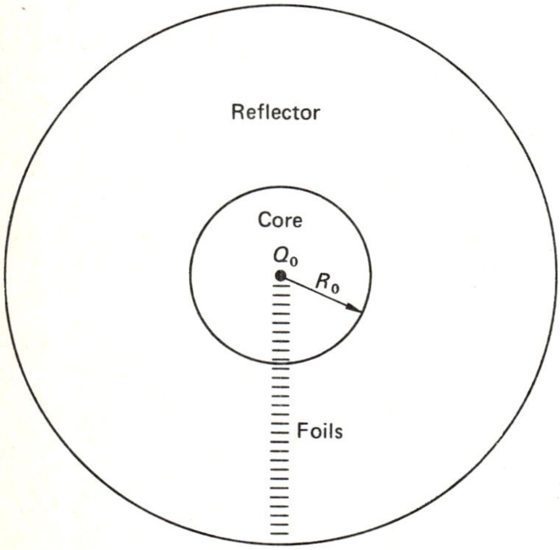

Fig. 8.2 French intermediate experiment.

In a separate experiment, the same radial thermal-flux plot was made, except that the core was removed. If you like, the second experiment can be considered the same as the first in the limit $R_0 \to 0$. (We shall see presently the importance of the second experiment.)

Let us use two-group theory to analyze this experiment. We shall denote by superscripts c and R the core and reflector regions, respectively, and by subscripts 1 and 2 the fast and thermal groups. Then the neutron flux in the core is described by the two equations

$$\nabla \cdot \mathbf{J}_1{}^c + \Sigma_{R1}{}^c\phi_1{}^c = \nu\Sigma_f{}^c\phi_2{}^c + Q_0\,\delta(\mathbf{r}), \tag{8.71}$$

$$\nabla \cdot \mathbf{J}_2{}^c + \Sigma_{a2}{}^c\phi_z{}^c = \Sigma_{12}{}^c\phi_1{}^c. \tag{8.72}$$

In writing these two equations, certain reasonable assumptions have been made:

1. All fission neutrons are born in the fast group.
2. The extraneous source (strength Q_0) emits fast neutrons.
3. We have ignored fast fission.

In the reflector, there is no uranium, so that $\Sigma_f{}^R = 0$. Since pure water has essentially a zero-capture cross section in the fast group, $\Sigma_{a1}{}^R = 0$ also. Then the neutron flux in the reflector is described by

$$\nabla \cdot \mathbf{J}_1{}^R + \Sigma_{12}{}^R\phi_1{}^R = 0, \tag{8.73a}$$

$$\nabla \cdot \mathbf{J}_2{}^R + \Sigma_{a2}{}^R\phi_2{}^R = \Sigma_{12}{}^R\phi_1{}^R. \tag{8.73b}$$

Now we integrate Eq. (8.73*a*) over r from 0 to R_0, and we integrate Eq. (8.73*b*) from R_0 to ∞. Using Gauss' theorem, we obtain

$$R_0{}^2J_1(R_0) + \Sigma_{R1}{}^c\Phi_1{}^c = \nu\Sigma_f{}^c\Phi_2{}^c + Q_0, \tag{8.74a}$$

$$R_0{}^2J_2(R_0) + \Sigma_{a2}{}^c\Phi_2{}^c = \Sigma_{12}{}^c\Phi_1{}^c, \tag{8.74b}$$

$$-R_0{}^2J_1(R_0) + \Sigma_{12}{}^R\Phi_1{}^R = 0, \tag{8.74c}$$

$$-R_0{}^{2J}{}_2(R_0) + \Sigma_{a2}{}^R\Phi_2{}^R = \Sigma_{12}{}^F\Phi_1{}^R. \tag{8.74d}$$

Here the superscript c or R on $J_1(R_0)$ or $J_2(R_0)$ is not necessary because $J_i{}^R(R_0) = J_i{}^c(R_0)$ (continuity of current).[1] Also, we have introduced the notation

$$\Phi_1{}^c = \int_0^{R_0} r^2\,dr\,\phi_i{}^c(r), \qquad i = 1, 2; \tag{8.75a}$$

$$\Phi_i{}^R = \int_{R_0}^{\infty} r^2\,dr\,\phi_i{}^R(r), \qquad i = 1, 2. \tag{8.75b}$$

[1] Also, the current has only a radial component, so we exhibit J as a scalar rather than a vector.

Equation (8.74c) can be used to eliminate $J_1(R_0)$ from (8.74a). If $\Phi_1{}^c$ and $\Phi_1{}^R$ are then eliminated from the remaining equations, we find

$$\left(\nu\Sigma_f{}^c - \frac{\Sigma_{R1}{}^c\Sigma_{a2}{}^c}{\Sigma_{12}{}^c}\right)\Phi_2{}^c = -Q_0 + \Sigma_{a2}{}^R\Phi_2{}^R + \left(\frac{\Sigma_{R1}{}^c}{\Sigma_{12}{}^c} - 1\right)R_0{}^2J_2(R_0). \tag{8.76}$$

To determine Q_0, we refer to the experiment already described in which there is no core, or, if you like, $R_0 \to 0$. Then, since any neutron emitted by the source must be absorbed as a thermal neutron, we have

$$Q_0 = \Sigma_{a2}{}^R\Phi_2{}^1, \tag{8.77}$$

where we have denoted by $\Phi_2{}^1$ the integrated thermal flux in this portion of the experiment

$$\Phi_2{}^1 = \int_0^\infty r^2\,dr\,\phi_2(r). \tag{8.78}$$

Now we recall the expressions for the resonance escape probability and k_∞,

$$p = \frac{\Sigma_{12}{}^c}{\Sigma_{R1}{}^c}, \tag{8.79}$$

$$\eta f = \frac{\nu\Sigma_f{}^c}{\Sigma_{a2}{}^c}, \tag{8.80}$$

$$k_\infty = \eta f p, \tag{8.81}$$

$$k_\infty = \frac{\nu\Sigma_f{}^c}{\Sigma_{a2}{}^c}\frac{\Sigma_{12}{}^c}{\Sigma_{R1}{}^c}. \tag{8.82}$$

Then Eq. (8.76) becomes, by using (8.77) and (8.82),

$$(k_\infty - 1)\Phi_2{}^c = \frac{\Sigma_{a2}{}^R}{\Sigma_{a2}{}^c}\,p(\Phi_2{}^R - \Phi_2{}^1) + \frac{1-p}{\Sigma_{a2}{}^c}\,R_0{}^2J_2(R_0). \tag{8.83}$$

The original experiments described here, and the analysis, was done by a French group in 1939 to 1940. This group was Frederic Joliot-Curie, Hans H. von Halban, and Lew Kowarski. In their analysis, they neglected $R_0{}^2J_2(R_0)$, that is, they assumed there was no transport of thermal neutrons between the core and the reflector. Then they noted they had only to determine the quantities $\Phi_2{}^R$ and $\Phi_2{}^1$ by integrating the activities of their foils. If $\Phi_2{}^R > \Phi_2{}^1$, it followed that $k_\infty > 1$, and vice versa.

We might question the neglect of $J_2(R_0)$. From our analysis of the reflected reactor, we expect that $J_2(R_0) < 0$, that is, there should be a net flow of neutrons from the reflector into the core. Then, since $k < 1$,

$$\Phi_2{}^R - \Phi_2{}^1 \approx (k_\infty - 1) + \delta \tag{8.84}$$

with $\delta > 0$. Suppose the experiments showed $\Phi_2{}^R - \Phi_2{}^1 < 0$; then *certainly* $k_\infty < 1$. However, if experimentally $\Phi_2{}^R - \Phi_2{}^1 > 0$, it would still be possible that $k_\infty < 1$. Thus a sufficient, but not a necessary, condition that a given reactor core configuration could never lead to a self-sustaining chain reaction is $\Phi_2{}^1 < \Phi_2{}^R$.

Since the French group, in those pre-World War II and early war days, were anxious to develop a guaranteed scheme for a nuclear reactor, they concentrated on finding a system for which $\Phi_2{}^R > \Phi_2{}^1$. After conducting experiments with natural uranium and H_2O in a homogeneous mixture and heterogeneous arrays of natural uranium and paraffin,[1] they concluded that there was no way to make $\Phi_2{}^R > \Phi_2{}^1$, and so it was very unlikely to make a natural uranium, hydrogen-moderated reactor operate. (We know today that this conclusion was correct. However, if the uranium is enriched, from its natural U^{235} abundance of 0.7 percent to about 0.9 percent, then a self-sustaining chain reaction *is* possible in a heterogeneous system.)

The French group next set up experiments with natural uranium and D_2O, reasoning that although deuterium was not so good a moderator as hydrogen, its thermal absorption cross section was so much less (compared with $\frac{1}{3}$ b) that k_∞ should be larger in the D_2O system. This reasoning proved to be correct. In December 1940, experiments in England (since the Germans had occupied Paris, it was necessary to move the experiments elsewhere) indicated $\Phi_2{}^R > \Phi_2{}^1$ for at least two different concentrations of natural uranium homogeneously mixed with D_2O. Rough estimation of the quantities appearing in Eq. (8.83), that is, $\Sigma_{a2}{}^R$, $\Sigma_{a2}{}^c$, and p, and measurements of $\Phi_2{}^c$ [they still assumed $J_2(R_0) = 0$] gave numerical values of k_∞ very close to those known today to be correct. The December 1940 test was the first proof that, in fact, a self-sustaining chain reaction was actually possible (although at the time, the problem of obtaining sufficient D_2O was still unsolved).

b FERMI'S EXPONENTIAL EXPERIMENT

We recall, from part a of this section, that the French were setting up and conducting their *intermediate* experiments between the spring of 1939 and December 1940. We also recall that they began studying assemblies consisting of natural uranium and H_2O and they turned to D_2O as a moderator after concluding that the use of H_2O would not lead to a critical system if fueled with natural uranium. The Americans, under the inspirational guidance of Enrico Fermi, had also begun to consider constructing reactors at this time and had also concluded that light water was not a suitable moderator. They decided to pin their hopes on graphite rather than heavy

[1] In Chap. 9 we explain why heterogeneous systems can be made to possess a higher k than correspondingly homogeneous mixtures.

water. It is important to note that the American decision was *not* made on the basis of any measurements of in light-water-moderated systems but was rather based on Fermi's genius intuition. This decision to use graphite as a moderator, rather than D_2O, was based on its much greater availability.[1] Before D_2O was available to construct a reactor, the Americans reasoned, an entire separations technology would have to be developed and implemented.

In the early winter months of 1941, before the news of the French intermediate experiment had reached the United States, Fermi was considering methods which he could use to measure k_∞ in his graphite assemblies. His first idea was to construct a number of different assemblies, identical in composition, but of varying sizes. He then proposed to measure the multiplication of these (see Sec. 4.9) as a function of B^2. For $k_\infty < 1$, the plot of M versus B^2 would asymptotically approach the value $1/(1 - k_\infty)$. For $k_\infty > 1$, on the other hand, the multiplication would increase without limit as B^2 decreased, i.e., as the size increased (see Fig. 8.3).

This method suffered from a number of deficiencies, for example, the need to build several different assemblies for each measurement and, as in the case of the French intermediate experiment, the impossibility of measuring k_∞ absolutely. (Actually, if M could be measured absolutely instead of only relatively, then k_∞ could be measured for values less than unity.)

Sometime in the late winter or spring of 1941, Fermi conceived of the *exponential experiment*. This had one similarity to the French intermediate experiment, that only a single assembly was needed to test k_∞. However, it had the advantage that it was much easier to obtain numerical values for k_∞ since the number of parameters which had to be known was vastly reduced from that in the French method [see Eq. (8.83)]. Also, only a single measurement and a single assembly were needed. The French, we recall, needed

[1] Again, later events bore out the correctness of this decision. The first graphite pile went critical in December 1942. In fact, graphite reactors had produced enough plutonium for the construction of an atomic bomb before the first D_2O reactor was even critical.

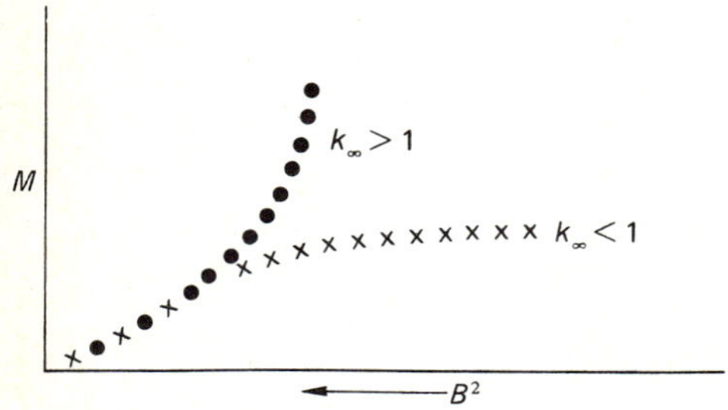

Fig. 8.3 M versus B^2 (note the different curvature for $k_\infty > 1$ and $k_\infty < 1$.

measurements with and without the active core in place. It is not known definitely whether learning of the French work gave Fermi the idea for the exponential pile. What *is* known is that he learned of the French work between the time he proposed the method described above and the time he invented the exponential rule, a period of only a few months.

The exponential experiment is quite similar to the measurement of the diffusion length described in Sec. 2.11, except that a portion of the reactor core actually becomes the medium in which an axial thermal-flux traverse is made. Today, the source of neutrons might come from the thermal column of a reactor—in those days, of course, there *were* no reactors. Neither were there any other sources of thermal neutrons. For this reason, it was necessary to use a vast neutron source, such as Ra-α-Be. However, the source was imbedded in a block of graphite so that effectively the core region was irradiated with thermal neutrons. A drawing of the assembly is seen in Fig. 8.4.

Again, we might use two-group theory to analyze this experiment.

$$-D_1 \nabla^2\phi_1 + \Sigma_{R1}\phi_1 = \nu\Sigma_f\phi_2 \tag{8.85a}$$

$$-D_2\nabla^2\phi_2 + \Sigma_{a2}\phi_2 = \Sigma_{12}\phi_1 + Q_0 \cos\alpha x \cos\beta y\, \delta(z), \tag{8.85b}$$

with

$$\alpha = \frac{\pi}{2a'} \tag{8.86a}$$

and

$$\beta = \frac{\pi}{2b'}. \tag{8.86b}$$

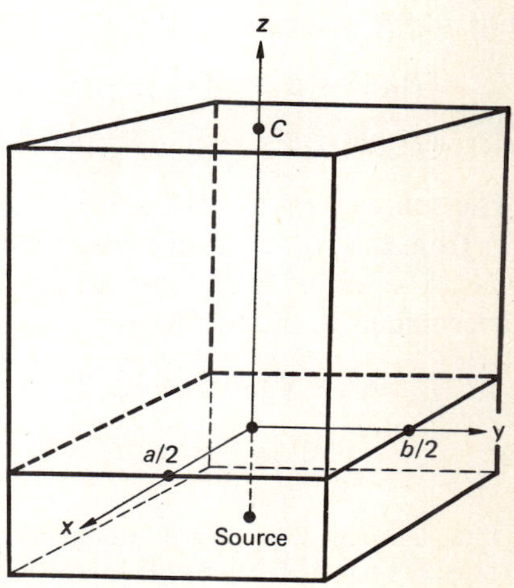

Fig. 8.4 The exponential pile.

The assumption in writing Eqs. (8.85) is that the source neutrons not only have *thermalized* before reaching the core region, but also have assumed a *fundamental* mode. This is identical with the assumption made in the analysis of the diffusion length measurements referred to above, the only difference being that because of the presence of fissionable material in the exponential pile, two-group theory (at least) is required.

An *ansatz* of the form

$$\phi_i(x,y,z) = Z_i(z) \cos \alpha x \cos \beta y, \qquad i = 1, 2, \tag{8.87}$$

reduces Eqs. (8.85) to

$$D_1\left(\alpha^2 + \beta^2 - \frac{\partial^2}{\partial z^2}\right)Z_1(z) + \Sigma_{R1}Z_1(z) = \nu\Sigma_f Z_2(z), \tag{8.88a}$$

$$D_2\left(\alpha^2 + \beta^2 - \frac{\partial^2}{\partial z^2}\right)Z_2(z) + \Sigma_{12}Z_2 = \Sigma_{12}Z_1 + Q_0\,\delta(z). \tag{8.88b}$$

A complete solution of these equations is not necessary for our purposes. We are only concerned with the *exponential* decay of the flux in the interior and near the upper edge of the core region. We seek solutions of the form

$$Z_1(z) = K_1 \sinh \gamma(c' - z), \tag{8.89a}$$

$$Z_2(z) = K_2 \sinh \gamma(c' - z), \tag{8.89b}$$

where K_1 and K_2 are constants to be determined. The form $\sinh \gamma(c' - z)$ was chosen because then $Z_1(c') = Z_2(c') = 0$, as must be true.[1] Since

$$\frac{\partial^2}{\partial z^2} \sinh \gamma(c' - z) = \gamma^2 \sinh \gamma(c' - z), \tag{8.90}$$

Eqs. (8.89) become

$$[D_1(\alpha^2 + \beta^2 - \gamma^2) + \Sigma_{R1}]K_1 - \nu\Sigma_f K_2 = 0, \tag{8.91a}$$

$$[D_2(\alpha^2 + \beta^2 - \gamma^2) + \Sigma_{a2}]K_2 - \Sigma_{12} K_1 = 0. \tag{8.91b}$$

(The source term is missing in (8.91*b*) because we are obtaining the solution far from the bottom of the core assembly.)

The set of homogeneous equations (8.91) has a solution only if the determinant of the coefficients vanishes, i.e.,

$$[D_1(\alpha^2 + \beta^2 - \gamma^2) + \Sigma_{R1}][D_2(\alpha^2 + \beta^2 - \gamma^2) + \Sigma_{a2}] - \nu\Sigma_f\Sigma_{12} = 0. \tag{8.92}$$

[1] If the assembly were critical, we should expect combinations of imaginary exponentials, that is, sines, rather than real exponentials, sinh's.

By dividing through by $\Sigma_{a2}\Sigma_{R1}$ and recalling the definitions

$$\frac{D_1}{\Sigma_{R1}} = L_1^2, \tag{8.93a}$$

$$\frac{D_2}{\Sigma_{a2}} = L_2^2, \tag{8.93b}$$

and

$$\frac{\nu\Sigma_{f2} + \Sigma_{12}}{\Sigma_{a2}\Sigma_{R2}} = \eta f p = k_\infty, \tag{8.93c}$$

(8.92) becomes

$$k_\infty = [1 + L_1^2(\alpha^2 + \beta^2 - \gamma^2)][1 + L_2^2(\alpha^2 + \beta^2 - \gamma^2)]. \tag{8.94}$$

Thus, if γ is known, k_∞ can be determined [the diffusion lengths are assumed known, and the bucklings α^2 and β^2 are determined from the dimensions, Eq. (8.80)].

The quantity γ is determined by making an axial thermal-flux plot (for example, by activation of foils placed along the z axis of the core assembly) and fitting the plot to express Eq. (8.89b). The term *exponential assembly* comes from the fact that for $z \ll c$ (that is, not too near the upper edge of the assembly),

$$\sinh \gamma(c' - z) = \tfrac{1}{2}\,(e^{\gamma(c'-z)} - e^{-\gamma(c'-z)}] \tag{8.95a}$$

$$\approx \tfrac{1}{2}e^{\gamma(c'-z)} \tag{8.95b}$$

$$= Ce^{-\gamma z}, \tag{8.95c}$$

where C is a constant.

Thus, in a certain region of the assembly, not too near either boundary, γ is simply the slope of the axial flux when plotted on semilog paper.

The above analysis could be refined in several ways, for example, by including fast fission in Eqs. (8.88). The reader may be interested in showing that this adds to the expression for k_∞ as given in Eq. (8.93c) a term $k_\infty^{(1)}$ corresponding to the contribution from fast fission.

Refined mathematical analysis of this experiment can also be found in a number of more advanced texts in reactor physics. The above analysis, however, is adequate not only for physical understanding, but also for most practical analyses. This type of experiment, as ancient as it is, is still in common use today.

8.8 EFFECTIVE DELAYED-NEUTRON FRACTION

In Chap. 4, we pointed out that the energy spectrum for prompt fission neutrons is considerably harder than that for delayed neutrons. Thus, we can expect that the effect of the delayed neutrons will be different from the prompt neutrons. This will require some modification of the results which were obtained in Chap. 4. To see this, let us consider the time-dependent multigroup equations. To obtain the time-dependent equation, the term $(1/u)(\partial\phi/\partial t)$ must be added to Eq. (8.14). It is clear that the $1/u$ factor will appear averaged over the kth group,

$$\frac{1}{v_k} = \frac{\int_{E_{k+1}}^{E_k} dE\,(1/u)\phi(\mathbf{r},E)}{\int_{E_{k+1}}^{E_k} dE\,\phi(\mathbf{r},E)}. \tag{8.96}$$

This will result in an additional term of the form $(1/v_k)(\partial/\partial t)\phi_k(\mathbf{r},t)$ in the multigroup equations.

In order to discuss the effect of the different energy distribution of delayed neutrons it will be necessary to consider at least three energy groups, the thermal group, an epithermal group which contains the delayed neutron source, and a fast group which contains the prompt neutron source. Of course, we must also include the delayed-neutron precursors. Then the kinetics equations with one delayed-neutron precursor, whose density we denote by $C(\mathbf{r},t)$, and three energy groups are as follows:[1]

$$\frac{1}{v_3}\frac{\partial}{\partial t}\phi_3(\mathbf{r},t) - D_3\nabla^2\phi_3(\mathbf{r},t) + \Sigma_{a3}\phi_3(\mathbf{r},t) = \Sigma_{23}\phi_2(\mathbf{r},t). \tag{8.97a}$$

As was pointed out, we assume that the delayed source is present only in the epithermal group. This gives

$$\frac{1}{v_2}\frac{\partial}{\partial t}\phi_2(\mathbf{r},t) - D_2\nabla^2\phi_2(\mathbf{r},t) + \Sigma_{R2}\phi_2(\mathbf{r},t) = \Sigma_{12}\phi_1(\mathbf{r},t) + \lambda C(\mathbf{r},t). \tag{8.97b}$$

We assume that the prompt-neutron source is present only in the fast group and that there is no fast fission to obtain,

$$\frac{1}{v_1}\frac{\partial}{\partial t}\phi_1(\mathbf{r},t) - D_1\nabla^2\phi_1(\mathbf{r},t) + \Sigma_{R1}\phi_1(\mathbf{r},t) = (1-\beta)\nu[\Sigma_{f3}\phi_3(\mathbf{r},t) + \Sigma_{f2}\phi_2(\mathbf{r},t)]. \tag{8.97c}$$

Finally the precursor concentration is given by

$$\frac{\partial}{\partial t}C(\mathbf{r},t) = -\lambda C(\mathbf{r},t) + \beta\nu[\Sigma_{f3}\phi_3(\mathbf{r},t) + \Sigma_{f2}\phi_2(\mathbf{r},t)]. \tag{8.97d}$$

We assume that only the lowest spatial mode contributes and that we can replace ∇^2 with $-B^2$. Furthermore, we shall look for a solution of the form

$$\phi_k(\mathbf{r},t) = \bar{\phi}_k e^{\omega t}, \qquad k = 1, 2, 3, \tag{8.98a}$$

$$C(\mathbf{r},t) = \bar{C}e^{\omega t}. \tag{8.98b}$$

[1] We assume that no neutrons are slowed down from group 1 to group 3, that is, $\Sigma_{13} = 0$.

Inserting Eqs. (8.98) into the multigroup Eqs. (8.97) and replacing ∇^2 with $-B^2$, we can rewrite the multigroup equations as a set of algebraic equations

$$\left(\frac{\omega}{v_3} + D_3 B^2 + \Sigma_{f3}\right)\bar{\Phi}_3 = \Sigma_{23}\bar{\Phi}_2, \tag{8.99a}$$

$$\left(\frac{\omega}{v_2} + D_2 B^2 + \Sigma_{R2}\right)\bar{\Phi}_2 = \Sigma_{12}\bar{\varphi}_1 + \lambda\bar{C}, \tag{8.99b}$$

$$\left(\frac{\omega}{v_1} + D_1 B^2 + \Sigma_{R1}\right)\bar{\Phi}_1 = (1-\beta)\nu(\Sigma_{f3}\bar{\Phi}_3 + \Sigma_{f2}\bar{\Phi}_2), \tag{8.99c}$$

$$\omega\bar{C} = -\lambda\bar{C} + \beta\nu(\Sigma_{f3}\bar{\Phi}_3 + \Sigma_{f2}\bar{\Phi}_2]. \tag{8.99d}$$

Since v_1 is the velocity in the fast group, we shall neglect the factor ω/v_1. These equations can be written in matrix form as

$$\begin{pmatrix} D_1B^2 + \Sigma_{R1} & -(1-\beta)\nu\Sigma_{f2} & -(1-\beta)\nu\Sigma_{f3} & 0 \\ -\Sigma_{12} & \dfrac{\omega}{v_2} + D_2 B^2 + \Sigma_{R2} & 0 & -\lambda \\ 0 & -\Sigma_{23} & \dfrac{\omega}{v_3} + D_3 B^2 + \Sigma_{a3} & 0 \\ 0 & -\beta\nu\Sigma_{f2} & -\beta\nu\Sigma_{f3} & \omega + \lambda \end{pmatrix} \begin{pmatrix} \bar{\Phi}_1 \\ \bar{\Phi}_2 \\ \bar{\Phi}_3 \\ \bar{C} \end{pmatrix} = 0. \tag{8.100}$$

This equation will have a solution only if the determinant of the coefficient matrix vanishes. After some algebra, this condition can be written as

$$(1+\omega l_2)(1+\omega l_3) - (1-\beta)k_2(1+\omega l_3) - (1-\beta)k_3$$
$$= \frac{\lambda}{\omega+\lambda}\left[\beta k_3 \frac{1+L_1{}^2B^2}{p_1} + (1+\omega l_3)\beta k_2 \frac{1+L_1{}^2B^2}{p_1}\right], \tag{8.101}$$

where we have introduced the definition of group multiplication factors k_2 and k_3,

$$k_2 = \frac{\nu\Sigma_{f2}}{\Sigma_{R2}} p_1 \frac{1}{(1+L_1{}^2B^2)(1+L_2{}^2B^2)}, \tag{8.102a}$$

$$k_3 = \frac{\nu\Sigma_{f3}}{\Sigma_{R3}} p_1 p_2 \frac{1}{(1+L_1{}^2B^2)(1+L_2{}^2B^2)(1+L_3{}^2B^2)}. \tag{8.102b}$$

We have also introduced the definitions of the group resonance escape probabilities and mean neutron lifetimes p_1, p_2, l_2, and l_3,

$$p_1 = \frac{\Sigma_{12}}{\Sigma_{R1}}, \tag{8.103a}$$

$$p_2 = \frac{\Sigma_{23}}{\Sigma_{R2}}, \tag{8.103b}$$

$$l_2 = \frac{1}{v_2 \Sigma_{R2}(1 + L_2{}^2 B^2)}, \tag{8.104a}$$

$$l_3 = \frac{1}{v_3 \Sigma_{a3}(1 + L_3{}^2 B^2)}. \tag{8.104b}$$

If we set $\omega = 0$ in Eq. (8.101), we merely obtain a three-group criticality equation involving k_2 and k_3 [cf. Eq. (8.58)].

Let us now apply Eq. (8.101) to thermal reactors. We have already made the assumption in deriving (8.101) that l_1, the effective lifetime of neutrons in the fast group, can be ignored. For thermal reactors, we shall set $k_2 = l_2 = 0$, as well. Then Eq. (8.101) reduces to (by setting $k_3 = k$ and $l_3 = l$)

$$(1 + \omega l) - (1 - \beta)k = \frac{\lambda}{\omega + \lambda}\left[\beta k \frac{1 + L_1{}^2 B^2}{p_1}\right]. \tag{8.105}$$

This equation is to be compared with Eq. (4.26*a*). It differs in that βk, the right-hand side, is multiplied by the factor $(1 + L_1{}^2 B^2)/p_1$. This reflects the fact that delayed neutrons are inserted into the reactor with an energy lower than that of the prompt neutrons, i.e., in group 2 rather than group 1 in our simple model. Thus, the delayed neutrons automatically escape capture and leakage in group 1. It is customary to define

$$\beta_{\text{eff}} = \beta \frac{1 + L_1{}^2 B^2}{p_1}, \tag{8.106}$$

that is, β_{eff} is the delayed-neutron fraction multiplied by the ratio of the probability that a prompt neutron will be absorbed as a thermal neutron to the probability that a delayed neutron will be likewise absorbed. We see $\beta_{\text{eff}} \geq \beta$. A simple, but not quite accurate, method of including β_{eff} in kinetic calculations is to replace β with β_{eff} wherever it appears in the kinetic equations.

We cannot directly apply (8.105) and (8.106) to fast reactors. For fast reactors, our neglect of k_2 and l_2 is not justified. Thus, a correct analysis requires that we return to (8.101). We can still put this equation into the form of (4.26*a*) by neglecting the terms of second order in the k's and l's (specifically we set $l_2 l_3$ and $k_2 l_3$ equal to zero). In this approximation, (8.101) becomes

$$1 + \omega(l_2 + l_3) - (1 - \beta)(k_2 + k_3) = \frac{\lambda}{\omega + \lambda}\beta(k_2 + k_3)\frac{1 + L_1{}^2 B^2}{p_1}. \tag{8.107}$$

This equation is of the form of (8.105) with k replaced by $k_2 + k_3$ and l by $l_2 + l_3$. The major difference between (8.107) and (8.105) lies in the size of the effective lifetimes l and the effective delayed-neutron fractions β_{eff}. For fast reactors, the effective lifetime is much shorter than for thermal reactors. The effective delayed-neutron fraction is smaller in fast reactors than in thermal reactors. Both of these effects combine to make fast reactors more difficult to control than thermal reactors.

PROBLEMS

8.1. Show that the inclusion of the terms $k' \neq k$ in (8.25*b*) requires the introduction of a diffusion matrix. Derive an expression for this diffusion matrix.

8.2. Compute χ_1 and χ_2 [Eq. (8.36)] for U^{235} with $E_2 = 1$ eV, and show that $\chi_2 \ll \chi_1$.

8.3. Some rather typical values for the group constants for a two-group model PWR (pressurized water reactor) are:

$\Sigma_{R1} = 0.02\ \text{cm}^{-1}$; $\nu\Sigma_{f1} = 0.006\ \text{cm}^{-1}$; $D_1 = 1.5$ cm; $D_2 = 0.5$ cm; $\nu\Sigma_{f2} = 0.13\ \text{cm}^{-1}$; $\Sigma_{12} = 0.01\ \text{cm}^{-1}$.

Suppose the reactor is an infinite slab 100 cm thick. What value of Σ_{a2} is required for criticality?

8.4. For the reactor in Prob. 8.3, what are the thermal and fast nonescape probabilities?

8.5. For the reactor in Prob. 8.3, what is the resonance escape probability?

8.6. For the reactor in Prob. 8.3, what is the fast fission factor?

8.7. For the reactor in Prob. 8.3, compute the effective one-group constants.

8.8. Suppose the critical bare reactor described in Prob. 8.3 is sandwiched between two slabs, each 100 cm thick and each having the following group constants:

$\Sigma_{R1}{}^R = 0.03\ \text{cm}^{-1}$; $\nu\Sigma_{f1}{}^R = \nu\Sigma_{f2}{}^R = 0$; $D_1{}^R = 1.3$ cm; $D_2{}^R = 0.4$ cm; $\Sigma_{12}{}^R = 0.02\ \text{cm}^{-1}$; $\Sigma_{a2}{}^R = 0.04\ \text{cm}^{-1}$.

What is the reactivity increase?

8.9. How much absorption must be added to the core in the reflected reactor in Prob. 8.8 to make it critical?

9
Slowing-down Calculations

9.1 INTRODUCTION

In the present chapter we consider some very special problems concerning neutron slowing-down in an infinite medium. That is, spatial effects are ignored. The rationale for this approach is based on the assumption that to a certain approximation the neutron flux in a reactor is separable in space and energy; we have seen in Sec. 8.6 that this separability does in fact hold in a bare, homogeneous reactor. A similar separability may be expected to hold in nonhomogeneous, reflected reactors sufficiently far from the boundaries between different regions.

The procedure typically followed, then, is to calculate energy flux averages of various cross sections of interest on the basis of these infinite-medium calculations, and then to use these averages in a multigroup scheme, as described in Chap. 8, to obtain spatial effects. While this procedure may not seem logically justifiable, it is in fact used as a matter of routine in almost all practical reactor calculations. This procedure, called the *spectral method*, can be refined by introducing a buckling B^2 into the calculation.[1] Then we calculate the neutron spectrum in a bare, homogeneous reactor with a geometrical buckling equal to the materials buckling for this region. The bucklings used in the calculation vary. For each core region of different composition, the materials buckling in that region should be computed, and a different spectrum calculated for each region. For nonfuel regions (e.g., reflector) it is customary to take $B^2 = 0$.

The actual procedures used to generate cross sections for multigroup calculations (called *group constants*) are described in detail later in this chapter.

In an infinite medium, the current **J** and the flux ϕ are independent of position. Then, the $\nabla \cdot \mathbf{J}$ term in Eq. (8.14) vanishes, and we can write

$$\Sigma_T(E)\phi(E) = S(E) + \int_0^\infty dE' \, \Sigma_s^0(E' \to E)\phi(E'). \tag{9.1}$$

Recall that this equation was derived directly from the transport equation without using Fick's rule. Thus it does not contain the approximations of diffusion theory. (Of course it does not describe spatial transport of neutrons either.)

We shall find it convenient to assume that slowing-down is isotropic, that is, $\Sigma_s^1(E' \to E) = 0$. In this case, as we determined in Chap. 8, an energy-dependent diffusion equation can be written, in which the diffusion coefficient is

$$D(E) = \frac{1}{3\Sigma_T(E)}, \tag{9.2}$$

and the diffusion equation is

$$-\nabla \cdot D(E)\nabla\phi(\mathbf{r},E) + \Sigma_T(E)\phi(\mathbf{r},E) = S(E) + \int_0^\infty dE' \, \Sigma_s^0(E' \to E)\phi(\mathbf{r},E'). \tag{9.3}$$

The integral of $\Sigma_s^0(E' \to E)$ over E gives the total scattering cross section $\Sigma_s(E')$. Thus, it is convenient to introduce the collision kernel $P(E' \to E)$, which describes the distribution of final energies for neutrons which suffer a scattering collision at the energy E'.

$$\Sigma_s^0(E' \to E) = \Sigma_s(E')P(E' \to E), \tag{9.4}$$

where $P(E' \to E)$ must be normalized such that

$$\int dE \, P(E' \to E) = 1. \tag{9.5}$$

Clearly the details of the moderation of neutrons will depend critically upon the detailed properties of $P(E' \to E)$.

[1] Note that the infinite-medium spectrum is a special case, corresponding to $B^2 = 0$.

9.2 THE COLLISION KERNEL: ELASTIC SCATTERING

We begin by studying the details of the collision process in which neutrons interact with a moderator nucleus. It is assumed that the reader has a knowledge of classical collision theory; so the results are presented here without derivation, but derivations may be found in Appendix E. (A number of definitions and results given there will be repeated here for the convenience of the reader who chooses to skip the Appendix.)

Suppose a neutron, whose mass we take equal to unity, makes an elastic collision with a nucleus of mass A (measured in neutron masses). Obviously $A \geq 1$.[1]

Let us define the following symbols (see Fig. 9.1):

E_i = initial neutron energy in the laboratory coordinate system (LCS).
E_f = final neutron energy in LCS.
μ_0 = cosine of the scattering angle in LCS.
μ_c = cosine of the scattering angle in the center-of-momentum coordinate system (CMCS).

Here we have referred to the laboratory and center-of-momentum coordinate systems. The laboratory system is used in reactor calculations. In this system the target nucleus is at rest before the collision. The center-of-momentum system is useful for theoretical analysis. It is defined by the

[1] Although, strictly speaking, A is slightly less than unity for collisions between a neutron and a proton. However, it is customary to read A to the nearest integer.

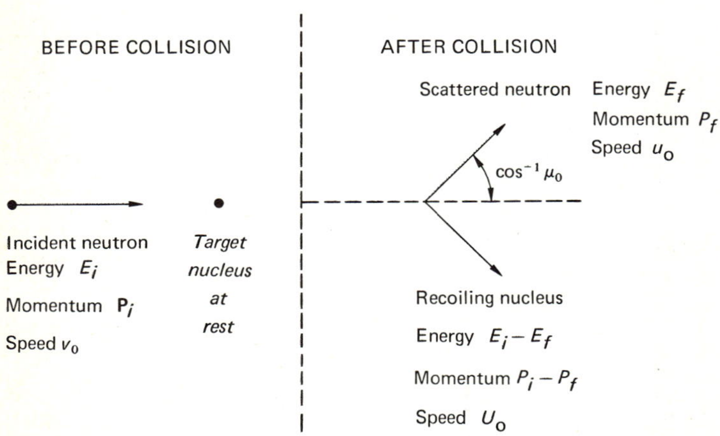

Fig. 9.1 Elastic scattering event in the laboratory coordinate system.

requirement that the total momentum (neutron plus target nucleus) be equal to zero.[1]

The following results are derived in Appendix E on the basis of energy and momentum conservation:

$$E_i \geq E_f \geq \alpha E_i, \tag{9.6}$$

$$\alpha = \frac{(A-1)^2}{(A+1)^2}. \tag{9.7}$$

This tells us that the energy of the neutron after collision is *less* than before the collision, but that the maximum energy loss of the neutron is an amount $(1-\alpha)E_i$. Physically, since the target nucleus is initially at rest, energy can be transferred only from the neutron *to* the target. Thus the neutron must *lose* energy in a collision, that is, $E_f \leq E_i$.[2] However, consideration of momentum conservation restricts the maximum energy loss to the amount $(1-\alpha)E_i$. For $A = 1$, that is, moderation by hydrogen, $\alpha = 0$, so that a neutron can in fact lose all its energy in a single collision. As A increases, α increases and approaches unity so that the maximum energy loss approaches zero; for this reason, light elements are better moderators than heavy elements.

The ratio E_i/E_f and the cosine of the scattering angle μ_0 (or μ_c) are related as follows:

$$\frac{E_f}{E_i} = \frac{A^2 + 2A\mu_c + 1}{(A+1)^2} \tag{9.8a}$$

$$= \frac{\gamma^2 + 2\gamma\mu_c + 1}{(\gamma+1)^2}, \tag{9.8b}$$

where

$$\gamma = \frac{1}{A}. \tag{9.9}$$

(By setting $\mu_c = \pm 1$ in Eq. (9.9), Eq. (9.6) follows immediately.) The cosine μ_0 and μ_c are related as follows:

$$\mu_0 = \frac{\mu_c + \gamma}{(\gamma^2 + 2\gamma\mu_c + 1)^{1/2}}. \tag{9.10}$$

[1] For nonrelativistic systems the center-of-momentum coordinate system is often referred to as the *center-of-mass coordinate system*. Incidentally, there is no change in neutron energy in an elastic collision as measured in the CMCS. Of course, the neutron energy as measured in the LCS does change.

[2] This is no longer true when the thermal energy of the moderator atom is of the order of neutron energies, i.e., for $E_i \approx kT$ (cf. Sec. 9.8).

It is convenient to introduce a new variable, the *lethargy*, defined by

$$u = \ln \frac{E_0}{E}, \qquad E = E_0 e^{-u}, \tag{9.11}$$

where E_0 is a conveniently chosen reference energy. We shall take $E_0 = 10$ MeV, so that $u = 0$ at $E = 10$ MeV. Our assumption that no fission neutrons are created with energy greater than 10 MeV means that negative values of the lethargy will never be encountered. (Note that as energy *decreases*, lethargy *increases*.)

In terms of the lethargy, we can write

$$\frac{E_f}{E_i} = e^{+(u_i - u_f)} = e^{-\Delta u}, \tag{9.12a}$$

where

$$\Delta u = u_f - u_i. \tag{9.12b}$$

That is, Δu is the lethargy change in a collision. From Eqs. (9.8*b*) and (9.10), we can derive an expression for $\mu_0(\Delta u)$ and $\mu_c(\Delta u)$, that is, the cosine of the scattering angle for a given lethargy change Δu:

$$\mu_c(\Delta u) = 1 - \frac{(1+\gamma)^2}{2\gamma}(1 - e^{-\Delta u}) \tag{9.13a}$$

and

$$\mu_0(\Delta u) = \frac{1}{2\gamma}[(1+\gamma)e^{-\Delta u/2} - (1-\gamma)e^{\Delta u/2}]. \tag{9.13b}$$

For many applications, it is adequate to assume the scattering to be *isotropic* in the CMCS.[1] Suppose we define $P(\mu_c)d\mu_c$ as the probability that a neutron will scatter through an angle whose cosine lies between μ_c and $\mu_c + d\mu_c$. Then, for isotropic scattering all *cosines* are equally probable, i.e.,

$$P(\mu_c) = \text{constant}. \tag{9.14}$$

This constant must be chosen such that $P(\mu_c)$ is normalized, i.e.,

$$\int_{-1}^{1} P(\mu_c)\, d\mu_c = 1. \tag{9.15a}$$

Clearly,

$$P(\mu_c) = \tfrac{1}{2}. \tag{9.15b}$$

[1] For a discussion of nonisotropic scattering, see Joel Ferziger and P. F. Zweifel: "The Theory of Neutron Slowing Down in Nuclear Reactors," Pergamon Press, New York, 1966.

We can now calculate the probability $P(\Delta u)\Delta u$ that in a collision the lethargy will change by an amount between Δu and $u + \Delta u$. Since $P(\Delta u)$ and $P(\mu_c)$ describe the same quantity in terms of uniquely related variables, the following must hold:[1]

$$P(\Delta u)|d(\Delta u)| = P(\mu_c)|d\mu_c|, \tag{9.16}$$

where, from (9.13a)

$$\frac{d\mu_c}{d(\Delta u)} = \frac{(1+\gamma)^2 e^{-\Delta u}}{2\gamma}. \tag{9.17a}$$

Thus we can write, using $\gamma = 1/A$ and Eq. (9.7),

$$P(\Delta u) = \frac{e^{-\Delta u}}{1-\alpha}. \tag{9.17b}$$

We note that Eq. (9.6) can be rewritten in terms of the lethargy variable,

$$0 \le \Delta u \le \ln\frac{1}{\alpha}. \tag{9.18}$$

[The lethargy always increases in a collision. Equation (9.17*b*) tells us the probability that it increases by a given amount Δu.] Note that $P(\Delta u) = 0$ for $\Delta u > \ln 1/\alpha$, from Eq. (9.18), so that Eq. (9.17) only applies for $0 \le \Delta u \le \ln 1/\alpha$. In Fig. 9.2, we plot $P(\Delta u)$ versus Δu.

[1] The two sides of (9.16) refer to the same probability, which cannot be changed merely because the independent variable is transformed. The absolute value signs express the fact that probabilities must be nonnegative.

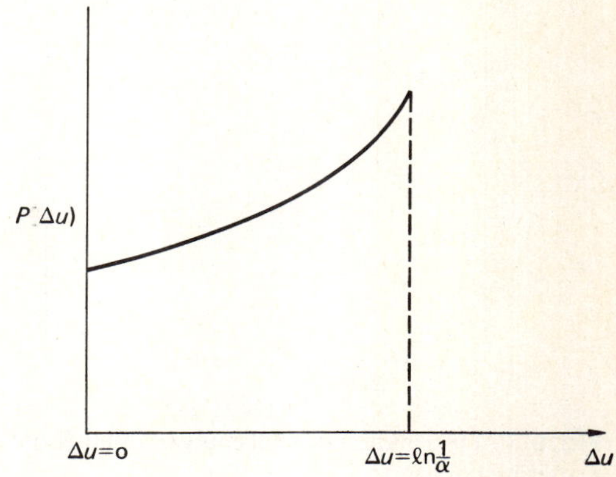

Fig. 9.2 Distribution of lethargy change in an elastic collision.

One can easily verify that

$$\int_0^{\ln 1/\alpha} P(\Delta u)\, d(\Delta u) = 1. \tag{9.19}$$

The quantity $P(E_i \to E_f)$ can also be calculated, e.g., from Eq. (9.17), by using the relation

$$P(E_i \to E_f) = P(\Delta u)\left|\frac{d(\Delta u)}{dE_f}\right|. \tag{9.20}$$

We find, since

$$E_f = E_i e^{-\Delta u}, \tag{9.21a}$$

that

$$P(E_i \to E_f) = \begin{cases} \dfrac{1}{(1-\alpha)E_i}, & E_i \geq E_f > \alpha E_i, \quad (9.21b) \\ 0, & \text{otherwise.} \quad (9.21c) \end{cases}$$

We note that in a collision the final energy of the neutron varies, with equal probability, over the range $E_i \geq E_f > \alpha E_i$ (see Fig. 9.3). This is a consequence, of course, of the assumption of isotropic scattering in the CMCS. For nonisotropic scattering, the distribution in final energies will be different.

The distribution functions which we have given here can be used to compute a number of average quantities. For example, the average cosine of the scattering angle in the CMCS is simply

$$\bar{\mu}_c = \int_{-1}^{1} \mu_c P(\mu_c)\, d\mu_c \tag{9.22a}$$

$$= 0. \tag{9.22b}$$

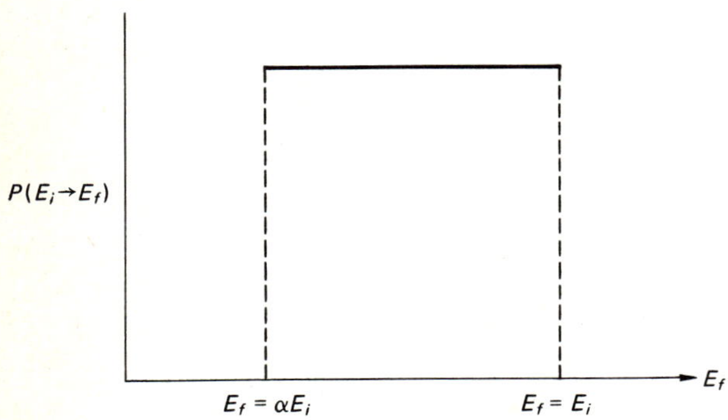

Fig. 9.3 Distribution of final energies in an (isotropic) elastic scattering collision.

In the LCS the scattering is *not* isotropic. Thus,

$$\bar{\mu}_0 = \int_{-1}^{1} \mu_0(\mu_c) P(\mu_c)\, d\mu_c, \tag{9.23}$$

where $\mu_0(\mu_c)$ is given in Eq. (9.10). Alternatively, one could write

$$\bar{\mu}_0 = \int_0^{\ln 1/\alpha} \mu_0(\Delta u) P(\Delta u)\, d(\Delta u), \tag{9.24}$$

using Eq. (9.13*b*) for $\mu_0(\Delta u)$ and (9.17*b*) for $P(\Delta u)$. In any event, one easily finds

$$\bar{\mu}_0 = \frac{2\gamma}{3} = \frac{2}{3A}. \tag{9.25}$$

For large A, that is, heavy scattering nuclei, $\bar{\mu}_0 \to 0$; so the scattering becomes isotropic in the LCS. For hydrogen, on the other hand, $A = 1$ and $\bar{\mu}_0 = 2/3$, which means the scattering is highly anisotropic (peaked in the forward direction). As $A \to \infty$, the CMCS and the LCS coincide, which explains why the scattering in the LCS tends to become isotropic.

Also of interest is the average energy change in a collision. Actually we shall have more occasion to use the average lethargy change $\overline{\Delta u}$, a quantity commonly denoted by ξ.

$$\xi = \int_0^{\ln 1/\alpha} \Delta u\, P(\Delta u)\, d(\Delta u) \tag{9.26a}$$

$$= 1 - \frac{\alpha}{1-\alpha} \ln \alpha. \tag{9.26b}$$

A formula valid for large A can be obtained as follows:

$$\ln \alpha = \ln \left[\left(\frac{1-\gamma}{1+\gamma} \right)^2 \right] \tag{9.27a}$$

$$= 2[\ln(1-\gamma) - \ln(1+\gamma)]. \tag{9.27b}$$

Now

$$\ln(1 \pm \varepsilon) = \pm \varepsilon - \frac{\varepsilon^2}{2} + \cdots. \tag{9.28}$$

Then,

$$\ln \alpha = 4\gamma, \tag{9.29}$$

or

$$\xi = 2\gamma + 0(\gamma^2). \tag{9.30a}$$

A slightly more accurate formula is

$$\xi \approx \frac{2}{A + \frac{2}{3}}. \tag{9.30b}$$

9.3 THE COLLISION KERNEL: INELASTIC SCATTERING

We have considered in our discussion of elastic scattering those processes in which an energetic neutron imparts some of its energy to a target nucleus by increasing the kinetic energy of the target nucleus. If the neutron has sufficient energy, it can also excite an internal state of the nucleus.

Usually we assume that the total energy of the neutrons and protons inside the nucleus takes on its smallest possible value,[1] which we call the *ground state*. Of course the nucleus can exist in states in which the neutrons and protons have a larger total energy. Such a state is called an *excited state* of the nucleus.

The possible energies of excited nuclear states occur at discrete values, depending upon the specific nucleus. In Table 9.1 we list the first two excited

Table 9.1 Energies of the first two excited states of several nuclides

Nucleus	*Energy of first excited state MeV*	*Energy of second excited state MeV*
C^{12}	4.43	7.65
O^{16}	6.06	6.14
Na^{23}	0.45	2.0
Al^{27}	0.84	1.01
Fe^{56}	0.84	2.1
U^{238}	0.045	0.145

states for several isotopes. We see then, that C^{12} can exist in its ground state and in states in which the internal energy is 4.43 and 7.65 MeV larger than the ground state. Of course, C^{12} has many other excited states at higher energies. We have only listed the first two in Table 9.1. In Fig. 9.4, some of the higher excited states of the O^{18} nucleus are shown. We can see from this figure one of the important properties of the nuclear excited states: The spacing between the high-energy excited states is smaller than the spacing between the first few excited states. The energies of the nuclear excited states are so large that naturally occurring nuclei are in their ground states. Then, if the kinetic

[1] Here we are referring to the internal energy.

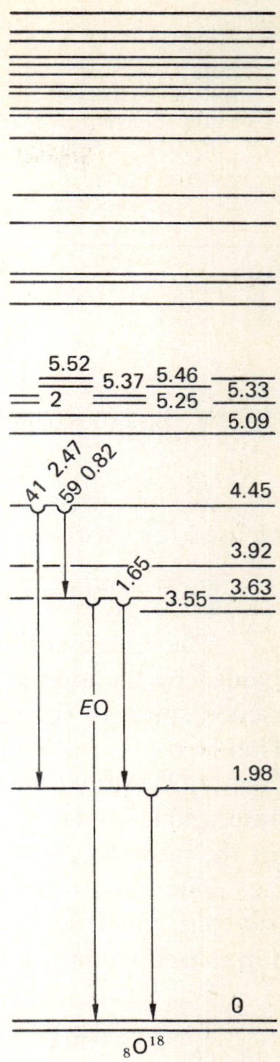

Fig. 9.4 Excited levels of O^{18}.

energy of a neutron is higher than the energy difference between the ground state and the first excited state of the nucleus, the neutron can leave the nucleus in an excited state. This process is referred to as *inelastic scattering* (see Fig. 9.5).

After the scattering, the nucleus eventually decays back to its ground state by emitting a high-energy photon (γ emission), and hence the process is denoted symbolically as

$$n + {}_ZX^A \rightarrow n' + \gamma + {}_ZX^A. \tag{9.31}$$

It is clear that inelastic processes can take place only for high-energy neutrons. Even for U^{238}, neutrons must have at least 45 keV for an inelastic event to take place (Table 9.1).

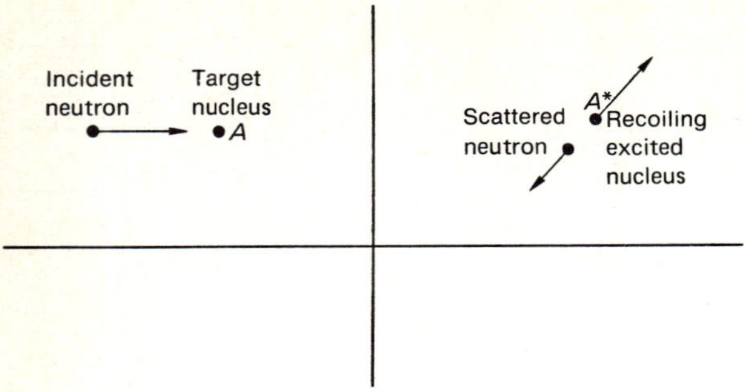

Fig. 9.5 Inelastic collision.

We can write the cross section for inelastic processes in the same form we used for elastic processes (9.4):

$$\Sigma_I(E_i \to E_f) = \Sigma_I(E_i)P_I(E_i \to E_f). \tag{9.32}$$

The microscopic, inelastic cross section $\sigma_I(E_i)$ is, as before, just Σ_I divided by the atomic density. Clearly, $\sigma_I(E_i)$ will display the fact that the excited states of the nucleus are discrete by exhibiting a threshold behavior. That is, $\sigma_I(E_i)$ will be zero for E_i less than the energy of the first excited state. Then $\sigma_I(E_i)$ should increase with E_i since it should be easier to excite the first state, and it will become possible to excite other, higher states. This behavior for the inelastic cross sections is shown for several typical nuclei in Fig. 9.6.

The factor $P_I(E_i \to E_f)$ is just the distribution of final energies E_f for neutrons which scatter inelastically at E_i. Of course $P_I(E_i \to E_f)$ must be normalized to unity,

$$\int_0^\infty dE_f P_I(E_i \to E_f) = 1. \tag{9.33}$$

In general, we do not have an analytic expression for $P_I(E_i \to E_f)$ as we did for elastic scattering, and so we must depend upon numerical data. However, consider the incident neutron energy E_i when it is much larger than the energy of the first excited state of the nucleus.[1] In this case, we think of the inelastic process as taking place in two steps. First, the neutron is captured by the nucleus to form a highly excited nucleus with mass number $A + 1$, called the *compound nucleus.* Second, the compound nucleus decays by emitting a neutron in much the same way evaporation of a liquid takes place. This method describes the compound nucleus by an *effective temperature*

$$k_B T_{\text{eff}} \equiv \theta = \frac{2\sqrt{E_i}}{B}, \qquad \text{MeV} \tag{9.34a}$$

[1] V. F. Weisskopf: *Phys. Rev.*, **52**: 294 (1937).

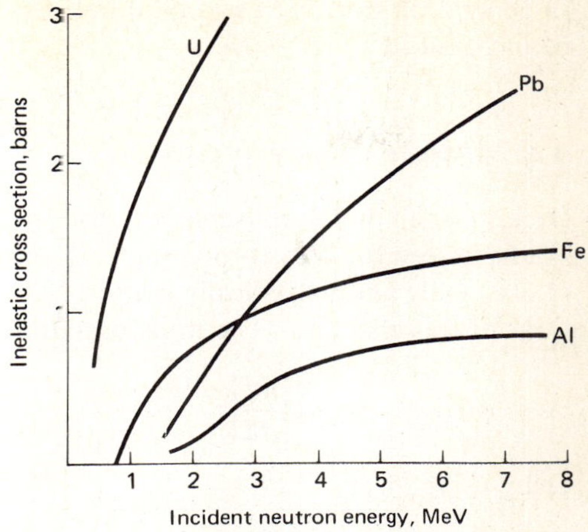

Fig. 9.6 Inelastic scattering cross sections for some reactor materials.

where

$$B \approx 0.62\sqrt{A} \tag{9.34b}$$

and of course A is the mass number of the target nucleus.

The detailed analysis of this model is beyond the scope of this book; however, the result is that

$$P_I(E_i \to E_f) = \frac{E_f}{\theta^2}\, e^{-E_f/\theta}. \tag{9.34c}$$

We must emphasize again that (9.34*c*) is completely invalid unless the incident neutron energies are much greater than the energy of the first excited state of the target nucleus; and even when this condition is fulfilled, it is only a fairly crude approximation.[1] Equations (9.34) provide us with an order-of-magnitude estimate of $P_I(E_i \to E_f)$ for inelastic events.

Equation (9.34*c*) can be readily converted to the corresponding expression using lethargy (9.11) as a variable,

$$P_I(u_i \to u_f) = P_I(E_i \to E_f)\left|\frac{dE_f}{du_f}\right| \tag{9.35a}$$

$$= \frac{{E_f}^2}{\theta^2}\, e^{-E_f/\theta}, \tag{9.35b}$$

$$= \frac{{E_0}^2}{\theta^2}\, e^{-2u_f} \exp\left(-\frac{E_0}{\theta}\, e^{-u_f}\right). \tag{9.35c}$$

[1] For this reason, it cannot generally be applied, in reactor calculations, to isotopes other than fuel materials—Pu, U—for which the first excited state lies low in energy (see Table 9.1). However, these fuel materials, along with iron, are by far the most important inelastic scatterers in a reactor. For other materials, such as iron, combinations of experimental data and more complicated theoretical models give estimates for the inelastic cross sections.

In general, in Eqs. (9.1) and (9.3) both elastic and inelastic processes should be included in $\Sigma_s{}^0(E' \to E)$.

9.4 FERMI AGE THEORY

Here we shall find it convenient to use lethargy instead of energy, Eq. (9.11). We have seen the transformation from energy to lethargy in the distributions of elastically and inelastically scattered neutrons in the second and third sections of this chapter. The transformation of the flux follows the usual rule,

$$\phi(\mathbf{r},E) = \phi(\mathbf{r},u)\left|\frac{du}{dE}\right|, \tag{9.36a}$$

or by using (9.11),

$$\phi(\mathbf{r},E) = \frac{1}{E}\,\phi(\mathbf{r},u). \tag{9.36b}$$

We can obtain the diffusion equation in terms of the lethargy variable by multiplying (9.3) by $|dE/du|$,

$$-\nabla \cdot D\,\nabla\phi(\mathbf{r},E)\left|\frac{dE}{du}\right| + \Sigma_T\,\phi(\mathbf{r},E)\left|\frac{dE}{du}\right| = S(E)\left|\frac{dE}{du}\right| + \int_0^\infty dE'\,\Sigma_s{}^0(E' \to E)\left|\frac{dE}{du}\right|\phi(\mathbf{r},E'). \tag{9.37}$$

Now we can make a change of variable in the E' integration,

$$dE' = \left|\frac{dE'}{du'}\right| du'. \tag{9.38}$$

Then using the results of the preceding two sections,

$$\Sigma_s{}^0(E' \to E)\left|\frac{dE}{du}\right| = \Sigma_s{}^0(u' \to u) \tag{9.39a}$$

$$= \Sigma_s(u')P_s(u' \to u), \tag{9.39b}$$

and (9.36*a*), and

$$S(E)\left|\frac{dE}{du}\right| = S(u), \tag{9.40}$$

we have for Eq. (9.37)

$$-\nabla \cdot D\,\nabla\phi(\mathbf{r},u) + \Sigma_T\,\phi(\mathbf{r},u) = S(u) + \int_0^\infty du'\,\Sigma_s{}^0(u' \to u)\phi(\mathbf{r},u'). \tag{9.41}$$

In converting the integral term in (9.37) into the integral term in (9.41), we have used the fact that $\phi(\mathbf{r},E')$ vanishes for $E' > E_0$.

Consider for a moment the integral term in (9.41). We shall assume that the scattering cross section $\Sigma_s{}^0(u' \to u)$ is just a sum of terms of the form discussed in Sec. 9.2. In particular, since there are a number of different types of atom, with different values of A, we represent $\Sigma_s{}^0(u' \to u)$ as a sum of terms of the form of (9.17*b*) with different values of A. We shall label the different atomic species by a subscript j. Then the integral term in (9.41) can be written

$$\int_0^\infty du'\, \Sigma_s{}^0(u' \to u)\phi(u') = \sum_j \int_{u-\ln 1/\alpha_j}^{u} \frac{e^{-(u-u')}}{1-\alpha_j} \Sigma_{sj}(u')\phi(u'). \tag{9.42}$$

If we substitute (9.42) in (9.41), we obtain an integro differential equation which is not particularly trivial to solve. However, suppose we restrict our attention to a fairly heavy moderator, like graphite. Then we note the interval $\Delta u = \ln 1/\alpha_j$ is small, and we may attempt a Taylor's series expansion of the product $\Sigma_{sj}(u')\phi(u')$ and retain only the first two terms:[1]

$$\Sigma_{sj}(u')\phi(u') = \Sigma_{sj}(u)\phi(u) - (u - u')\frac{\partial}{\partial u}[\Sigma_{sj}(u)\phi(u)] + \cdots. \tag{9.43}$$

When we insert this result into (9.42), we obtain

$$\begin{aligned}\int_0^\infty du'\, \Sigma_s{}^0(u' \to u)\phi(u') = \sum_j \Bigg\{\Sigma_{sj}(u)\phi(u) \int_{u-\ln 1/\alpha_j}^{u} \frac{e^{-(u-u')}}{1-\alpha_j}\, du' \\ - \int_{u-\ln 1/\alpha_j}^{u} (u-u') \frac{e^{-(u-u')}}{1-\alpha_j}\, du' \frac{\partial}{\partial u}[\Sigma_{sj}(u)\phi(u)]\Bigg\}.\end{aligned} \tag{9.44a}$$

The first integral has been normalized to unity, while the second is simply the definition of ξ_j [see (9.26)]. Thus when we insert (9.44) into (9.41), we obtain the following approximate equation for the flux:

$$-\nabla \cdot D\, \nabla\phi(\mathbf{r},u) + \Sigma_T\, \phi(\mathbf{r},u) = \Sigma_s\, \phi(\mathbf{r},u) - \xi \frac{\partial}{\partial u}[\Sigma_s\, \phi(\mathbf{r},u)] + S(u) \tag{9.44b}$$

or, noting that $\Sigma_T - \Sigma_s$ is just Σ_a,

$$-\nabla \cdot D\, \nabla\phi(\mathbf{r},u) + \Sigma_a\, \phi(\mathbf{r},u) = S(u) - \xi \frac{\partial}{\partial u}[\Sigma_s\, \phi(\mathbf{r},u)]. \tag{9.44c}$$

This equation is essentially the *Fermi age-diffusion equation*, although Fermi originally presented it in a somewhat different form. This form (9.44*b*) is

[1] This assumption is not valid for H or D, and so age theory is not valid for them either.

more convenient for numerical computation, however. Note that the ξ introduced into Eq. (9.44*b*) is an average over atomic species, i.e.,

$$\xi = \sum_j \frac{\xi_j \Sigma_{sj}}{\Sigma_s}, \tag{9.45a}$$

where, of course,

$$\Sigma_s = \sum_j \Sigma_{sj}. \tag{9.45b}$$

For historical interest, as well as for some practical applications, it is of interest to transform Eq. (9.44*c*) to its original form. First we define the slowing-down density $q(\mathbf{r},u)$ such that $q(\mathbf{r},u)\, d^3r$ represents the number of neutrons in d^3r at $\mathbf{r}$ which slow down, per unit time, past lethargy u. This represents contributions from all lethargies below u to all lethargies above u consistent with the maximum lethargy change condition. We see that

$$\frac{d^2q(u'' \to u')}{du''\, du'} = \sum_j \Sigma_{sj}(u'')\phi(u'')\, \frac{e^{-(u'-u'')}}{1-\alpha_j}. \tag{9.46}$$

The initial lethargy u'' varies between u and $u - \ln 1/\alpha_j$, while the final lethargy u' varies between u and $u'' + \ln 1/\alpha_j$. Thus (see Fig. 9.7),

$$q(u) = \sum_j \int_{u-\ln 1/\alpha_j}^{u} du'' \int_{u}^{u''+\ln 1/\alpha_j} du'\, \Sigma_{sj}(u'')\phi(u'')\, \frac{e^{-(u'-u'')}}{1-\alpha_j}. \tag{9.47}$$

The u' integration can be carried out to give

$$q(u) = \sum_j \int_{u-\ln 1/\alpha_j}^{u} du''\, \Sigma_{sj}(u'')\phi(u'')\, \frac{e^{-(u-u'')} - \alpha_j}{1-\alpha_j}. \tag{9.48}$$

This is the general expression for the density of neutrons slowing down past

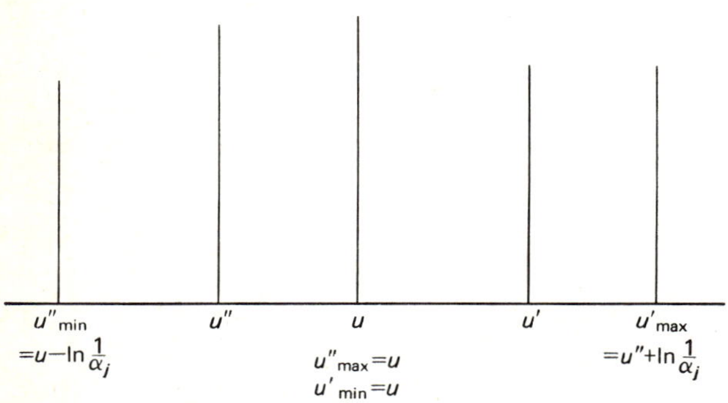

Fig. 9.7 Integration limits in the calculation of slowing-down density.

lethargy u in terms of the flux above. If we expand $\Sigma_s(u'')\phi(u'')$ in a Taylor's series and retain only a single term, we find the approximate expression

$$q(u) = \xi\Sigma_s\phi(u), \tag{9.49}$$

which is used in Fermi age theory.

Also, in age theory a new independent variable was introduced in place of the energy. It is called the *age* τ and is defined by

$$\tau(u) = \int_0^u \frac{D(u')}{\xi\Sigma_s(u')}\,du'. \tag{9.50}$$

From this we see that

$$\frac{\partial q(u)}{\partial u} = \frac{\partial q}{\partial \tau}\frac{\partial \tau}{\partial u} \tag{9.51a}$$

$$= \frac{D}{\xi\Sigma_s}\frac{\partial q}{\partial \tau}. \tag{9.51b}$$

In order for the present procedure to be the least bit useful, we must assume the cross sections are space independent. Furthermore, we assume $\Sigma_a = 0$, and that $S(u)$ in (9.44) is zero. Then we see from the above that Eq. (9.44) may be written in terms of q and τ as

$$\nabla^2 q = \frac{\partial q}{\partial \tau}. \tag{9.52}$$

This equation is very similar to the equation for heat conduction

$$\nabla^2 Q = \frac{\partial Q}{\partial t}. \tag{9.53}$$

We again refer to Eq. (9.44*c*) and assume constant cross sections (spatially). However, here we do not set Σ_a and S equal to zero. We have (using q as the dependent variable, rather than ϕ)

$$-\frac{D}{\xi\Sigma_s}\nabla^2 q + \frac{\Sigma_a}{\xi\Sigma_s}q = S - \frac{\partial q}{\partial u}. \tag{9.54}$$

For application to the critical equation, Sec. 8.6, we require the Fourier transform of the solution to this equation for a point source in an infinite medium. In such a case,

$$S = \delta(\mathbf{r})\chi(u), \tag{9.55}$$

and q obeys[1]

$$\left(\frac{\partial}{\partial u} + \frac{Dk^2}{\xi\Sigma_s} + \frac{\Sigma_a}{\xi\Sigma_s}\right)\bar{q}_s = \chi(u). \tag{9.56}$$

[1] Here, $\chi(u)$ is the fission spectrum, and k is the transform variable.

This equation is readily integrated to give

$$\bar{q}_s = \int_0^u \chi(u')e^{-k^2\tau(u'\to u)}p(u'\to u)\,du', \tag{9.57}$$

where, in analogy with Eq. (9.50), we have defined $\tau(u'\to u)$ as

$$\tau(u'\to u) = \int_{u'}^{u} \frac{D(u'')}{\xi\Sigma_s(u'')}\,du'', \tag{9.58a}$$

and we have also defined

$$p(u'\to u) = \exp\left[-\int_{u'}^{u} \frac{\Sigma_a(u'')}{\xi\Sigma_s(u'')}\,du''\right]. \tag{9.58b}$$

Since $\bar{q}_s$ has the physical significance of being the number of neutrons from a fission source that are slowing down past a given lethargy, we can readily interpret p as the resonance escape probability. Also, $e^{-k^2\tau}$ is a non-leakage probability, as we shall show below.

For $k^2(=B^2)$ small, Eq. (9.57) takes the simpler approximate form

$$\bar{q}_s = e^{-k^2\tau(u)}p(u), \tag{9.59}$$

where

$$\tau(u) = \int_0^u \chi(u')\tau(u'\to u)\,du', \tag{9.60}$$

with a similar expression for $p(u)$. That is, functions of averages have replaced averages of functions.

Inserting this result into the critical equation gives

$$\frac{\eta f p e^{-B^2\tau}}{1+L^2B^2} = 1, \tag{9.61}$$

which confirms our identification of $e^{-B^2\tau}$ as the fast nonleakage probability of a reactor of buckling B^2 [this can also be derived as in Sec. 3.6].

When we speak of the *resonance escape probability* and the *neutron age*, we shall generally mean their averages over a fission spectrum. We now note that we are in a position to compute τ and p directly from cross-section data. In particular, consider a lethargy group in a multigroup calculation which extends, say, from lethargy u_K to u_{K+1}. Using the above formulas, we can derive $\tau(u_K\to u_{K+1})$ and $p(u_K\to u_{K+1})$. Then we make the identifications (cf. Sec. 8.4),

$$p(u_K\to u_{K+1}) = \frac{\Sigma_{RK}-\Sigma_{aK}}{\Sigma_{RK}} \tag{9.62a}$$

and

$$\tau(u_K \to u_{K+1}) = L_K{}^2 = \frac{D_K}{\Sigma_{aK}}. \tag{9.62b}$$

Equation (9.58*b*) follows from noting that for small B^2,

$$e^{-B^2\tau} \approx \frac{1}{1 + B^2\tau}, \tag{9.63}$$

which gives just the form of the nonleakage probability in Eq. (8.45) if τ is replaced by $L_1{}^2$. This same identification can also be seen in another way, namely, by considering the physical meaning of the age. We note that Eq. (9.59) gives the Fourier transform of the slowing-down density from a point source. This result may be Fourier inverted to yield the spatial distribution[1]

$$q(r) = \frac{1}{(2\pi)^3} \int e^{i\mathbf{k}\cdot\mathbf{r}} e^{-k^2\tau}\, d^3k \tag{9.64a}$$

$$= \frac{e^{-r^2/4\tau}}{(4\pi\tau)^{3/2}}. \tag{9.64b}$$

We see that the spatial distribution of neutrons is gaussian. If we calculate

$$\overline{r^2} = \frac{\int r^2 q(r)\, d^3r}{\int q(r)\, d^3r}, \tag{9.65}$$

we find

$$\overline{r^2} = 6\tau. \tag{9.66}$$

Thus, in analogy with Eq. (2.111*c*), $\sqrt{\tau}$ is seen to be a diffusion length for fast neutrons.

For light moderators such as hydrogen, for which age theory is not applicable, τ is still defined by Eq. (9.50). It can be either measured or calculated numerically.[2] Once τ has been determined, the nonescape probability can be computed as $e^{-B^2\tau}$, but unfortunately this relation is not exact, even with the correct τ, for systems which do not obey the age equation. However, it may be used as a rough approximation.

The neutron spectrum in an infinite medium can be found by setting $B^2 = 0$. If we also consider no absorption, we have (for energies below the bottom of the fission spectrum)

$$q = 1 \tag{9.67a}$$

[1] See Appendix C.
[2] J. Ferziger and P. F. Zweifel: *op. cit.*

or

$$\phi(u) = \frac{1}{\xi\Sigma_s}. \tag{9.67b}$$

Of course this result is physically reasonable, because if there is no absorption and no leakage, then the slowing-down density must be constant, equal to the strength of the source (which was normalized to unity). Since

$$\phi(E) = \phi(u)\left|\frac{du}{dE}\right|, \tag{9.68a}$$

we find

$$\phi(E) = \frac{1}{\xi\Sigma_s E}, \tag{9.68b}$$

the famous $1/E$ spectrum. Although this result has been derived from age theory, it is actually valid asymptotically (i.e., far from sources or sinks[1]) for any moderator.[2]

Equations (9.62) can now be used to eliminate two of the group constants, for example, Σ_{RK} and D_K. The remaining (Σ_{fK} and Σ_{aK}) can be averaged over the spectrum given by (9.68). This spectral procedure is commonly used, although it is strictly correct only if the spectrum is the same everywhere in the reactor, i.e., if complete space-energy separability obtains. This, we know, is true only for a bare, homogeneous reactor.

Actually the use of the infinite-medium spectrum (9.68) is not valid. From (9.59) we see

$$\phi(E) = \frac{1}{\xi\Sigma_s E}\,p(E)e^{-B^2\tau(E)}. \tag{9.69}$$

This spectrum could be used to average constants with more accuracy than the simple $1/E$ spectrum. As was mentioned before, the materials buckling of the core (or of each core region) should be used for B^2. In the reflector, it is customary to take $B^2 = 0$.

More accurate spectra can be generated by other methods,[3] but the procedure of defining a group constant is identical. This spectral procedure is widely used in obtaining group constants for both fast and thermal reactors. Another approach sometimes used is to define a set of group cross sections

[1] Namely, resonances.

[2] See J. Ferziger and P. F. Zweifel: *op. cit.* Of course ξ is still the average over atomic species.

[3] See J. Ferziger and P. F. Zweifel: *ibid.* Also Chap. 10 of this text.

once and for all and use them for all reactors of a given type. This procedure is particularly popular in the design of fast reactors. However, it is clear that the spectral procedure must be more accurate and, in particular, permits the use of fewer groups.

9.5 INFINITE–MEDIUM RESONANCE ABSORPTION

In order to obtain Eq. (9.58*b*) for the resonance escape probability, we assumed that $\Sigma_s(u)\phi(u)$ was not a strongly varying function of u, Eq. (9.42). The name alone, resonance absorption, indicates that this supposed slow variation is not valid. The absorption cross sections for heavy nuclei such as U^{238} and Th^{232} have sharp resonance, as shown in Fig. 1.3. In the vicinity of these resonances the flux must be drastically depressed owing to the high-absorption cross section, and consequently, the flux varies rapidly.

We can account for this by modifying the form of (9.58*b*) somewhat. We note that the exponent in (9.58*b*) is just the integral of the absorption cross section times the asymptotic flux, (9.67*b*).

$$p(u' \to u) = \exp\left[-\int_{u'}^{u} \phi(u'')\Sigma_a(u'')\,du''\right], \tag{9.70a}$$

or in terms of energy,

$$p(E' \to E) = \exp\left[-\int_{E}^{E'} \phi(E'')\Sigma_a(E'')\,dE''\right]. \tag{9.70b}$$

We can account for the flux depression in the vicinity of resonances by using the exact flux in (9.70), rather than the asymptotic flux.

To compute the resonance escape probability then, we must find the flux in the resonance region (where $\chi(E)$ and $\Sigma_I(E)$ are both zero) normalized asymptotically as

$$\phi(u) \xrightarrow[u\ \text{small}]{} \frac{1}{\xi\Sigma_s} \tag{9.71a}$$

or

$$\phi(E) \xrightarrow[E\ \text{large}]{} \frac{1}{E\xi\Sigma_s}. \tag{9.71b}$$

In an infinite medium this means that we must solve Eq. (9.1) with $S(E) = 0$ and with $\Sigma_s{}^0(E' \to E)$ given by a sum over elastic scattering contributions from all the nuclear species. For the sake of simplicity, we shall only consider two types of nuclei, a heavy absorbing nucleus (such as U^{238} or Th^{232}) and a light, nonabsorbing moderator nucleus (such as hydrogen or carbon). We shall label these with F and M for the absorber and moderator, respectively. Then using (9.21) for the scattering cross section, we have

$$\phi(E)\Sigma_T(E) = \frac{1}{1-\alpha_M}\int_E^{E/\alpha_M} dE'\,\frac{\phi(E')\Sigma_{sM}(E')}{E'} + \frac{1}{1-\alpha_F}\int_E^{E/\alpha_F} dE'\,\frac{\phi(E')\Sigma_{sF}(E')}{E'}. \tag{9.72}$$

Before proceeding further with our analysis of Eq. (9.72), it will be necessary to specify in some detail the properties of the cross sections in the energy region of interest (that is, ≈ 1 eV to 1 keV). In this region the moderator scattering cross section is a constant function of energy and, except for the resonance, the absorber scattering cross section is also a constant function of energy. In the neighborhood of a resonance we can write the absorption cross section as[1]

$$\Sigma_{aF}(E) = \Sigma_0 \frac{\Gamma_\gamma}{\Gamma}\sqrt{\frac{E_r}{E}}\,\frac{1}{4(E-E_r)^2/\Gamma^2+1}; \tag{9.73}$$

E_r is called the *resonance energy*, Γ_γ is called the *radiation line width*, and Γ is called the *total line width*. The radiation line width Γ_γ is essentially constant for all of the resonances of a particular isotope. For example, in U^{238},

$$\Gamma_\gamma(U^{238}) = 0.0246 \text{ eV}, \tag{9.74a}$$

and for thorium,

$$\Gamma_\gamma(Th^{232}) = 0.030 \text{ eV} \tag{9.74b}$$

for all the resonances.

The absorber's scattering cross section also exhibits resonance behavior in the neighborhood of E and a constant value Σ_p away from the resonance. That is, the scattering cross section can be written

$$\Sigma_{sF}(E) = \Sigma_0 \frac{\Gamma_n}{\Gamma}\,\frac{1}{4(E-E_r)^2/\Gamma^2+1} + \Sigma_I\,\frac{2(E-E_r)/\Gamma}{4(E-E_r)^2/\Gamma^2+1} + \Sigma_p. \tag{9.75}$$

The first term in (9.75) is referred to as *resonance scattering*, the second term as *interference scattering*, and the third term as *potential scattering*. The behavior of $\Sigma_{sF}(E)$ is sketched in Fig. 9.8. The main peak is given by the

[1] See A. M. Weinberg and E. P. Wigner: "The Physical Theory of Neutron Chain Reactors," The University of Chicago Press, Chicago, 1958.

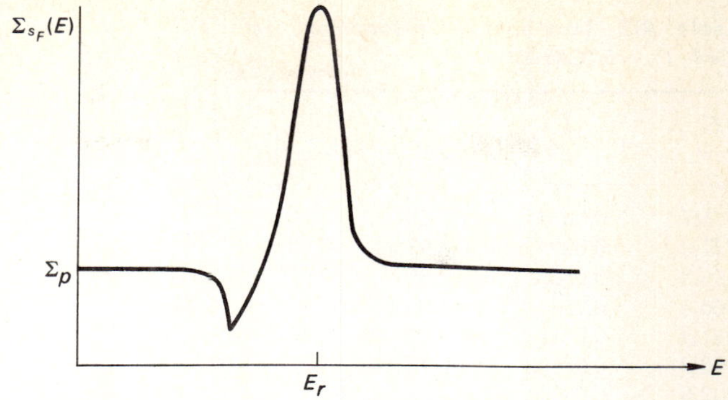

Fig. 9.8 Scattering cross section in the neighborhood of a resonance.

resonance scattering term, the constant *background* by Σ_p, and the dip in front of the resonance by the interference term. The factor Σ_I is related to the cross sections Σ_0 and Σ_p. However, there is a *statistical factor* which depends upon the spin of the compound nucleus in the formula relating these factors.[1] The factor Γ_n is the neutron line width. The total line width is the sum of the neutron and radiation line widths,

$$\Gamma = \Gamma_\gamma + \Gamma_n. \tag{9.76}$$

A measure of the *effective width* of the resonance is given by the practical line width,

$$\Gamma_p = \Gamma\sqrt{\frac{\Sigma_0}{\Sigma_p}}. \tag{9.77}$$

The practical line width is a measure of the amount one must move from the resonance energy before the absorption cross section is smaller than Σ_p. Since Σ_0 is usually much larger than Σ_p, then Γ_p is much larger than Γ. These parameters are listed for the first 12 resonances of U^{238} in Table 9.2. Here we also show the energy loss for a neutron with energy E_r in an elastic collision with U^{238} [that is, $(1 - \alpha_F)E_r$].

With this background, we may now return to Eq. (9.72). In the moderator integral, the energy interval $E/\alpha_M - E$ will be large compared with the effective width of the resonance Γ_p. Then over most of the integral the energy will be far from a resonance, and we shall not make too great an approximation by replacing $\phi(E')$ with its asymptotic value (9.71*b*) (and recall

[1] Weinberg and Wigner: *op. cit.*

Table 9.2 Low-energy resonances in U^{238}, $\Gamma_\gamma = 0.0246$ eV

E_r eV	Γ_n eV	Γ_p eV	$(1-\alpha_F)E_r$
6.7	0.00148	1.26	0.11*
21.0	0.009	1.95	0.35*
36.8	0.033	3.65	0.62*
66.3	0.023	2.26	1.11*
81.3	0.0021	0.76	1.36
90.0	0.00008	0.13	1.51
102.8	0.07	3.98	1.73*
117.0	0.018	1.32	1.97
146.0	0.0008	0.68	2.44
166.0	0.0035	0.98	2.78
190.0	0.135	5.66	3.21*
209.0	0.055	2.63	3.55

that $\xi\Sigma_s$ is constant).[1] This will be true even if the energy interval contains E_r.

$$\frac{1}{1-\alpha_M}\int_E^{E/\alpha_M} dE' \frac{\phi(E')\Sigma_{sM}}{E'} \approx \frac{1}{1-\alpha_M}\frac{\Sigma_{sM}}{\xi\Sigma_s}\int_E^{E/\alpha_M}\frac{dE'}{E'^2} \tag{9.78a}$$

$$= \frac{\Sigma_{sM}}{\xi\Sigma_s E}. \tag{9.78b}$$

Using this result in (9.72), we obtain

$$\phi(E)\Sigma_T(E) = \frac{1}{1-\alpha_F}\int_E^{E/\alpha_F}\frac{\phi(E')}{E'}\Sigma_{sF}(E') + \frac{1}{\xi\Sigma_s E}. \tag{9.78c}$$

Now consider Table 9.2. For those resonances (marked by an asterisk) for which $(1-\alpha_F)E_r$ is less than Γ_p, we shall use the same approximation as we used in the moderator integral, and replace $\phi\Sigma_s$ with its asymptotic value

$$\frac{1}{1-\alpha_F}\int_E^{E/\alpha_F} dE'\frac{\phi(E')}{E'}\Sigma_{sF}(E') \approx \frac{\Sigma_p}{\xi\Sigma_s E}. \tag{9.79}$$

Then, using (9.72), we have for the flux,

$$\phi_{NR}(E) = \frac{\Sigma_{sM}+\Sigma_p}{\xi\Sigma_s E\Sigma_T}. \tag{9.80}$$

[1] We recall that, for this case, Σ_s is just the sum of Σ_p and Σ_{sM}. Also, we note that Σ_s is constant.

For obvious reasons, this approximation is called the *narrow resonance approximation.* It is also obvious that (9.79) is considerably less accurate than (9.78).

With this approximation for the flux, the resonance escape probability for one resonance can be written as

$$p_{\mathrm{NR}} = e^{-N_F I_{\mathrm{NR}}/\xi\Sigma_s}, \tag{9.81a}$$

where

$$I_{\mathrm{NR}} = \int dE\, \frac{\sigma_{aF}(E)(\Sigma_{sM} + \Sigma_p)}{E\Sigma_T(E)}, \tag{9.81b}$$

and N_F is the number density of absorber nuclei. Of course, σ_{aF} is the microscopic absorption cross section. We have factored out the N_F so that our definition of the resonance integral would agree with standard notation.

For the other resonances in Table 9.7, for which Γ_p is less than $(1 - \alpha_F)E_r$, we do not expect the narrow resonance approximation to apply at all. In this case, we shall take the opposite extreme. That is, we assume that the energy interval $E/\alpha_F - E$ is so small that we can ignore the variation of $\phi\Sigma_{sF}$ over the energy interval:

$$\frac{1}{1-\alpha_F}\int_E^{E/\alpha_F} dE'\, \frac{\phi(E')}{E'}\, \Sigma_{sF}(E') \approx \frac{\phi(E)\Sigma_{sF}(E)}{1-\alpha_F}\int_E^{E/\alpha_F} \frac{dE'}{E'}. \tag{9.82}$$

The integral in (9.82) should be evaluated in the limit $\alpha_F = [(A-1)/(A+1)]^2 \to 1$,

$$\lim_{\alpha_F\to 1} \frac{1}{1-\alpha_F}\int_E^{E/\alpha_F} \frac{dE'}{E'} = 1. \tag{9.83}$$

Then (9.82) becomes

$$\frac{1}{1-\alpha_F}\int_E^{E/\alpha_F} dE'\, \frac{\phi(E')\Sigma_{sF}(E')}{E'} \approx \phi(E')\,\Sigma_{sF}(E'). \tag{9.84}$$

This, again for obvious reasons, is called the *infinite mass*, or IM, approximation.[1] Using (9.84) and (9.78*b*), we can solve (9.72) for the flux in the IM approximation

$$\phi_{\mathrm{IM}}(\Sigma_T - \Sigma_{sF}) = \frac{\Sigma_{sM}}{\xi\Sigma_s\, E}, \tag{9.85a}$$

or

$$\phi_{\mathrm{IM}}(E) = \frac{\Sigma_{sM}}{\xi\Sigma_s\, E(\Sigma_T - \Sigma_{sF})}. \tag{9.85b}$$

[1] Sometimes the result is referred to as the NRIM approximation since the moderator is treated in the NR approximation.

Here, the resonance escape probability is

$$p_{\text{IM}} = e^{-I_{\text{IM}}/\xi\Sigma_s}, \tag{9.86a}$$

where

$$I_{\text{IM}} = \int dE \, \frac{\sigma_{aF}\Sigma_{sM}}{E(\Sigma_T - \Sigma_{sF})}. \tag{9.86b}$$

We see that it is essentially identical with I_{NR} except that scattering in the fuel is ignored. We do not always expect the approximations made in (9.78), (9.79), or (9.84) to be terribly accurate since, in many cases, the energy intervals $E/\alpha_F - E$ and the line widths Γ_p are comparable, instead of one's being very much larger than the other. Then, modified methods may be used at the expense of more complicated calculations.[1]

Of some importance is the *infinitely dilute resonance integral* which can be obtained from either I_{NR} [Eq. (9.81*b*)] or I_{IM} [Eq. (9.86*b*)] in the limit $N_F \to 0$. In either case we find

$$I_\infty = \int \sigma_{aF} \frac{dE}{E}. \tag{9.86c}$$

The difference between the infinitely dilute resonance integral and the expression for I_{NR} or I_{IM} is due to flux depression within the resonance, and is sometimes called *energy self-shielding*. Note also that the use of I_∞ in (9.81*a*) reproduces the resonance escape probability as calculated by Fermi age theory, Eq. (9.58*b*).

9.6 RESONANCE ABSORPTION IN LUMPED ABSORBERS

Here we consider the situation in which the heavy absorber nuclei are contained in a lump surrounded by moderating material (as, e.g., a fuel pin surrounded by coolant, shown in Fig. 9.9.) We can see the effect of the size of the fuel region in Fig. 9.10.

As we did in Sec. 7.3, we shall designate the cross sections in regions A and B by the appropriate superscripts.

We have assumed that region B contains only moderator with a cross section $\Sigma_s{}^B$ and average lethargy change $\xi_M{}^B$. The asymptotic flux in the moderator region has the characteristic $1/E$ dependence,[2]

$$\phi_B(E) = \frac{1}{E}. \tag{9.87}$$

[1] See, e.g., R. Goldstein and E. R. Cohen: *Nucl. Sci. Eng.*, **13**: 132 (1962).

[2] Here we do not include the normalization factor $1/\xi\Sigma_s$. This factor must be appropriately averaged over moderator and absorber regions.

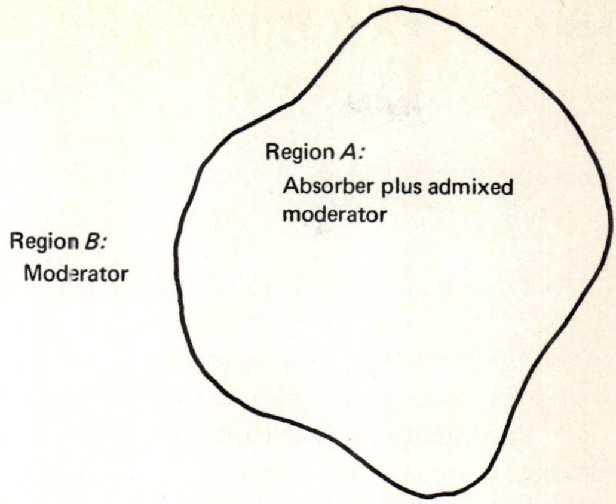

Fig. 9.9 Resonance capture in lumped absorber.

The flux in region A must satisfy Eq. (8.14) with $S = 0$,

$$\nabla \cdot \mathbf{J}_A(\mathbf{r},E) + \Sigma_T{}^A\phi_A(\mathbf{r},E) = \int dE'\, \Sigma_s{}^{A0}(E' \to E)\phi_A(\mathbf{r},E'). \qquad (9.88a)$$

Now multiply (9.88a) by d^3r, and integrate over region A:

$$\int_A d^3r\, \nabla \cdot \mathbf{J}_A(\mathbf{r},E) + \Sigma_T{}^A\bar{\phi}_A(E) = \int dE'\, \Sigma_s{}^{A0}(E' \to E)\bar{\phi}_A(E'), \qquad (9.88b)$$

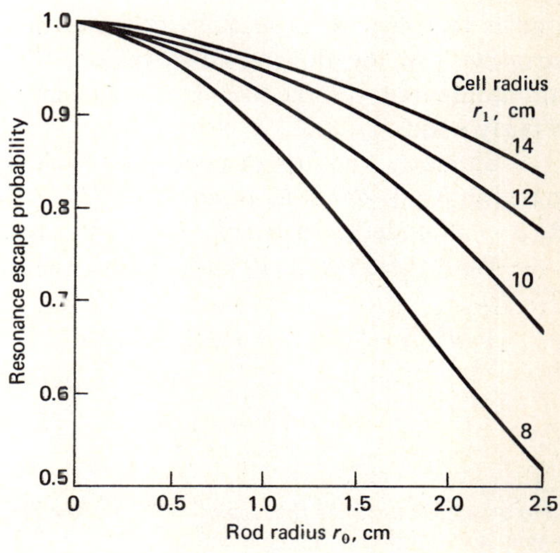

Fig. 9.10 Resonance escape probability in heterogeneous natural uranium—graphite systems.

where

$$\bar{\phi}_A(E) = \int_A d^3r\, \bar{\phi}_A(\mathbf{r},E). \tag{9.88c}$$

The first integral in (9.88*b*) can be evaluated with Gauss' theorem [see Eq. (2.20*b*)].

$$\int d^3r\, \nabla \cdot \mathbf{J}_A(\mathbf{r},E) = \int_A dS\, \mathbf{n} \cdot \mathbf{J}_A(\mathbf{r},E), \tag{9.89a}$$

where the second integral is to be carried out over the surface area of region A and $\mathbf{n}$ is the unit normal to the surface.

The integral in (9.89*a*), we see, is just the net number of neutrons leaving region A by passing through the bounding surface S.

There will be two contributions to this net current. Those neutrons which suffer an elastic collision in region A and then leave region A give a positive contribution.[1] Neutrons which suffer an elastic collision in region B and then undergo their next collisions in region A give a negative contribution. Neutrons which undergo absorption collisions or which enter and leave region A without undergoing a collision will not contribute to the net current leaving region A.

The flux of neutrons which have undergone an elastic collision in region A is $\int dE'\, \Sigma_s^{A0}(E' \to E)\phi_A(E')$. Of these neutrons, a fraction $P_0(A \to B)$ will undergo their next collisions in region B, where $P_0(A \to B)$ is the first-flight escape probability from region A into region B. This quantity has been defined in Sec. 7.4 for the case in which there is a uniform source of neutrons in region A. Then if $\int dE'\, \Sigma_s^{A0}(E' \to E)\phi_A(\mathbf{r},E')$ is approximately uniform over region A, we can use the results of Sec. 7.4. Since this approximation implies that $\phi_A(\mathbf{r},E)$ is spatially constant, it is referred to as the *flat flux assumption.* In the slowing-down region, the scattering mean free paths are large compared with the size of spatial heterogeneities, and this approximation is fairly good.

Similarly, the flux of neutrons which have undergone elastic collisions in region B is $\int_B d^3r \int dE'\, \phi_B(\mathbf{r},E')\Sigma_s^{B0}(E' \to E)$. Of these neutrons a fraction $P_0(B \to A)$ undergo their next collision in region A.

With these results (9.89*a*) can be written

$$\int_A d^3r\, \nabla \cdot \mathbf{J}_A(\mathbf{r},E) = P_0(A \to B) \int dE'\, \Sigma_s^{A0}(E' \to E)\bar{\phi}_A(E')$$
$$- P_0(B \to A) \int dE'\, \Sigma_s^{B0}(E' \to E) \int_B d^3r\, \phi_B(\mathbf{r},E'). \tag{9.89b}$$

[1] Here we have assumed that region A is convex so that a neutron cannot leave and reenter region A without undergoing a collision.

We can use the flat flux approximation and the asymptotic expression for ϕ_B, Eq. (9.87). Then using (9.21) for the scattering cross section in region B, we obtain

$$P_0(B \to A) \int dE'\, \Sigma_s^{B0}(E' \to E) \int_B d^3r\, \phi_B(\mathbf{r},E')$$

$$= P_0(B \to A)\Sigma_s^B \frac{V_B}{1-\alpha_M{}^B} \int_E^{E/\alpha_m{}^B} \frac{dE'}{E'^2} \tag{9.89c}$$

$$= P_0(B \to A) \frac{\Sigma_s^B V_B}{E}. \tag{9.89d}$$

We can use the reciprocity theorem which was derived in Chap. 7, Eq. (7.9). When we apply this to the first-flight escape probability, every event is an effective absorption and the reciprocity relation should contain the total cross section rather than those for absorption only.[1]

$$P_0(B \to A)\Sigma_s^B V_B = P_0(A \to B)\Sigma_T{}^A V_A. \tag{9.89e}$$

We can use these results in (9.89b) and abbreviate our notation by writing P_0 for $P_0(A \to B)$;

$$\int_A d^3r\, \nabla \cdot \mathbf{J}_A(\mathbf{r},E) = P_0 \int dE'\, \Sigma_s^{A0}(E' \to E)\bar{\phi}_A(E') - P_0 \frac{\Sigma_T{}^A V_A}{E}. \tag{9.89f}$$

If the flux in A were actually flat, we could write

$$\phi_A(\mathbf{r},E) = \frac{\phi_A}{V_A}. \tag{9.90}$$

Let us use this as a definition of $\phi_A(E)$. Then in terms of $\phi_A(E)$, Eq. (9.89b) can be rewritten as

$$\Sigma_T{}^A\phi_A(E) = (1-P_0) \int dE'\, \Sigma_s^{A0}(E' \to E)\phi_A(E') + P_0 \frac{\Sigma_T{}^A}{E}. \tag{9.91a}$$

If we divide both sides of (9.91a) by $1 - P_0$, we obtain

$$\frac{\Sigma_T{}^A}{1-P_0}\phi_A(E) = \int dE'\, \Sigma_s^{A0}(E' \to E)\phi_A(E') + \frac{P_0\Sigma_T{}^A}{1-P_0}\frac{1}{E}. \tag{9.91b}$$

The last term in (9.91b) has the form of a scattering integral in the NR approximation [see Eq. (9.78b)]. The effective-scattering cross section for this term is

$$\Sigma_s^{\text{eff}} = \frac{P_0\Sigma_T{}^A}{1-P_0}. \tag{9.92}$$

[1] The total cross section in region B is Σ_s^B.

In terms of Σ_s^{eff}, (9.91*b*) can be written as[1]

$$(\Sigma_T^A + \Sigma_s^{\text{eff}})\phi_A(E) = \int dE' \, \Sigma_s^{A0}(E' \to E)\phi_A(E') + \frac{\Sigma_s^{\text{eff}}}{E}. \tag{9.93}$$

Then the resonance integral is

$$I = \int dE \, \sigma_{aF}^A(E)\phi_A(E). \tag{9.94}$$

The resonance escape probability can be written in the form of (9.81*a*) and (9.86*a*). However, the product $\xi\Sigma_s$ must be appropriately averaged over the absorber cross sections. Then we write $\xi\Sigma_s$ as

$$\xi\Sigma_s = \frac{V_A \xi^A \Sigma_s^A + V_B \xi_M{}^B \Sigma_s^B}{V_A + V_B}, \tag{9.95}$$

and the resonance escape probability is

$$p = e^{-N_F I/\xi\Sigma_s}. \tag{9.96}$$

Now in solving the integral Eq. (9.93), we can use either the NR or IM approximation as we did in (9.80) and (9.85). Using these approximations, we obtain

$$\phi_{A\text{NR}} = \frac{\Sigma_{sM}^A + \Sigma_p^A + \Sigma_s^{\text{eff}}}{(\Sigma_T^A + \Sigma_s^{\text{eff}})E} \tag{9.97a}$$

and

$$\phi_{A\text{IM}} = \frac{\Sigma_{sM}^A + \Sigma_s^{\text{eff}}}{(\Sigma_T^A + \Sigma_s^{\text{eff}} - \Sigma_{sF}^A)E}. \tag{9.97b}$$

Then the expression for the resonance integral is obtained by inserting these expressions into (9.94). Thus, we obtain

$$I_{\text{NR}} = I_{\text{NR}}{}^V + I_{\text{NR}}{}^S, \tag{9.98a}$$

where

$$I_{\text{NR}}{}^V = \int \frac{dE}{E} \frac{\Sigma_{sM}^A + \Sigma_p^A}{\Sigma_T^A} \sigma_{aF}^A(E) \tag{9.98b}$$

and

$$I_{\text{NR}}{}^S = \int P_0 \frac{dE}{E} \frac{\Sigma_T^A - \Sigma_{sM}^A - \Sigma_p^A}{\Sigma_T^A} \sigma_{aF}^A(E). \tag{9.98c}$$

[1] This result is usually referred to as the *equivalence principle*. Stated simply, it says that lumping an absorber is equivalent to adding a moderator with the scattering cross section Σ_s^{eff} in the homogeneous case.

The superscripts V and S refer to the *volume* and *surface* contributions, respectively. In the limit that the absorber becomes optically thick, we can use Eq. (7.28*d*) for P_0,

$$P_0 = \frac{S_A}{4V_A \Sigma_T^{\ A}}. \tag{9.99}$$

Then $I_{NR}^{\ S} \approx S_A/V_A \approx S_A/M$, where M is the mass of the absorber body.

Analogous formulas for I_{IM} are

$$I_{IM} = I_{IM}^{\ S} + I_{IM}^{\ V}, \tag{9.100a}$$

where

$$I_{IM}^{\ V} = \int \frac{dE}{E} \frac{\Sigma_{sM}^{\ A}}{\Sigma_T^{\ A} - \Sigma_{sF}^{\ A}(1 - P_0)} \sigma_{aF}^{\ A}(E) \tag{9.100b}$$

and

$$I_{IM}^{\ S} = \int \frac{dE}{E} P_0 \frac{\Sigma_T^{\ A} - \Sigma_{sM}^{\ A}}{\Sigma_T^{\ A} - \Sigma_{sF}^{\ A}(1 - P_0)} \sigma_{aF}^{\ A}(E). \tag{9.100c}$$

The above formulas are valid for single lumps, as might be encountered in experimental determination of p. For reactors, analogous formulas must be obtained for lattices. One simply replaces P_0 above with P_0^*, the collision probability for lattices, described in Sec. 7.4.

9.7 TEMPERATURE EFFECTS

The formulas for the resonance integral which have been obtained thus far apply to zero absolute temperature, and thus the class of reactors for which these formulas are useful is somewhat limited.

The physical reason that the absorption rate changes with temperature is the following: We note that the formula, Eq. (9.73), for a resonance cross section depicts a function which varies rapidly with energy (see Fig. 7.3). The E in that formula is the relative energy between the neutron velocity, and the nucleus, i.e., proportional to $|v_n - v_N|^2$, where v_n is the neutron velocity, and v_N the nucleus velocity.

If the nucleus is at rest, as it is for temperature of absolute zero, this quantity degenerates to v_n, and E becomes the neutron energy. For higher temperatures, the nucleus will be in thermal motion, and the appropriate energy E is obtained by averaging over the thermal motions of the nucleus. The detailed analysis is beyond the scope of even this book; it may be found in a number of references.[1] The result is that the Breit-Wigner *line shape*, i.e., Eq. (9.73), is replaced by a *Doppler broadened line shape*, as shown in

[1] See, e.g., Ferziger and Zweifel: *op. cit.*

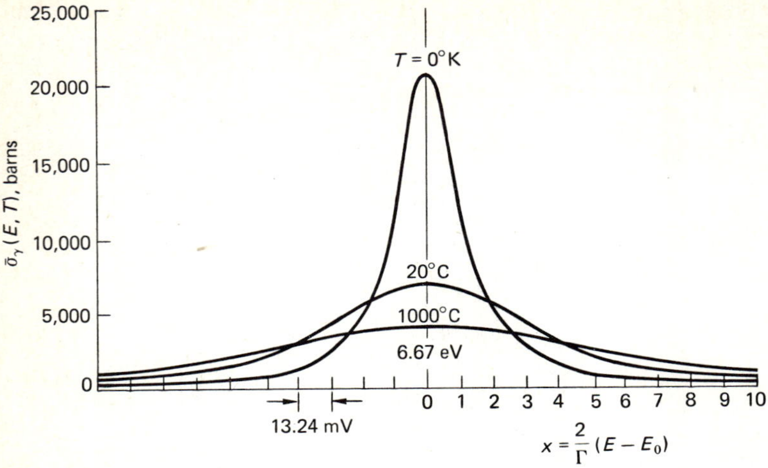

Fig. 9.11 Doppler broadening of the capture cross section of U^{238} at the 6.67-eV resonance.

Fig. 9.11. We see that the effect of increasing the temperature is to lower the center of the resonance line, and to spread it out. To an excellent approximation, the area under the curves remains the same, independent of temperature. One would then guess that the absorption or the resonance escape probability would not change. However, we recall that the center of the resonance line is strongly self-shielded. Put another way, the peak resonance absorption is so high that every neutron near the resonant energy is gobbled up anyway. Lowering the peak even by a factor of 10, does not appreciably affect the absorption rate in the center of the resonance. But the higher the temperature, the more absorption there is in the "wings" of the resonance. For this reason, absorption increases with increasing temperature, see Fig. 9.12.

The expression that replaces $\Sigma_a(E)$, namely, Eq. (9.73), for nonzero temperature is

$$\bar{\Sigma}_a(E) = \frac{\Gamma_\gamma}{\Gamma}\sqrt{\frac{E_r}{E}}\,\Sigma_0\psi(\zeta,x), \tag{9.101a}$$

where

$$\zeta = \frac{\Gamma}{\Gamma_D}, \tag{9.101b}$$

$$\Gamma_D = \sqrt{\frac{4E_r kT}{A}}, \tag{9.101c}$$

$$x = 2\,\frac{E - E_r}{\Gamma}, \tag{9.101d}$$

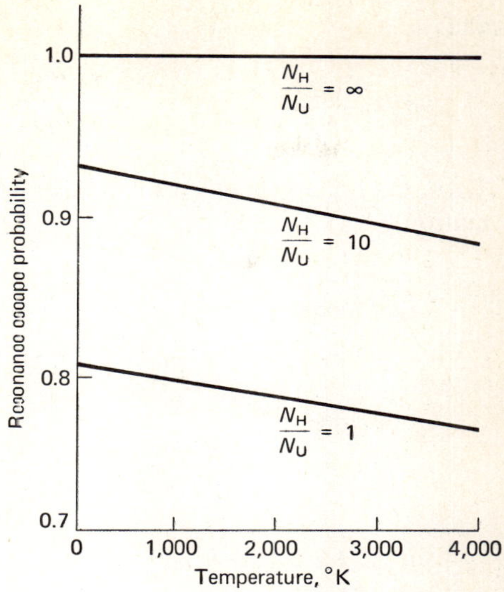

Fig. 9.12 The resonance escape probability for the 6.67-eV resonance of U^{238} as a function of temperature for various concentrations of uranium in an infinite hydrogen moderator.

and

$$\psi(\zeta,x) = \frac{\zeta}{\sqrt{4\pi}} \int_{-\infty}^{\infty} dy \frac{1}{1+y^2} e^{-\zeta^2(x-y)^2/4}. \tag{9.101e}$$

Furthermore, the scattering cross section in the absorber shows this broadening effect due to the thermal motions of the nuclei.

$$\overline{\Sigma}_s(E) = \frac{\Gamma_n}{\Gamma} \Sigma_0 \psi(\zeta,x) + \Sigma_i \chi(\zeta,x) + \Sigma_p, \tag{9.102a}$$

where

$$\chi(\zeta,x) = \frac{\zeta}{\sqrt{4\pi}} \int_{-\infty}^{\infty} 2y\, dy \frac{e^{-\zeta^2(x-y)^2/4}}{1+y^2}. \tag{9.102b}$$

To obtain expressions for I_{NR} and I_{IM}, the formulas (9.98) and (9.100) are utilized, except that $\sigma_{aF}{}^A$ and $\Sigma_{sF}{}^A$ are replaced by $\overline{\sigma}_{aF}{}^A$ and $\overline{\Sigma}_{sF}{}^A$ above. Numerical quadrature is required except for I_∞, the infinitely dilute resonance integral,

$$I_\infty = \frac{\pi}{2} \frac{\sigma_0}{E_r} \Gamma_\gamma. \tag{9.103}$$

It is convenient to introduce the following notation:

$$\sigma_p' = \frac{\Sigma_p{}^A + \Sigma_{sM}{}^A}{N_F} \tag{9.104a}$$

and

$$\sigma_p'' = \frac{\Sigma_{sM}{}^A}{N_F}. \tag{9.104b}$$

Then, the homogeneous resonance integrals are given by, in the NR approximation,

$$I_{\text{NR}} = \frac{\sigma_p' \Gamma_\gamma}{E_r} J(\zeta,\beta), \tag{9.105a}$$

where

$$J(\zeta,\beta) = \int_0^\infty \frac{\psi(\zeta,x)}{\psi(\zeta,x) + \beta}\, dx, \tag{9.105b}$$

$$\beta = \frac{\sigma_p'}{\sigma_0}. \tag{9.105c}$$

Here J is a tabulated function.[1] In obtaining this result, interference scattering has been neglected, as is true throughout the present section.

For IM, a similar calculation gives

$$I_{\text{IM}} = \frac{\sigma_p'' \Gamma_\gamma}{E_r} J(\zeta,\beta'), \tag{9.106a}$$

$$\beta' = \frac{\sigma_p'' \Gamma_\gamma}{\sigma_0 \Gamma}. \tag{9.106b}$$

In the heterogeneous case, the following formulas hold:

$$I_{\text{NR}}{}^V = \frac{\sigma_p' \Gamma_\gamma}{E_r} J(\zeta,\beta), \tag{9.107a}$$

$$I_{\text{NR}}{}^S = \frac{\sigma_0 \Gamma_\gamma}{E_r} L(t,\zeta,\beta). \tag{9.107b}$$

Here $L(t,\zeta,\beta)$ is another tabulated function;[2] also,

$$t = \frac{4V_A}{S_A} (\Sigma_p{}^A + \Sigma_{sM}{}^A). \tag{9.107c}$$

[1] F. T. Adler and L. Nordheim: *GA* 377 (1958).
[2] Adler and Nordheim: *ibid.*

Similarly,

$$I_{\text{IM}}{}^V = \frac{\sigma_p''}{E_r}\Gamma_\gamma J(\zeta,\beta'), \tag{9.108a}$$

$$I_{\text{IM}}{}^S = \frac{\sigma_p''}{\beta'}\frac{\Gamma_\gamma}{E_r}L(t',\zeta,\beta'), \tag{9.108b}$$

$$t' = \frac{4V_A\Sigma_{sM}{}^A}{S_A}. \tag{9.108c}$$

When no moderator is mixed with the absorber, $\beta' = \sigma_p'' = 0$ and $I_{\text{IM}}{}^V \to 0$. However, this is clearly incorrect as some neutrons will always be slowed down into the resonance. We can make an approximate correction for this by assuming that all neutrons degraded into the resonance are captured. Then $I_{\text{IM}}{}^V$ becomes simply the slowing-down density and A = fuel mass,

$$I_{\text{IM}}{}^V = \frac{2\Sigma_p{}^A}{AN_F}. \tag{9.109}$$

When we apply the Wigner rational approximation in the expression for $\Sigma_s{}^{\text{eff}}$, namely, (9.92), we obtain

$$\Sigma_s{}^{\text{eff}} = \frac{S_A}{4V_A}. \tag{9.110}$$

Inserting this into the expression for I_{NR}, namely, Eq. (9.98), we find

$$I_{\text{NR}} = \int\frac{dE}{E}\sigma_{aF}{}^A(E)\frac{\Sigma_p{}^A + \Sigma_{sM}{}^A + S_A/4V_A}{\Sigma_T{}^A + S_A/4V_A}. \tag{9.111}$$

On comparing this result with Eq. (9.81*b*), we see that it is identical with the resonance integral for a homogeneous system if in the lattice, Σ_p were replaced by $\Sigma_p{}^A + S_A/4V_A$. A similar result[1] is easily seen to apply to I_{IM}. Thus, to the accuracy of the rational approximation, homogeneous calculations may be used.

Dancoff corrections, for lattices, can be estimated in the same way by replacing S_A in the formulas above with[2]

$$S_{\text{eff}} = \frac{S_A}{1 + S_AV_A/\Sigma_{sM}{}^BV_B}. \tag{9.112}$$

These results have been experimentally verified, and with them, it is possible to do such things as calculate the resonance integrals of oxides from those of metals, and vice versa. However, as one would expect, the resonance

[1] These results were derived by J. Cernick and A. Vernon: *Nucl. Sci. Eng.*, **4**: 649 (1958).
[2] This was derived by G. I. Bell: *Nucl. Sci. Eng.*, **5**: 75 (1959).

integrals of heterogeneous assemblies calculated with (9.111) are about 10 percent higher than those of equivalent homogeneous mixtures, because of the error in rational approximation.

As we apply the equivalence relation, it is important to remember that if two systems have the same resonance integrals, it does not mean that they have the same resonance escape probability. This is true because, for heterogeneous assemblies, the slowing-down power depends not only on the admixed moderator but also on the surrounding moderator. Thus, by lumping an absorber, it is possible to increase the slowing-down power, without as large an increase in the resonance integral, and bring about a great reduction in the resonance capture. This procedure is essential for natural uranium reactors.

9.8 THERMAL GROUP CONSTANTS

As we have indicated previously, the so-called spectral assumption is generally used to obtain multigroup constants for reactor calculations. That is, one assumes the flux within some region of a reactor has a particular energy distribution, and averages cross sections over that distribution. For the fast group, various methods of calculating the energy distribution are commonly used, many of them amounting to nothing more than numerical integration of the energy-dependent transport equation. The age theory method described in this chapter may be used for certain classes of reactors, particularly graphite reactors. Since nobody builds graphite reactors today, the method has more historical than practical interest. Some of the computer methods currently used for fast reactors and light-water-moderated reactors are briefly discussed in Chap. 10.

The thermal spectrum, over which cross-section averages are taken to obtain thermal group constants, must be calculated in a different fashion. In either case, one has an equation of the form of (9.1). For the fast spectrum, $\Sigma_s{}^0(E' \to E)$ represents elastic and inelastic slowing-down (Secs. 9.2 and 9.3). For the thermal group, however, the thermal energy of the moderator is of the order of the neutron energy (this is the definition of the thermal group). Thus the cross section $\Sigma_s{}^0(E' \to E)$ represents collisions between two moving objects. In particular, the neutron may either gain or lose energy in a collision, depending upon the energy of the nucleus with which it collides.

The derivation of the appropriate form of $\Sigma_s{}^0(E' \to E)$ for various moderators is a complicated procedure beyond the scope of this book. Once the scattering cross section has been determined, the spectral equation is generally solved numerically. We discuss this further in Chap. 10.

Another simpler approach is to assume a spectrum of some sort and perform direct averages. The simplest assumption is that the thermal neutron spectrum has a maxwellian distribution with an absolute temperature T,

$$\phi_{\text{th}}(E) \approx Ee^{-E/kT}, \tag{9.113}$$

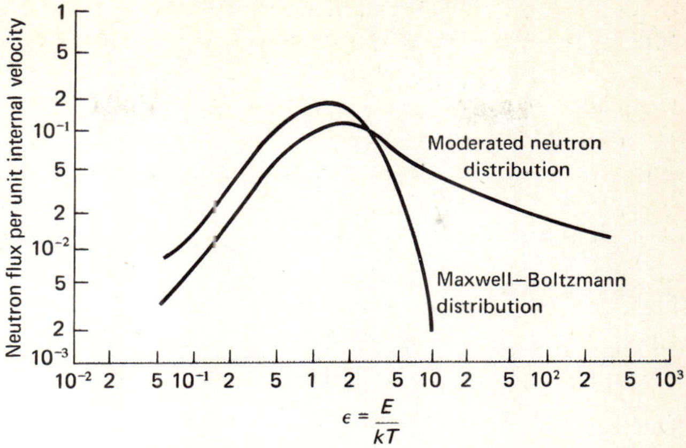

Fig. 9.13 Thermal neutron distribution. [*Source:* *R.R. Coveyou, R.R. Bate, and R.K. Osborn, J. Nucl. Energy,* **2:** 133 (1956).]

where k is Boltzmann's constant. Actually this assumption would be valid if the neutrons came into thermal equilibrium with the moderator. However, they only come into some sort of quasi-equilibrium because neutrons are constantly being created by fission and lost by absorption and leakage. It can be shown (see Fig. 9.13) by comparison with exact thermalization calculations that a fair approximation to the thermal spectrum in a reactor is a maxwellian at an effective temperature T' which is higher than the physical temperature T of the system (the spectrum is said to be *hardened*).

Coveyou, Bate, and Osborn[1] suggest a formula of the type

$$T' = T\left[1 + 0.46\,\frac{\Sigma_a(kT)}{\xi\Sigma_s}\right], \tag{9.114}$$

where $\Sigma_a(kT)$ is the absorption cross section at the energy kT ($= 0.025$ eV for $T = 20°\text{C}$), and Σ_s is the scattering cross section at energies above thermal.[2]

One defines a thermal group constant, then, by a relation such as

$$\Sigma_{\text{th}} = \frac{\int \phi_{\text{th}}(E)\Sigma(E)\,dE}{\int \phi_{\text{th}}(E)\,dE}, \tag{9.115}$$

where for ϕ_{th} one may use either a maxwellian (at temperature T') or a better-calculated or measured spectrum if one is available.

[1] *J. Nucl. Energy*, **2**: 133 (1956).

[2] Above thermal energies, the scattering cross section tends to approach a constant value, denoted by Σ_s. In the thermal region chemical binding effects tend to increase the cross section, particularly for the lighter elements.

For a $1/v$ cross section, the integration in Eq. (9.115) can be carried out explicitly. We have then

$$\Sigma = \frac{K}{\sqrt{E}}, \tag{9.116a}$$

with

$$K = \Sigma(kT')\sqrt{kT'}. \tag{9.116b}$$

Then[1]

$$\Sigma_{\text{th}} = \Sigma(kT')\sqrt{kT'}\,\frac{\int_0^\infty E^{3/2}e^{-E/kT'}\,dE}{\int_0^\infty Ee^{-E/kT'}\,dE}. \tag{9.117}$$

One easily finds

$$\Sigma_{\text{th}} = \frac{\sqrt{\pi}}{2}\,\Sigma(kT'). \tag{9.118}$$

9.9 VARIATIONAL PRINCIPLES AND PERTURBATION THEORY IN THE MULTIGROUP FORMALISM

Variational principles and perturbation theory have been discussed in Chap. 5. The applications were made in the one-speed approximation. In multigroup theory the energy dependence is included by writing the flux as an N-component vector. Of course, each component is a function of position. This form for the fluxes is shown in Eq. (8.6).

Then in place of the definitions which we used in Chap. 5, and in particular the definition of the scalar product equation (5.23), we must use a scalar product which must account for the multigroup column vectors. That is, **f** is given as

$$\mathbf{f} = \begin{pmatrix} f_1(\mathbf{r}) \\ f_2(\mathbf{r}) \\ \vdots \\ f_N(\mathbf{r}) \end{pmatrix}. \tag{9.119}$$

The scalar product is taken to be

$$(f,g) = \sum_j \int d^3r\, f_j^*(r) g_j(r). \tag{9.120a}$$

[1] The upper limit of the integral can be extended to infinity because the maxwellian factor falls off so rapidly with increasing energy. The upper limit of the thermal group should actually be chosen as some energy above which there is no appreciable amount of upscattering. This is about 0.6 eV in practice.

The operators which appear in multigroup diffusion theory are matrices of differential operators, as, for example, in Eq. (8.7). For two groups, this is

$$\mathbf{M} = \begin{pmatrix} -\nabla \cdot D_1\nabla + \Sigma_{R1} & 0 \\ \Sigma_{12} & -\nabla \cdot D_2\nabla + \Sigma_{R2} \end{pmatrix}. \tag{9.120b}$$

We see that in general, for N groups, the operator $\mathbf{M}$ acting on a vector $\mathbf{f}$ gives another vector $\mathbf{c}$,

$$\mathbf{Mf} = \mathbf{c}, \tag{9.121a}$$

where the components c_k of $\mathbf{c}$ are given by

$$\sum_{u} M_{R\mu} f_\mu = c_k. \tag{9.121b}$$

The $M_{R\mu}$'s are of the form shown in (9.120). That is, they still represent differential operators or functions of $\mathbf{r}$. The adjoint of $\mathbf{M}$, namely, $\mathbf{M}^\dagger$, is defined in the same way as in Chap. 5, Eq. (5.22).

$$(\mathbf{M}^\dagger\mathbf{g},\mathbf{f}) = (\mathbf{g},\mathbf{Mf}). \tag{9.122}$$

The matrix multiplication is discussed in Appendix A. There we show that the adjoint of a matrix is the complex conjugate of the transposed matrix. Here the elements of $\mathbf{M}$ are real. However, there are differential operators of the form $\nabla \cdot D\nabla$ appearing in the elements of $\mathbf{M}$, and the adjoint of these differential operators must also be taken. We have shown in Chap. 5 that operators of this form are self-adjoint. Thus, the elements of the adjoint of $\mathbf{M}$ are

$$(M^\dagger)_{ji} = M_{ij}. \tag{9.123}$$

Thus, referring to our example, Eq. (9.120),

$$\mathbf{M}^\dagger = \begin{pmatrix} -\nabla \cdot D_1\nabla + \Sigma_{R1} & \Sigma_{12} \\ 0 & -\nabla \cdot D_2\nabla + \Sigma_{R2} \end{pmatrix}. \tag{9.124}$$

The expressions for variational principles and perturbation theory are similar to those given in Chap. 5, except that matrix multiplication is also required. For example, consider the calculation of v_c made in Sec. 5.3. In two-group theory, this becomes the following [see Eq. (8.5)]:

$$\mathbf{M}\boldsymbol{\phi} = v_c\mathbf{F}\boldsymbol{\phi}. \tag{9.125}$$

Taking the scalar product with $\boldsymbol{\phi}^\dagger$ gives

$$v_c = \frac{(\boldsymbol{\phi}^\dagger,\mathbf{M}\boldsymbol{\phi})}{(\boldsymbol{\phi}^\dagger,\mathbf{F}\boldsymbol{\phi})}. \tag{9.126}$$

This expression corresponds to Eq. (5.40), and like it, is variationally stationary. However, since the operators are not self-adjoint, the property expressed by Eq. (5.55) does not hold. Thus, the variational principle loses much of its appeal.

Perturbation theory, however, can still be obtained for Eq. (9.126) by replacing $\mathbf{M}$ with $\mathbf{M}_0 + \delta\mathbf{M}$ and $\mathbf{F}$ with $\mathbf{F}_0 + \delta\mathbf{F}$ and using $\boldsymbol{\phi}_0$ and $\boldsymbol{\phi}_0^\dagger$ for trial functions:

$$\mathbf{M}_0 \boldsymbol{\phi}_0 = \nu_c{}^0 \mathbf{F}_0 \boldsymbol{\phi}_0 \tag{9.127a}$$

and

$$\mathbf{M}_0^\dagger \boldsymbol{\phi}_0 = \nu_c \mathbf{F}_0^\dagger \boldsymbol{\phi}_0^\dagger. \tag{9.127b}$$

Then, using (9.126), we have

$$\nu_c = \frac{(\boldsymbol{\phi}_0^\dagger, \mathbf{M}_0 \boldsymbol{\phi}_0) + (\boldsymbol{\phi}_0^\dagger, \delta\mathbf{M}\boldsymbol{\phi}_0)}{(\boldsymbol{\phi}_0^\dagger, \mathbf{F}_0 \boldsymbol{\phi}_0) + (\boldsymbol{\phi}_0^\dagger, \delta\mathbf{F}\boldsymbol{\phi}_0)}. \tag{9.128}$$

Dividing numerator and denominator by $(\boldsymbol{\phi}_0^\dagger, \mathbf{F}_0 \boldsymbol{\phi}_0)$, expanding the denominator, and noting that

$$\delta\nu = \nu_c - \nu_c{}^0 \tag{9.129a}$$

$$= \nu_c - \frac{(\boldsymbol{\phi}_0^\dagger, \mathbf{M}_0 \boldsymbol{\phi}_0)}{(\boldsymbol{\phi}_0^\dagger, \mathbf{F}_0 \boldsymbol{\phi}_0)}, \tag{9.129b}$$

we see that

$$\delta\nu = \frac{(\boldsymbol{\phi}_0^\dagger, \delta\mathbf{M}\boldsymbol{\phi}_0)}{(\boldsymbol{\phi}_0^\dagger, \mathbf{F}_0 \boldsymbol{\phi}_0)} - \nu_c{}^0 \frac{(\boldsymbol{\phi}_0^\dagger, \delta\mathbf{F}\boldsymbol{\phi}_0)}{(\boldsymbol{\phi}_0, \mathbf{F}_0 \boldsymbol{\phi}_0)}. \tag{9.130}$$

This is general for any number of groups. For our two-group matrix, Eqs. (9.124), with $\mathbf{F}$ given by

$$\mathbf{F} = \begin{pmatrix} 0 & \Sigma_f \\ 0 & 0 \end{pmatrix} \tag{9.131}$$

(i.e., ignoring fast fission), Eq. (9.130) reduces to

$$\delta\nu = \left\{ \int d^3r\, \phi_{01}^\dagger(\mathbf{r})(-\nabla \cdot \delta D_1 \nabla + \delta\Sigma_{R1})\phi_{01}(\mathbf{r}) + \int d^3r\, \phi_{02}^\dagger(\mathbf{r})(\Sigma_{12}\phi_{01} - \nabla \cdot \delta D_2 \nabla\phi_{02} + \delta\Sigma_{a2}\phi_{02}) \right\} \Big/ \int d^3r\, \phi_{01}\Sigma_f \phi_{02}. \tag{9.132}$$

PROBLEMS

9.1. Calculate E_f/E_i, and compare with e^{ξ}. What do you conclude from this?

9.2. Using E_f/E_i, calculate the number of collisions required to slow a neutron down from 2 MeV to 0.025 eV in hydrogen; in graphite. Compare with the same calculation based on ξ.

9.3. Assuming the scattering law in the CMCS to be not isotropic, but rather of the form

$$P(\mu_c) = A + B\mu_c,$$

where A and B are constant, calculate:

(*a*) $\bar{\mu}_0$,
(*b*) ξ.

9.4. Derive a formula for $\overline{\Delta u \mu_0}$.

9.5. Compute the probability that a 5-MeV neutron which scatters inelastically from U^{238} will end up in the energy interval 0 to 1 eV using the Weisskopf model, Sec. 9.3.

9.6. Suppose that there is a resonance at 1 MeV which can be characterized by the parameters $\Gamma = 0.12$ MeV, $\sigma_0 = 10^4$ b, and $\Gamma_\gamma = 0.02$ MeV. Compute the infinite-dilution resonance integral for this resonance.

10
Selected Special Topics

10.1 INTRODUCTION

In this chapter we shall discuss several special problems actually encountered in the design and operation of reactors. We have chosen these topics to illustrate the calculational techniques which were developed earlier in the text. We shall not attempt to present a comprehensive list of the problems encountered in reactor physics.[1] A number of the examples have been chosen to illustrate problems which must be treated by other, more advanced techniques.

10.2 SOME PROBLEMS ENCOUNTERED IN THE DESIGN OF LIGHT–WATER REACTORS

a GENERAL REMARKS

Light-water power reactors fall into two general classes, pressurized (PWR) and boiling (BWR). These types have many special problems in common,

[1] There are several excellent compilations of this type, e.g., *ANL* 5800.

and a few particular problems. A major design problem is the large energy loss suffered by a neutron when it collides with a hydrogen nucleus. Since it is desirable to have the group width sufficiently large that a neutron can be assumed to suffer many collisions in each group, multigroup calculations are generally replaced by "few-group" calculations. The number of groups chosen is generally about four.

The determination of the group constants for a few-group scheme proceeds as described in Chap. 9. However, the appropriate spectrum cannot be deduced from Fermi age theory, since age theory is only valid for heavy moderators (like graphite). We discuss below the methods used to obtain the fast spectrum.

Another problem is the determination of the thermal spectrum. The *effective temperature* method described in the previous chapter is too crude, particularly if there is any appreciable amount of plutonium present (plutonium has a large resonance in the thermal group). Since light-water reactors have enrichments of about 2 to 3 percent, it is clear that a great amount of U^{238} is present. This means that after a certain amount of operating time, Pu^{239} will in fact be built up, and its low-lying resonance distorts the thermal spectrum.

This raises another issue of importance. As the U^{235} is depleted and Pu is built up, the composition of the fuel elements (or fuel *rods*) changes. Eventually, old fuel elements must be replaced by new. It is not customary simply to take every old element out of the reactor and put in new elements. The reason for this is that the fuel elements in portions of the reactor core where the flux is high will be more *burned out* than those elements near the edges, say, where the flux is lower. The search for an optimum, economical method of replacing fuel elements includes the important and analytically difficult area of *fuel management.*

A problem peculiar to BWRs is the interplay between thermal hydraulics and *neutronics*. Basically, the problem is the following. In a BWR, boiling occurs inside the core, and a given region of the core will have a certain bubble fraction, depending upon the power level in that region. But the power level depends upon the amount of hydrogen present, i.e., the bubble fraction. Thus, the calculation of criticality alone is not enough; the criticality calculation must be made consistent with bubble distribution throughout the core.

Below, we discuss some of these problems in little more detail.

b FAST SPECTRUM

We recall, from Sec. 9.4 that age theory was obtained from an integro-differential equation by expanding the neutron flux in a Taylor's series and carrying out certain integrations. This procedure, as we have pointed out, is not valid for hydrogen. Let us refer to Eqs. (8.14) and (8.20), and see what might be done. First, we rewrite these in terms of the lethargy, and make the generalization that the scattering integral must be summed over all moderators present.

$$\nabla \cdot \mathbf{J}(\mathbf{r},u) + \Sigma_T(\mathbf{r},u)\phi(\mathbf{r},u)$$
$$= S(\mathbf{r},u) + \sum_{i=1}^{n} \int_{u-\ln 1/\alpha_i}^{u} du'\, \Sigma_{si}(\mathbf{r},u')P_i{}^0(u' \to u)\phi(\mathbf{r},u'); \qquad (10.1a)$$

$$\tfrac{1}{3}\nabla\phi(\mathbf{r},u) + \Sigma_T\phi = \sum_{i=1}^{n} \int_{u-\ln 1/\alpha_i}^{u} du'\, \mathbf{J}(\mathbf{r},u')\Sigma_{si}{}^1(\mathbf{r},u')P_i{}^1(u' \to u). \qquad (10.1b)$$

The summation is taken over hydrogen, oxygen, iron, uranium, etc.—whatever materials may be present in the reactor. Also $P^0(u' \to u)$ was defined in (9.21). Here $P^1(u' \to u)$ is the analogous kernel for linearly anisotropic scattering [cf. Eq. (8.18*b*)]. Age theory is obtained if in Eq. (10.1*b*) we make the approximation

$$\Sigma_{si}(u')\mathbf{J}(u') \approx \Sigma_{si}(u)\mathbf{J}(u), \qquad (10.2a)$$

and in (10.1*a*)

$$\Sigma_{si}(u')\phi(u') \approx \Sigma_{si}(u)\phi(u) + (u' - u)\frac{\partial}{\partial u}[\Sigma_{si}(u)\phi(u)]. \qquad (10.2b)$$

Under these approximations it is easy to show that

$$\int_{u-\ln 1/\alpha_i}^{u} du'\, \mathbf{J}(u')\Sigma_{si}(u')P_i{}^1(u' \to u) = \mathbf{J}(u)\Sigma_{si}(u)\bar{\mu}_{0i}, \qquad (10.3a)$$

where $\bar{\mu}_0$ is the average of the cosine of the scattering angle from a nucleus of mass A_i ($= 2/3A_i$), and

$$\int_{u-\ln 1/\alpha'}^{u} du'\, \phi(u')\Sigma_{si}(u')P^0(u' \to u) = \phi(u)\Sigma_{si}(u) - \xi_i\frac{\partial}{\partial u}[\phi(u)\Sigma_{si}(u)]. \qquad (10.3b)$$

Suppose these approximations are made for every element except hydrogen.[1] Then the P_1 equations become

$$\nabla \cdot \mathbf{J} + (\Sigma_a + \Sigma_{sH})\phi = S + \int_0^u du'\, e^{-(u-u')}\Sigma_{sH}(u')\phi(u')\,du'$$
$$- \frac{\partial}{\partial u}\sum_{i \neq H}\xi_i\Sigma_{si}\phi(u) \qquad (10.4a)$$

and

$$\tfrac{1}{3}\nabla\phi + (\Sigma_{tr} + \tfrac{2}{3}\Sigma_{sH})\mathbf{J} = \int_0^u e^{-3(u'-u)/2}\Sigma_{sH}(u')\mathbf{J}(u')\,du'. \qquad (10.4b)$$

[1] And deuterium, if any is present. See J. Ferziger and P. F. Zweifel: "The Theory of Neutron Slowing Down in Nuclear Reactors," Pergamon Press, New York, 1966.

Here the notation is self-explanatory, and we have explicitly introduced the functions for hydrogen,

$$P_H{}^0(u' \to u) = e^{-(u-u')}, \tag{10.5a}$$

$$P_H{}^1(u' \to u) = e^{-3(u-u')/2}. \tag{10.5b}$$

If the buckling approximation is made (that is, ϕ and $\mathbf{J}$ are replaced by $\hat{\phi}e^{iB\cdot r}$ and $\hat{\mathbf{J}}e^{iB\cdot r}$), the result is two coupled, linear integral equations for $\hat{\phi}$ and $\hat{\mathbf{J}}$.

In practice, Eqs. (10.4*a*) and (10.4*b*) are solved by noticing that the integral terms

$$\hat{I}_0{}^H = \int_0^u du'\, e^{-(u-u')}\Sigma_{sH}(u')\hat{\phi}(u') \tag{10.6a}$$

and

$$\hat{I}_1{}^H = \int_0^u du'\, e^{-3(u-u')/2}\Sigma_{sH}(u')\hat{J}(u') \tag{10.6b}$$

satisfy simple differential equations in u,

$$\hat{I}_0{}^H(u) + \frac{\partial \hat{I}_0{}^H}{\partial u} = \Sigma_{sH}(u)\hat{\phi}(u) \tag{10.7a}$$

and

$$\hat{I}_1{}^H(u) + \frac{\partial \hat{I}_1{}^H(u)}{\partial u} = \Sigma_{sH}(u)\hat{J}(u). \tag{10.7b}$$

If we insert the definitions (10.6*a*) and (10.6*b*) into (10.4*a*) and (10.4*b*), we obtain

$$iB\hat{J} + (\Sigma_a + \Sigma_{sH})\hat{\phi} = S + \hat{I}_0{}^H - \frac{\partial}{\partial u}\left[\sum_i \xi_i \Sigma_{si}\hat{\phi}(u)\right] \tag{10.7c}$$

and

$$\frac{iB}{3}\hat{\phi} + (\Sigma_{tr} + \tfrac{2}{3}\Sigma_{sH})\hat{J} = \hat{I}_1{}^H(u). \tag{10.7d}$$

Equations (10.7) are solved numerically, and the calculated fluxes are used to obtain constants. The source S is, of course, taken to be a fission spectrum, and the buckling B is chosen as described in Chap. 9.

c RESONANCE CAPTURE

The method of calculating resonance escape probability, described in Chap. 9, is highly accurate if all the resonance data are available. Since they generally are not, empirical information is used to provide proper resonance integral information. Hellstrand[1] and various other workers have measured resonance integrals in homogeneous mixtures of U^{238} and moderators as well as U^{238} rods surrounded by moderating material (see Figs. 10.1). Similar measurements have been made for thorium.

One obvious way in which we could test the accuracy of resonance absorption calculations would be simply to perform the computations for the experiments for which the measurements have been taken. There is a more useful technique with the so-called metal oxide correlation. The experimental results indicated that the resonance integrals for all these cases could be written as a simple function of the geometry and composition. Namely, we can define a variable

$$x = \frac{\Sigma_S{}^A}{N_F(1 - P_0^*)} \tag{10.8a}$$

where P_0^* is the first-flight escape probability which was defined in Sec. 7.4. It was shown that the U^{238} resonance data could be fitted by

$$I^{28} = 2.16x + 2.56 + (0.0279x - 0.0537)\sqrt{T_{\text{eff}}}\,. \tag{10.8b}$$

Equation (10.8*b*) gives the total absorption in U^{238}, including 1.08 b for the $1/v$ absorptions. In order to apply (10.8) we must still settle two questions. We must decide what to use for T_{eff}, and we must choose appropriate averages for the cross sections appearing in Eq. (10.8*a*).

[1] For a fairly complete list of references and a compilation of these data, the reader should see the article by L. E. Strawbridge and R. F. Barry, *Nucl. Sci. Eng.*, **23**: 58 (1965).

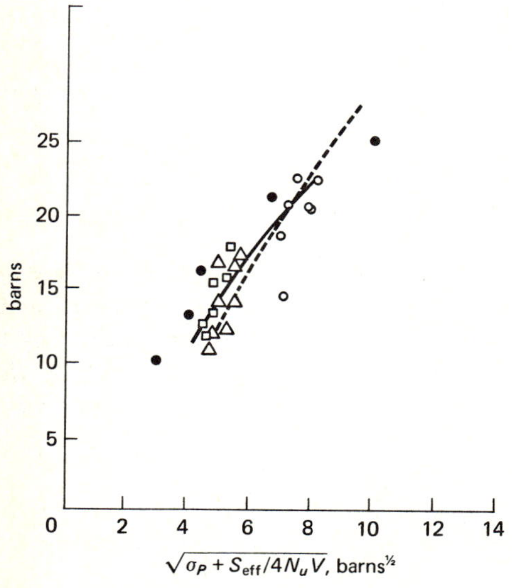

Fig. 10.1a U^{238} resonance integral. The experimental data are from the following sources: ● *Homogeneous systems; A. M. Weinberg and E. P. Wigner, op. cit., p.* 313. □ *Metal rods; H. Kouts et al., PICG P*/1841 (1958). △ *Metal rods; D. Klein et al., Nucl. Sci. Eng.*, **3**: 403 (1958). ○ *Oxide rods; ibid.* --- *Metal rods; E. Hellstrand, J. Appl. Phys.*, **28**: 1443 (1957). — *Oxide rods; ibid.*

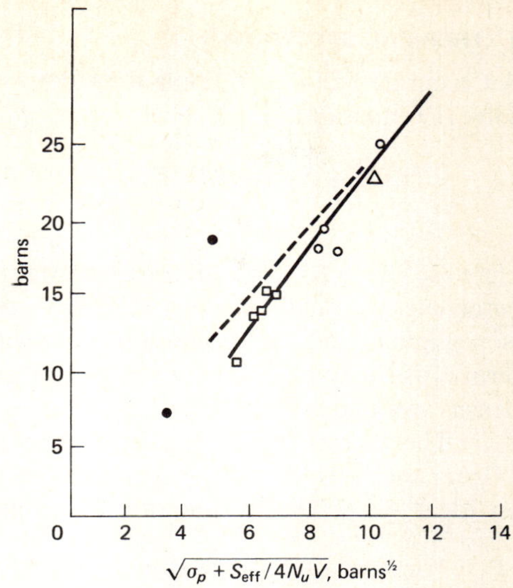

Fig. 10.1b Thorium resonance integral. The experimental data are from the following sources: ● *Homogeneous systems; A. M. Weinberg and E. P. Wigner, op. cit., p.* 313. □ *Metal rods; D. Klein et al., Nucl. Sci. Eng.*, **3**, 403 (1958). △ *Metal rods; M. C. Edlund et al., PICG P*/2405 (1958). ○ *Oxide rods; ibid.* --- *Metal rods; W. G. Pettus, BAW-TM*-203 (1959). — *Oxide rods; ibid.*

The second problem is easily settled. For the moderators the scattering cross section in the resonance region is constant, and those values are used. For example, the oxygen cross section is 3.8 b. The appropriate average cross section to be used for U^{238} is 10.7 b.

The choice of T_{eff} is somewhat harder to make. The temperature in a rod containing U^{238} will vary rather drastically from the centerline to the outside. In this case, then, one must take some average of the centerline and surface temperatures of the rod, which we denote by T_V and T_S, respectively,

$$T_{\text{eff}} = \frac{\alpha T_S + \beta T_V}{\alpha + \beta}. \tag{10.8c}$$

One possible choice of α and β is to calculate, by the techniques described in Chap. 9, the surface volume contributions to the resonance integrals I_S and I_V, and to weight T_S and T_V by these numbers.

In any event, the resonance integral is calculated according to Chap. 9;[1] the total resonance integral is then normalized to Eq. (10.8*b*). The calculated resonance integral, renormalized in this way, is used in each group of the spectrum calculation previously described as an absorption cross section. The ratio of the homogeneously calculated resonance integral to its renormalized value is called the *l* factor.

[1] In some cases a simple homogeneous calculation is used; all the geometrical effects are introduced through the correlation equations (10.8). In other codes the geometry is included in the calculation of *I*.

d THERMAL SPECTRUM

An accurate method of computing the thermal spectrum in a reactor is to solve the equation

$$[D(E)B^2 + \Sigma_T(E)]\phi(B^2,E) = \int_0^{E_c} \Sigma_S{}^0(E' \to E)\phi(B^2,E')\,dE' + q_T(B^2). \tag{10.9}$$

Here $\Sigma_S{}^0(E' \to E)$ is the thermal scattering kernel, and E_c is the thermal cutoff energy, usually chosen as 0.625 eV, since above that, energy upscattering is negligible and ordinary slowing-down theory may be used. The last term q_T is the source of neutrons to the thermal group, i.e., the slowing-down density evaluated at $E = E_c$.

The thermal scattering kernel most frequently used was derived by Wigner and Wilkens[1] on the assumption that the moderator atoms constituted an ideal gas. This assumption is certainly physically incorrect, but the

[1] E. P. Wigner and J. E. Wilkens, Jr.: *Rept. AECD* 2275, 1944.

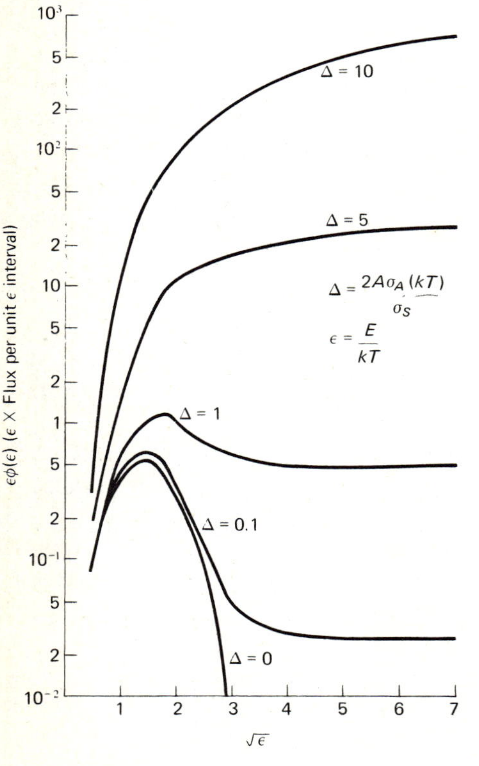

Fig. 10.2 Energy distribution of neutrons thermalized in a heavy gaseous moderator (large A), $1/r$ absorber. [*Source: H. Hurwitz, Jr., M. S. Nelkin, and G. J. Habetler, Nucl. Sci. Eng.*, **1**: 280 (1956).]

thermal spectra calculated from (10.9) with an ideal gas kernel turn out to give reasonably accurate results.

Clearly, the kernel must depend upon the mass of the moderator A. For $A = 1$, such as applies to H_2O reactors, Eq. (10.9) can be reduced to a differential equation,[1] numerical solutions can be obtained on a computer, and cross sections can be averaged to obtain thermal group constants. Examples of computed neutron densities are shown in Fig. 10.2.

For a heavy moderator $A \gg 1$, a differential equation can also be obtained.[2] For other moderators, numerical solution of Eq. (10.9) is necessary.

e FUEL MANAGEMENT

A number of different reloading schemes have been proposed for H_2O reactors. One, known as the *out-in* scheme, divides the core of the reactor into three concentric regions. At the end of each cycle, the fuel in the innermost region is removed from the reactor, fuel from the middle region goes to the center, fuel from the outside region goes to the middle, and new fuel goes into the outside. This simple scheme is designed to flatten the power distribution by putting more depleted fuel, i.e., fuel with a higher burnup, in the center of the core, where the flux is higher.

A more advanced scheme shuffles individual fuel elements, and is called the *scatter-load* strategy. The idea behind this strategy is to keep the isotopic composition of the fuel as uniformly mixed as possible, and to remove fuel that has the maximum burnup. The replacement of fuel assemblies is by no means random, although it would appear so to the casual observer. For example, in boiling-water reactors a *unit cell* consists of four fuel assemblies clustered about a cruciform control rod. The assembly of the four with the maximum burnup is always replaced with a fresh assembly, and the other three remain in the same position until the next reloading.

In any event, proper analysis of fuel management problems requires the following program of calculations. First, a criticality and flux calculation is performed for the *clean* reactor. The uranium depletion and plutonium and fission-product buildup are computed for a short time step (perhaps 20 days) on the assumption that the flux and composition are constant over that period of time. (This is discussed in more detail in Chap. 7.) At the end of the time step, a discontinuous change in reactor composition is made, new fluxes are calculated, and depletion and buildup are computed for a second time step. This procedure is continued until the *end of life*—the point at which the reactor can no longer operate, usually because $k < 1$. Next, a reloading

[1] M. M. R. Williams: "The Slowing Down and Thermalization of Neutrons," John Wiley & Sons, Inc., New York, 1966.

[2] Williams: *ibid.*

pattern is assumed and tested by computation through several reloading cycles. If several different schemes are tested in this way, an optimum fuel management strategy can be deduced.

Calculations such as those described here are very lengthy and quite costly in computer time.

10.3 FAST REACTORS

Fast reactors are constructed with almost no moderating material. Thus the energy spectrum of the neutrons is peaked at considerably higher energy (the terminology *hard spectrum* is used) than in H_2O-, D_2O-, or graphite-moderated reactors (see Fig. 10.3). A representative number which is used as a measure of the hardness of the spectrum is the average energy at which a neutron causes a fission.[1] In fast reactors the mean fission energy ranges from a few kilovolts to a few hundred kilovolts depending upon the reactor type.

Specifically, reactors with metal fuel elements have mean fission energies in the 100-kV region. However, oxide and carbide fuel elements are more resistant to radiation damage, and hence are more likely to be used in power reactors. In that case, the oxygen or carbon (or both) degrades the spectrum, through elastic scattering, to much lower energies. The liquid sodium, generally used as a coolant, also helps to degrade the spectrum somewhat. Actually, the energy degradation is caused primarily by inelastic

[1] Usually referred to as the *mean fission energy* in spite of the possible confusion with the average energy of the fission spectrum, $N(E)$.

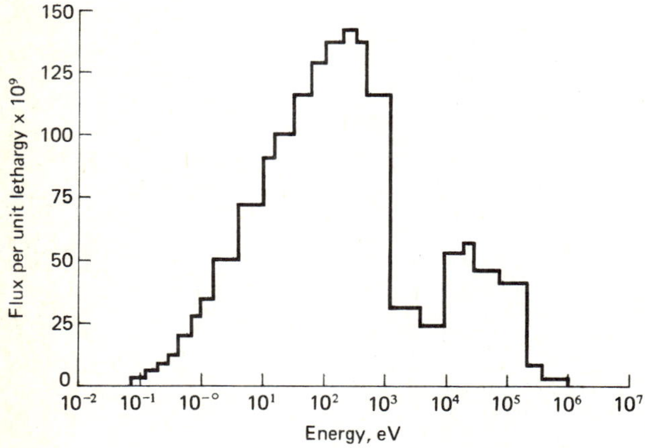

Fig. 10.3 Spectrum at inner surface of vessel wall of fast testing reactor.

scattering with the heavy nuclei, an effect which predominates over the elastic scattering mentioned above. However, the low-energy *tail*, particularly in oxide and carbide cores, is largely determined by elastic scattering.

We recall the fission cross section decreases with increasing energy. Thus the hard spectrum in fast reactors requires that they contain considerably more fissionable material than thermal reactors. This increase is brought about by increasing the enrichment of the fuel and its concentration in the reactor core. Thus, in an H_2O reactor the fertile material is found within the core, while in a fast reactor it is generally found in an external low-enrichment breeding blanket as well as in the core.

These large concentrations of fissionable material create extra problems in ensuring that the reactor is safe. In particular, a meltdown accident could be considerably more hazardous in a fast reactor because the fissile enrichment in the fast reactor is large enough that it can become supercritical if formed into a reasonably coherent compacted mass of pure fuel after the coolant has boiled away. A "credible accident" would involve the gravitational collapse and coalescence of the fuel material near the center and top of the core after the coolant has boiled away and this portion of the fuel has become molten.

Another safety problem is caused by the fact that the prompt-neutron lifetime l_{eff} is very short in a fast reactor. In a thermal reactor this quantity may vary between 100 μs for a highly enriched system to 1 ms for a graphite pile. In a fast reactor, on the other hand, $l_{\text{eff}} \approx 10$ μs. Thus for large reactivity increments, of about $1, the response of a fast reactor is much more rapid than is that of a thermal reactor. [The kinetic behavior associated with smaller reactivity changes, of course, is controlled by the delayed-neutron lifetime (see Chap. 4), and hence is similar for fast and thermal reactors.]

Still another safety problem has to do with the value of the dollar. In thermal ractors, $\beta_{\text{eff}} \approx 0.008$.[1] In Pu^{239} and U^{233}-fueled fast reactors, because β is much less for Pu^{239} than for U^{235}, β_{eff} is much smaller. Typically $\beta_{\text{eff}} \approx 0.003$ in Pu breeders. Clearly, the smaller β_{eff} is, the more difficult it is to control a reactor. (For U^{233} fueled reactors, β_{eff} is also about 0.003, and similar conclusions obtain.)

In another area, the control of a fast reactor is not so simple for still a different reason, namely, the lack of materials which strongly absorb fast neutrons. We can see this effect in the flux profile of Fig. 10.4.

Another problem concerns a possible loss of coolant. In H_2O reactors, the hydrogen has such a strong positive reactivity effect (because of its moderating properties) that any loss of coolant will shut the reactor down.[2] Then,

[1] See Sec. 8.8.

[2] Except when a large amount of boron is dissolved in the water, as, e.g., in a PWR near the beginning of life.

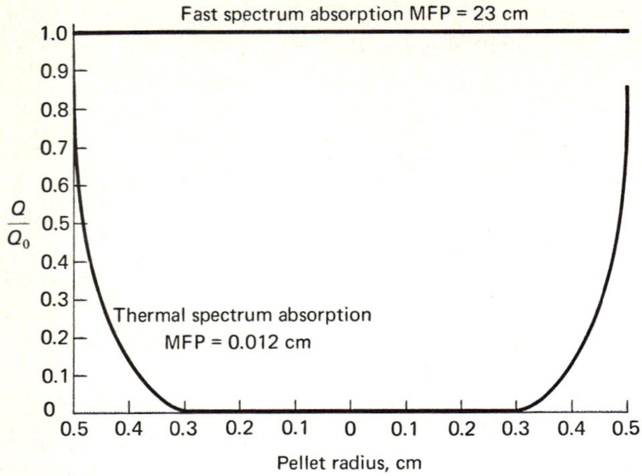

Fig. 10.4 Flux depression in boron carbide pellets.

fuel elements in the thermal reactor might still be expected to melt because of residual heating, but no supercritical excursion would take place. In a fast reactor, on the other hand, loss of sodium coolant can very easily lead to an increase in reactivity. The reason for this is somewhat complicated, but we shall attempt to sketch it below.

The only important physical effects of sodium in a reactor are (1) scattering and (2) energy degradation which accompanies scattering. The effect of scattering is to prevent leakage from the reactor core; i.e., the sodium acts as a sort of internal reflector. Thus, when coolant is lost, this effect decreases reactivity—more neutrons leak from the core. On the other hand, the energy degradation may provide a positive reactivity component as coolant is lost and the neutron spectrum accordingly hardens. The reason is that the capture-to-fission ratio of the fissile material decreases as the neutron energy increases (in the energy range appropriate to fast reactors) and, to a lesser extent, the fission cross section of the *fertile* material increases. Depending upon the size (and shape) of the fast reactor core the net effect can be positive or negative; for example, a core with high leakage would tend to have a net negative sodium-void coefficient, and vice versa.

All these difficulties are reflected as computational problems in fast-reactor design. Very careful analysis of reactivity coefficients is required, for example, to satisfy pertinent safety criteria. For water reactors the safety analysis is, for reasons pointed out above, not quite so difficult.

As a general rule, the behavior of a thermal system is not terribly sensitive to the details of the fast spectrum. A reasonable calculation will usually bring home the bacon. Thus, effort can be concentrated on the thermal region where nuclear data are better, cross-section knowledge greater, and computational techniques more advanced.[1] (As an example, we note that MUFT, a popular fast-spectrum code used in H_2O reactor design, has only 54 energy groups.)

In a fast reactor, on the other hand, if the fast spectrum is unknown, there is nothing known! So all the difficulties which do *not* occur for thermal reactors do occur for fast reactors. Questionable calculational methods must be used in energy regions where the cross sections are poorly known—and this is particularly true of the oxide and carbide cores, with their softer spectra. There are many computational problems, but one of the most severe is the fact that multigroup theory may just not be adequate unless many groups are used, in which case machine time requirements may preclude the use of more than one dimension. In fact, because of the large diffusion length compared with the dimensions of the core for fast assemblies, it is sometimes found that diffusion theory is inadequate, and transport theoretical calculations are required. (This difficulty is manifested in thermal reactors, however, in the frequent need for transport theoretical cell calculations. The reason is the extreme sensitivity of thermal reactors to the details of the spatial distribution of fission sources. However, transport theory is certainly easier to apply to cells than to larger regions of a reactor.) Group-constant codes for fast reactors sometimes contain thousands of groups. This is necessitated by the factors already outlined in the previous paragraph. Until a few years ago, spectral codes were not in general used in fast-reactor design; standard sets of group constants were used instead. However, today the spectral codes are used almost universally.

Because of many of the problems listed above, but particularly the safety problem, the commercial fast reactor is at least a decade away.

10.4 THE THORIUM CONVERTER

As was explained in Chap. 1, the primary interest in fast reactors is that breeding plutonium is possible because of the high value of η (≈ 2.7 or 2.8) for fast neutrons incident in Pu^{239}. For thermal neutrons, η of both U^{235} and Pu^{238} is too low to permit breeding[2] (see Figs. 1.1, 1.2, 1.3, and 1.4).

We also pointed out in Chap. 1 that breeding U^{233} from Th^{232} in thermal reactors was in principle possible because $\eta(U^{233}) \approx 2.28$ for thermal neutrons.[3] Thus $\eta - 2 = 0.28$, which is just marginal for building a reactor which produces more fissionable material than it consumes.

[1] Since thermal neutrons do not diffuse so far as fast neutrons, the spectral assumption is more nearly correct.

[2] We recall that $n - 2 > 0$ is a minimal condition. For practical purposes $\eta - 2$ should be considerably greater than zero; see Sec. 1.2.

[3] Fast reactors also work well with thorium.

The reactions appropriate to this type of reactor are

$$\begin{aligned}
&Th^{232} + n \rightarrow Th^{233} + \gamma \\
&Th^{233} \rightarrow \beta^- + Pa^{233} + \bar{\nu} \qquad 23.5 \text{ min} \\
&Pa^{233} \rightarrow \beta^- + U^{233} + \bar{\nu} \qquad 27.4 \text{ days} \\
&U^{233} \rightarrow Th^{229} + \alpha \qquad 1.0 \times 10^5 \text{ years.}
\end{aligned} \tag{10.10}$$

Many of the physics problems associated with thorium converters have to do with our lack of knowledge of cross sections. There are also some engineering problems which have slowed the development of the thorium converter, but perhaps the major reason that the fast reactor has achieved more attention is that the relatively higher value of η in a fast reactor makes it a more promising source of fissile material.

The safety problems associated with the thorium converter, on the other hand, are probably not so severe as are those peculiar to the fast reactor.

10.5 HEAVY–WATER REACTORS

Heavy-water reactors are currently very popular in Canada, but except for some government-owned and -operated production reactors, there are not too many D_2O reactors in the United States.

One important fact of D_2O reactors is the negligible absorption cross section of deuterium compared with the fairly significant cross section ($\approx \frac{1}{3}$ b at 0.025 eV) of ordinary hydrogen. For this reason, D_2O reactors require lower critical mass and a lower Σ_f. Since at a given power level

$$\text{Power density} \approx \Sigma_f \phi, \tag{10.11}$$

and since Σ_f is lower in a D_2O than in an H_2O reactor, ϕ will be higher—substantially higher. For this reason D_2O-moderated reactors are good research reactors, i.e., they serve as copious sources of thermal neutrons.[1] On the other hand, there is a need to shield the high neutron flux. This is not so much of a problem as one might at first think, however, since thermal neutrons are not so difficult to shield. The fast neutron and gamma doses are proportional to the fission rate, and hence to the power

Because deuterium undergoes photo disintegration with a 2.24 MeV threshold,

$$\gamma + d \rightarrow n + p; \tag{10.12}$$

and because delayed *gammas* are emitted by fission products just as are delayed neutrons, an operating D_2O reactor has more delayed-neutron groups

[1] These remarks do not even remotely apply to D_2O power reactors, for which the fissile inventory is usually determined by heat transfer conditions.

than an H_2O reactor has. This complicates some of the kinetic analysis,[1] even to the possibility of having negative effective delayed-neutron fractions (see Sec. 4.4).

One of the major analytical difficulties of treating D_2O reactors is that of calculating the fast spectrum. Deuterium ($A = 2$) is not sufficiently massive that age theory is valid. However, the transport theoretical treatment, as described in Sec. 10.2, is more complicated when applied to deuterium than when applied to hydrogen. For example, the limits of the slowing-down integrals are no longer $(0,u)$ but $(u - \ln 1/\alpha_D, u)$, where $\alpha_D = [(A-1)/(A+1)]^2 = \frac{1}{9}$. This makes numerical computations somewhat more complicated. In particular, the set of two coupled, integral equations cannot be converted to the more easily treated coupled, differential equations (10.7).

However, Greuling and Goertzel[2] invented an approximate scheme by which this conversion could be done. As long as the range of integration of the slowing-down integrals can be extended downward to zero, a differential equation can be written. Greuling and Goertzel constructed a slowing-down kernel such that its integral over the range $(0,u)$ was equal to the integral of the exact slowing-down kernel over the range $(u - \ln 1/\alpha, u)$. The net result is that the fast flux and current can be obtained as a solution of the four simultaneous differential equations,[3]

$$iB\hat{J} + \Sigma_a \hat{\phi} = S_0 - \frac{\partial \hat{I}_D{}^0}{\partial u}, \tag{10.13a}$$

$$\Sigma_{tr} \hat{J} + iB\hat{\phi} = \frac{\partial \hat{I}_D{}^1}{\partial u}, \tag{10.13b}$$

$$I_D{}^0 + \lambda_0 \frac{\partial \hat{I}_D{}^0}{\partial u} = \hat{\phi}\xi\Sigma_{SD}, \tag{10.13c}$$

$$\hat{I}_D{}^1 + \lambda_1 \frac{\partial \hat{I}_D{}^1}{\partial u} = \hat{J}_1 \Sigma \xi_{SD}. \tag{10.13d}$$

Here

$$\lambda_0 = \frac{1 - \alpha[1 + \ln 1/\alpha + \frac{1}{2}\ln^2 1/\alpha]}{1 - \alpha(1 + \ln 1/\alpha)} = 0.593. \tag{10.14a}$$

$$\begin{aligned} \lambda_1 = &-\left(\frac{(1+\gamma)^2}{4\gamma}\left\{\frac{(1+\gamma)}{3}\left[\frac{8}{9} - \alpha^{3/2}\left(\ln^2\frac{1}{\alpha} - \frac{4}{3}\ln\frac{1}{\alpha} + \frac{8}{9}\right)\right]\right\}\right. \\ &\left. - (1-\gamma)\left[8 - \alpha^{1/2}\left(\ln^2\frac{1}{\alpha} - 4\ln\frac{1}{\alpha} + 8\right)\right]\right)\Big/\zeta \\ = &-693.6. \end{aligned} \tag{10.14b}$$

[1] This is also true of beryllium-moderated reactors because of the reaction $Be^9 + \gamma \to n + 2\alpha$.
[2] E. Greuling and G. Goertzel: *Nucl. Sci. Eng.*, **7**: 69 (1960).
[3] See Ferziger and Zweifel: *op. cit.*

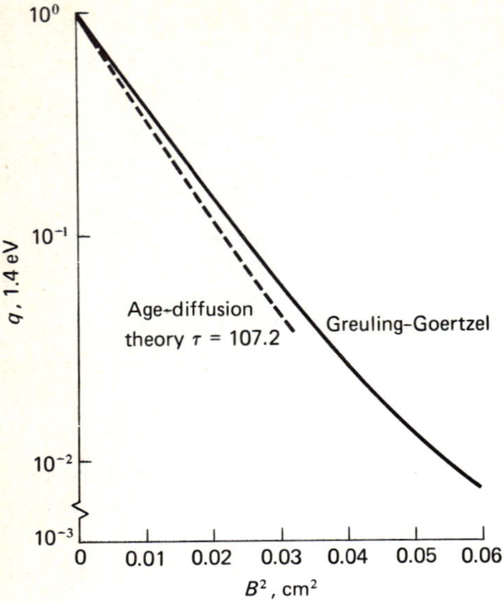

Fig. 10.5 The calculated slowing down density (non leakage probability) in 99.8% D_2O. [*Source: M. M. Levine, K. E. Roach, D. B. Wehmeyer, and P. F. Zweifel, Nucl. Sci. Eng.*, **7**: 14 (1960).]

$$\zeta = \frac{(1+\gamma)^2}{\gamma^2}\left\{\frac{1+\gamma}{9}\left[1-\alpha^{3/2}\left(\frac{3}{2}\ln\frac{1}{\alpha}+1\right)\right]\right\}$$

$$-(1-\gamma)\left[1-\alpha^{1/2}\left(\frac{1}{2}\ln\frac{1}{\alpha}+1\right)\right]$$

$$= -0.011. \qquad (10.14c)$$

The numerical values above are for $A = 2$. A representative solution of these equations is shown in Fig. 10.5.

These equations turn out to be more accurate than Fermi's age theory even for elements such as oxygen, graphite, etc., and so might be employed profitably as a more accurate slowing-down model than Fermi age theory. However, the major application has been to calculations in D_2O reactors.

There are many engineering problems in the design and operation of D_2O reactors, having to do with the presence of large quantities of such an exotic material as heavy water in a radiation environment. In accordance with our mission specified in Sec. 1.1, we do not discuss these engineering problems here.

10.6 GRAPHITE PILES

Graphite reactors today are only of historical importance, so our discussion will be brief. When the American scientists, in 1941 to 1942, were attempting to build a reactor, only natural uranium was available as a fuel, and D_2O was not available in sufficient quantity to construct a reactor.

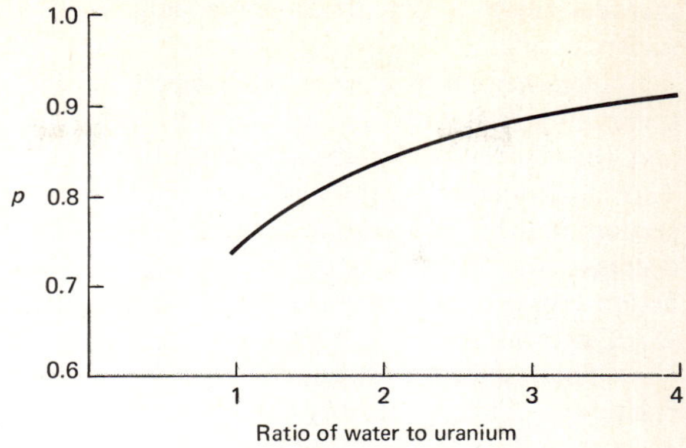

Fig. 10.6 Resonance escape probability for uranium-water lattices. (*The data for this curve are from H. Kouts and R. Sher, BNL-486, September,* 1957.)

From exponential experiments[1] it was determined that $k_\infty < 1$ for any homogeneous mixture of graphite and natural uranium. However, Fermi soon concluded that a lumping of the uranium and graphite, and "piling" them together, might raise k_∞ for the following reason: Self-shielding effects would reduce f and increase p, which would leave, of course, the other factors in the four-factor formula essentially unchanged.[2] If, however, f was decreased less than p was increased, the net effect would be to increase the product of pf, and actually increase $k_\infty = \eta\varepsilon pf$ above unity.

This actually turned out to be the case. The factors p and f for any heterogeneous system can be calculated from the formulas given for p (in Chap. 8) and for f, say in Sec. 2.10. Typical values for p are plotted in Fig. 10.6.

10.7 TEMPERATURE COEFFICIENTS

As the power and hence the temperature of a reactor are increased, the physical properties of the reactor change. These changes take several forms.

The most important contribution for water reactors is caused by the fact that the volume coefficient of thermal expansion for water is so large. Thus, a given temperature rise will decrease the effective density of water in the core (which will increase leakage) and will decrease the ratio of hydrogen to fuel in the core. The latter effect, in a low-enrichment water reactor, increases resonance absorption more than it increases thermal utilization, and

[1] See Sec. 8.7.
[2] Except for a slight increase in ε.

provides the most important component of negative reactivity feedback in such reactors.[1]

Another effect is that when temperature increases, the thermal neutron spectrum gets slightly harder. For example, the T in Eq. (9.114) increases, and this increases the effective temperature of the thermal spectrum T'. This hardening of the thermal spectrum causes changes in the average cross sections in the thermal group. For example, the $1/v$ absorption cross sections decrease, which causes a net decrease in reactivity. The reason is that the fission cross section decreases more rapidly than $1/v$ does, and so the ratio $\nu\Sigma_f/\Sigma_a$ decreases.

These effects are treated by defining the temperature coefficient of reactivity. That is, quantity $C(T)$ is defined as

$$C(T) = \frac{\delta\rho}{\delta T}. \tag{10.15}$$

If $C(T)$ is positive, an increase in temperature will cause an increase in reactivity. This will then lead to an increase in power (and temperature), which will further increase the reactivity. This is a clearly unstable situation, and will eventually lead to the melting of the core.

A negative temperature coefficient means that an increase in power leads to a decrease in reactivity. This would seem to be an inherently stable situation.

We must be careful, however; for even with net negative temperature coefficients, the reactor can be effectively unstable. For example, oscillations can be introduced through feedback mechanisms in which the peak temperatures eventually cause melting. Also there may be two effects giving rise to temperature coefficients. These effects can occur on different time scales. For example, the hardening of the thermal spectrum should take place simultaneously with the temperature increase. On the other hand, those effects which arise from expansion of reactor supporting structures do not occur until after the fuel temperatures have increased. If a reactor has a prompt positive temperature coefficient and a slower negative temperature coefficient, the initial increase in the temperature could produce melting before the slower negative coefficient could take effect.

There are a number of effects which will produce negative temperature coefficients.[2] We have discussed the effects of temperature on resonance

[1] By *feedback*, we mean the further reactivity effect produced by a given reactivity change. For example, suppose positive reactivity is introduced into a water reactor. The power and hence the temperature will begin to rise. This effect will introduce negative reactivity, causing the power and temperature rises to diminish. This is negative feedback.

[2] At the very least, the reactor should be designed so that melting produces a decrease in reactivity. Still, nobody knows at present how it can be accomplished for a fast reactor, which explains why the commercial fast reactor is still some distance in the future.

absorption in some detail in Sec. 9.7 and again in Sec. 10.2. In particular, we have noted that the temperature dependence of the resonance integral is $\approx \sqrt{T}$. We discussed this contribution extensively in Sec. 9.7, where we noted that the resonance integral increases with temperature. This of course causes the resonance escape probability to decrease with temperature. Thus the resonance absorption temperature reactivity coefficient, called the *Doppler coefficient*, is negative. This effect is particularly important in fast reactors, since the resonance absorption is particularly important for the harder, fast spectrum. It is also important for any low-enrichment reactor, such as PWRs or BWRs.

It is certainly clear that an accurate calculation of the temperature coefficients for a reactor are of paramount importance in determining whether or not the reactor is safe. In fact, temperature coefficients were referred to as *danger coefficients* prior to one of the campaigns to improve the public image of nuclear power.

We have only outlined the kinds of problems faced in these temperature feedback effects. Serious reactor designers will obviously need to refer to the more advanced texts devoted specifically to these topics.[1]

[1] A. Z. Akcasu, G. S. Lellouche, L. M. Shotkin: "Mathematical Methods in Nuclear Reactor Dynamics," Academic Press, Inc., New York, 1971.

appendix A
Matrices

Matrix notation is commonly used in reactor physics and should be thoroughly understood before Chap. 5 and subsequent chapters are read.

Consider the set of linear algebraic equations

$$a_1 x + b_1 y = c_1, \tag{A.1a}$$

$$a_2 x + b_2 y = c_2 . \tag{A.1b}$$

If we define a matrix **M** as

$$\mathbf{M} = \begin{pmatrix} a_1 & b_1 \\ a_2 & b_2 \end{pmatrix}, \tag{A.2}$$

and column vectors $\boldsymbol{\phi}$ and $\boldsymbol{\psi}$ by

$$\boldsymbol{\phi} = \begin{pmatrix} x \\ y \end{pmatrix}, \qquad \boldsymbol{\psi} = \begin{pmatrix} c_1 \\ c_2 \end{pmatrix}, \tag{A.3}$$

then Eqs. (A.1) may be rewritten in compact notation as

$$\mathbf{M}\boldsymbol{\phi} = \boldsymbol{\psi}. \tag{A.4}$$

Here we must understand the definition of matrix multiplication. If we define the element of a matrix **M** which appears in the ith row and jth column as M_{ij}, then the product of two matrices **MN** is

$$(\mathbf{MN})_{ij} = \sum_k M_{ik} N_{kj}. \tag{A.5}$$

Note that the product **MN** is another matrix, i.e., the product of two matrices is a matrix. The matrix **M** given by (A.2) is a special case—it is a square matrix of rank 2. In (A.4) the matrices **M** and **N** need not be square, but it is required that the number of columns of **M** equal the number of rows of **N**. Otherwise the matrix product is not defined.

The column vectors $\boldsymbol{\phi}$ and $\boldsymbol{\psi}$ defined in Eqs. (A.3) are really matrices. We refer to them as 2×1 matrices (two rows and one column). The product $\mathbf{A}\boldsymbol{\psi}$ is readily seen, by utilizing definition (A.5), to be another 2×1 matrix. In fact, if **M** is an $m_1 \times m_2$ matrix and **N** is an $m_2 \times n_2$ matrix, then **MN** is an $m_1 \times n_2$ matrix.

Some matrices have an inverse, i.e., if **M** is a matrix, then there may exist a matrix $\mathbf{M}^{-1}$ with the properties

$$\mathbf{M}\mathbf{M}^{-1} = \mathbf{M}^{-1}\mathbf{M} = \mathbf{I}, \tag{A.6}$$

where **I** is a unit matrix, i.e., a matrix with unity on the diagonal and zero elsewhere. In terms of the Kronecker delta symbol defined in Chap. 5, the matrix elements of **I** may be written as

$$(\mathbf{I})_{ij} = \delta_{ij}. \tag{A.7}$$

The solution of (A.3) is, in terms of $\mathbf{M}^{-1}$,

$$\boldsymbol{\phi} = \mathbf{M}^{-1}\boldsymbol{\psi}.$$

Recalling Kramer's rule, the high school approach to solving systems such as (A.1), we see that $\mathbf{M}^{-1}$ exists if the determinant of the matrix **M** does not vanish, i.e., if

$$\det \mathbf{M} = \begin{vmatrix} a_1 & b_1 \\ a_2 & b_2 \end{vmatrix} \neq 0. \tag{A.8}$$

Kramer's rule is, in fact, nothing other than a method for calculating $\mathbf{M}^{-1}$.

In the same sense that the linear operators, discussed in Chap. 5, have eigenfunctions and eigenvalues, so do matrices. A square matrix of rank N has, in fact, N eigenvalues (some of which may be the same). Eigenvalues and eigenfunctions are calculated in a straightforward manner. If **M** is a matrix, then the N eigenfunctions will obey an equation of the form

$$\mathbf{M}\boldsymbol{\psi}_\lambda = \lambda\boldsymbol{\psi}_\lambda, \qquad \lambda = 1, 2, \ldots, N. \tag{A.9}$$

Rewriting this equation as

$$(\mathbf{M} - \lambda\mathbf{I})\boldsymbol{\psi}_\lambda = 0, \tag{A.10}$$

we see, again from Kramer's rule, that this *homogeneous* set has no solution unless the determinant of the coefficients *does* vanish, i.e.,

$$0 = \det|\mathbf{M} - \lambda\mathbf{I}| = \begin{vmatrix} M_{11} - \lambda & M_{12} & M_{13} & \cdots & M_{IN} \\ M_{21} & M_{22} - \lambda & \cdots & \cdots & M_{2N} \\ \cdots & \cdots & \cdots & \cdots & \cdots \\ M_{N1} & \cdots & \cdots & \cdots & M_{NN} - \lambda \end{vmatrix}. \tag{A.11}$$

[Since the above determinental equation is of degree N, it is clear why there are N eigenvalues, that is, N roots of (A.11)]. Equation (A.11) is called the *secular equation.*

Once the N eigenvalues $\lambda_1, \lambda_2, \ldots, \lambda_N$ are obtained, one readily finds the N eigenvectors[1] $\boldsymbol{\psi}_1, \boldsymbol{\psi}_2, \ldots, \boldsymbol{\psi}_N$ as follows: Referring to (A.9) and inserting λi, say, we get

$$\mathbf{M}\boldsymbol{\psi}_{\lambda i} = \lambda_i \psi_{\lambda i}, \tag{A.12}$$

or writing it out,

$$M_{11}(\psi_{\lambda i})_1 + M_{12}(\psi_{\lambda i})_2 + \cdots + M_{1N}(\psi_{\lambda i})_N = \lambda_i(\psi_{\lambda i})_1, \tag{A.13a}$$

$$M_{21}(\psi_{\lambda i})_1 + M_{22}(\psi_{\lambda i})_2 + \cdots + M_{2N}(\psi_{\lambda i})_N = \lambda_i(\psi_{\lambda i})_2, \tag{A.13b}$$

$$\cdots\cdots\cdots\cdots\cdots\cdots\cdots\cdots\cdots\cdots$$

$$M_{N1}(\psi_{\lambda i})_N + \quad \cdots \quad + \cdots + M_{NN}(\psi_{\lambda i})_N = \lambda_i(\psi_{\lambda i})_N. \tag{A.13c}$$

It is not generally true that this set of simultaneous equations has a solution, indeed (A.11) is a necessary but not sufficient condition that there be a solution to (A.10). Let us assume that (A.10) does in fact leave a solution. If the N equations (A.10) are solved, the N unknowns $(\psi_{\lambda i})_1, (\psi_{\lambda i})_2, \ldots, (\psi_{\lambda i})_N$ are the N entries of the eigenvector $\psi_{\lambda i}$. Because the set is homogeneous, one of the $(\psi_{\lambda i})_K$ will be arbitrary. Sometimes it is chosen by requiring that the eigenvector be normal, i.e.,

$$\boldsymbol{\psi}^T\boldsymbol{\psi} = 1, \tag{A.14}$$

where the transpose matrix, denoted by the superscript T, is obtained by interchanging rows and columns. It is left as an exercise to the reader to show that the matrix obtained by multiplying ψ by ψ^T is a square matrix of rank N.

[1] In some cases a matrix may be *deficient*, i.e., some eigenvalues may not have eigenvectors. Consider, for example, the matrix

$$\begin{pmatrix} 1 & 1 \\ 0 & 1 \end{pmatrix}.$$

We observe, for example from Prob. A.1, that the product of matrices in general depends upon the order in which the product is formed, i.e., if **A** and **B** are two matrices, in general

$$\mathbf{AB} \neq \mathbf{BA}. \tag{A.15}$$

Finally, we simply mention that the analogs of the self-adjoint operators discussed in Chap. 5 are hermitean matrices, i.e., matrices whose elements obey

$$A_{ij} = A_{ji}^*. \tag{A.16}$$

(Since in reactor physics, most matrices which occur are real, Eq. (A.16) becomes $A_{ij} = A_{ji}$, or $\mathbf{A} = \mathbf{A}^T$. A real hermitean matrix is said to be *symmetric*).

Scalar products in matrix notation are simply defined as matrix products, i.e.,

$$(\boldsymbol{\phi},\boldsymbol{\psi}) = \boldsymbol{\phi}^T\boldsymbol{\psi} \tag{A.17a}$$

and

$$(\boldsymbol{\phi},\mathbf{A}\boldsymbol{\psi}) = \boldsymbol{\phi}^T\mathbf{A}\boldsymbol{\psi} \tag{A.17b}$$

$$= (\boldsymbol{\phi}^T)(\mathbf{A}\boldsymbol{\psi}) \tag{A.17c}$$

$$= (\boldsymbol{\phi}^T\mathbf{A})\boldsymbol{\psi}, \tag{A.17d}$$

where $\boldsymbol{\phi}$ and $\boldsymbol{\psi}$ are $1 \times N$ matrices (column vectors), and **A** is an $N \times N$ matrix. On occasion, the elements of the matrix **A** will be differential operators. Then the scalar product will be a combination of the matrix product and integration, i.e.,

$$(\boldsymbol{\phi},\boldsymbol{\psi}) = \int dw\, \boldsymbol{\phi}^T\boldsymbol{\psi}, \tag{A.18a}$$

$$(\boldsymbol{\phi},\mathbf{A}\boldsymbol{\psi}) = \int dw\, \boldsymbol{\phi}^T\mathbf{A}\boldsymbol{\psi}, \tag{A.18b}$$

etc.

PROBLEMS

A.1. Show $\boldsymbol{\phi}\boldsymbol{\phi}^T = N \times N$ matrix if $\boldsymbol{\phi}$ is a column vector with N elements.

A.2. Calculate eigenvalues and eigenfunctions of some matrix.

A.3. Using Eq. (5.22), which defines the adjoint operator, prove that the adjoint matrix to **A** has elements which obey (A.16).

appendix **B**

The Delta Distribution[1]

Throughout this book, we have considered neutron distributions generated by sources restricted to points, lines, etc. These physical idealizations have been represented by a mathematical idealization, the Dirac delta distribution.[2]

Originally, Dirac proposed a function $\delta(x)$ with the following properties:

$$\delta(x) = 0, \qquad x \neq 0. \tag{B.1a}$$

$$\int_a^b \delta(x)\, dx = 1, \qquad 0\varepsilon(a,b), = 0 \text{ otherwise.} \tag{B.1b}$$

[The notation $0\varepsilon(a,b)$ means that point $x = 0$ is a member of the open set (a,b), that is, $a < 0 < b$]. Equations (B.1) clearly express the physical idealization of a point source located at $x = 0$ [Eq. (B.1a)], with a finite "strength" [Eq. (B.1b)]. [Incidentally, if $0 = a$ or $0 = b$, the integral in (B.1b) is defined as equal to 1/2. See Prob. B.2].

[1] The author is indebted to Dr. Richard Carmichael for help in the preparation of this appendix.

[2] Often mistakenly called the "delta function."

No function with the properties (B.1) exists! However, it *is* possible to define $\delta(x)$ as a linear *functional* on a space of test functions with certain properties.[1] Thus, consider the set of all functions whose derivatives of every order exist (mathematicians call this the space C_∞), defined on the real line (i.e., functions of the variable x). We require further that these functions have *compact support*, i.e., that they vanish identically for $|x|$ greater than some finite value. Such a function space has a name; it is called D.

We now consider linear functionals on this space D, that is, we associate with each function in D, call it ϕ_n, a complex number, call it $T\phi_n$. Linear means, as for operators,[2] that if α_1 and α_2 are complex numbers, then

$$T(\alpha_1\phi_1 + \alpha_2\phi_2) = \alpha_1 T\phi_1 + \alpha_2 T\phi_2. \tag{B.2}$$

The *distributions*, as they are called, then are defined as linear functionals on an appropriate space of test functions, say, D. The delta "function" is really a delta distribution, i.e., a linear functional[3] which associates with a function $\phi_n(x)$ the value $\phi_n(0)$. One writes

$$\delta(x)\phi_n(x) = \phi_n(0). \tag{B.3}$$

From (B.3) we see

$$\delta(x - x_0)\phi_n(x) = \phi_n(x_0). \tag{B.4}$$

The derivative of a distribution is defined, in analogy with the formula for partial integration, as

$$T'\phi_n = -T\phi_n'. \tag{B.5}$$

Another example of a distribution is the Heaviside distribution[4] H, defined by

$$H\phi_n(x) = \int_0^\infty \phi_n(x)\,dx. \tag{B.6}$$

We see, from (B.5), that

$$H'\phi_n(x) = -H\phi_n'(x) \tag{B.7a}$$

$$= -\int_0^\infty \phi_n'(x)\,dx \tag{B.7b}$$

$$= \phi_n(0), \tag{B.7c}$$

[1] See, e.g., John W. Dettman: "Mathematical Methods in Physics and Engineering," 2d ed., McGraw-Hill Book Company, New York, 1969.
[2] See Chapter 5.
[3] Note the difference between a functional and an operator. A functional associates a complex number with a function. An operator associates another function with a function.
[4] Also called the *step function*. That is, $H\phi_n(x) = \int_{-\infty}^{\infty} S(x)\phi_n(x)\,dx$, where $S(x)$ is the usual step function.

since $\phi_n(\infty) = 0$ by the assumption of compact support. Therefore we say

$$H' = \delta. \tag{B.8}$$

The usual hocus-pocus mathematics which physicists and engineers apply to delta and Heaviside distributions can be justified rigorously with the theory outlined above. Thus, we shall frequently write formulas like

$$\int f(x)\, \delta(x - x_0)\, dx = f(x_0), \tag{B.9}$$

understanding that they are symbolic for relations like (B.4).

A point source is represented as a product of delta functionals,

$$\delta(r - r_0) \equiv \delta(x - x_0)\, \delta(y - y_0)\, \delta(z - z_0). \tag{B.10}$$

These three functionals operate in three different spaces (i.e., the x, y, and z axes), and so their product is well defined. (A similar relation for line sources is left as an exercise to the reader.)

In differentiating a discontinuous function, a little care is required. For example, consider $f(x) = |x|$. We see $f'(x) = 1$ with $x > 0$, while $f'(x) = -1$ with $x < 0$. Thus $f''(x)$ must be a delta distribution, that is,

$$f''(x) = C\, \delta(x), \tag{B.11}$$

where C is a constant to be determined. *Integrating* Eq. (B.11) from $-\varepsilon$ to $+\varepsilon$ ($\varepsilon > 0$), we find

$$f'(+\varepsilon) - f'(-\varepsilon) = C, \tag{B.12a}$$

so that

$$C = 2, \tag{B.12b}$$

i.e., the coefficient of the delta distribution is just the amplitude of the "jump."

Finally, we note that a delta distribution has a Fourier representation.[1] We write, symbolically,

$$\delta(x) = \int \bar{f}(k) e^{ikx}\, dk, \tag{B.13a}$$

or applying the Fourier inversion formula,

$$\bar{f}(k) = \frac{1}{2\pi} \int f(x)^{-ikx}\, dk, \tag{B.13b}$$

$$f(k) = \frac{1}{2\pi}. \tag{B.13c}$$

Thus, the Fourier representation of $\delta(x)$ is seen to be

$$\delta(x) = \frac{1}{2\pi} \int e^{ikx}\, dk. \tag{B.14}$$

[1] See Appendix C.

appendix C
Fourier and Laplace Transforms

The method of integral transforms is a powerful tool for solution of certain differential and integral equations. For a more complete discussion, see Dettman.[1]

The Fourier transform $\bar{f}(k)$ of a function $f(x)$ is defined by the relation

$$\bar{f}(k) = \int_{-\infty}^{\infty} f(x)e^{ikx}\,dx. \tag{C.1}$$

The inverse relation is

$$f(x) = \frac{1}{2\pi}\int \bar{f}(k)e^{-ikx}\,dk. \tag{C.2}$$

Also

$$\overline{\frac{\partial f}{\partial x}} = \int_{-\infty}^{\infty} \frac{\partial f}{\partial x}\,e^{ikx}\,dx \tag{C.3a}$$

$$= -ik\bar{f}(k), \tag{C.3b}$$

[1] John W. Dettman: "Mathematical Methods in Physics and Engineering," 2d ed., McGraw-Hill Book Company, New York, 1969.

where the last line has been obtained by partial integration, under the assumption that $f(\pm\infty) = 0$ (as is the case for essentially all functions of physical interest).

The purpose of the transform method is that certain differential (and integral) equations can be converted to algebraic equations and solved trivially. Then, the inversion formula (C.2) can be applied to obtain the solution.

As an example, consider the equation describing neutron diffusion from a plane source in an infinite homogeneous medium,

$$-\frac{\partial^2\phi}{\partial x^2} + \kappa^2\phi = \frac{Q\,\delta(x)}{D}. \tag{C.4}$$

Applying Eq. (C.3) twice, we see that

$$\overline{\frac{\partial^2\phi}{\partial x^2}} = -k^2\bar{\phi}(k). \tag{C.5}$$

Also (see Appendix B),

$$\bar{\delta}(x) = 1. \tag{C.6}$$

Thus, the Fourier transform of (C.4) is

$$k^2\bar{\phi} + \kappa^2\bar{\phi} = \frac{Q}{D}, \tag{C.7a}$$

or

$$\bar{\phi}(k) = \frac{Q}{D}\frac{1}{\kappa^2 + k^2}. \tag{C.7b}$$

Then

$$\phi(x) = \frac{Q}{2\pi D}\int_{-\infty}^{\infty} e^{-ikx}\frac{1}{\kappa^2 + k^2}\,dk. \tag{C.8}$$

The integral in Eq. (C.8) may be evaluated by complex integration techniques.[1] The idea is to extend the range of integration from the real axis (in the complex k plane) to a closed contour which includes the real axis as one segment. This can only be done if the integrand vanishes on the added segments. Consider first the case $x < 0$ ($x > 0$ will be considered separately). Writing

$$k = R(\cos\theta + i\sin\theta), \tag{C.9}$$

we see that

$$e^{ikx} \approx e^{Rx\sin\theta}e^{iRx\cos\theta}. \tag{C.10}$$

[1] See Dettman, *op. cit.*

For $0 < \theta < \pi$ (that is, in quadrants I and II of the complex k plane), $\sin\theta > 0$ and $e^{ikx} \to 0$ as $R \to \infty$ (recall that $x < 0$). Thus we can add to the right side of Eq. (C.8) the expression

$$\frac{1}{2\pi}\int_C e^{-ikx}\frac{1}{\kappa^2 + k^2}\,dk, \tag{C.11}$$

which is identically equal to zero. (The contour is shown in Fig. C.1). Thus

$$\phi(x) = \frac{1}{2\pi}\oint e^{-ikx}\frac{1}{\kappa^2 + k^2}\,dk \tag{C.12a}$$

$$= \frac{1}{2\pi}\oint e^{-ikx}\frac{1}{k - i\kappa}\frac{1}{k + i\kappa}\,dk. \tag{C.12b}$$

The symbol $\oint$ indicates integration about a closed contour, i.e., the real axis plus C.

The residue theorem tells us that the value of the above integral is equal to $2\pi i$ times the residue at the pole(s) inside the contour. The only pole is at $k = +i\kappa$, and the residue is the coefficient of $(k - i\kappa)^{-1}$ evaluated at $k = i\kappa$,

$$\text{Residue} = \frac{Qe^{+\kappa x}}{4\pi i\kappa D}, \tag{C.13}$$

and

$$\phi(x) = \frac{Qe^{\kappa x}}{2\kappa D}, \qquad x < 0. \tag{C.14}$$

For $x > 0$, the calculation is analogous, except that the contour must be closed in the lower half-plane. One easily finds

$$\phi(x) = \frac{Qe^{-\kappa x}}{2\kappa D}, \qquad x > 0. \tag{C.15}$$

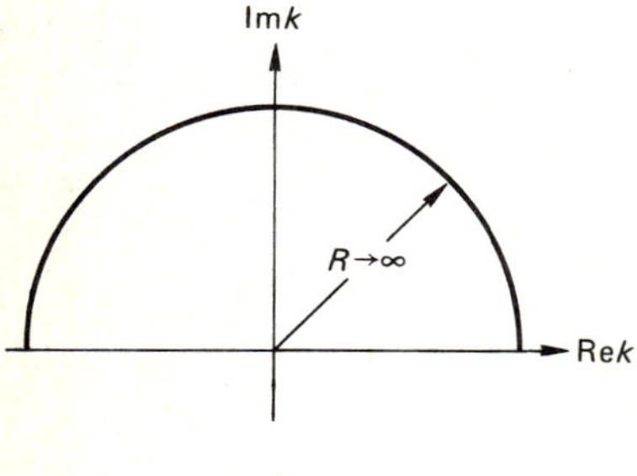

Fig. C.1 The contour C in Eq. (C-11).

Of considerable use is the convolution theorem. A convolution of two functions f and g is defined as

$$f * g = \int_{-\infty}^{\infty} f(x')g(x - x')\, dx'. \tag{C.16}$$

Then

$$\overline{f * g} = \int_{-\infty}^{\infty} e^{ikx} \int_{-\infty}^{\infty} f(x')g(x' - x)\, dx'\, dx \tag{C.17a}$$

$$= \int_{-\infty}^{\infty} dx \int_{-\infty}^{\infty} dx'\, e^{ik(x-x')}e^{ikx'}f(x')g(x - x'). \tag{C.17b}$$

If we change variables from x, x' to $x', y = x - x'$, we find

$$\overline{f * g} = \int_{-\infty}^{\infty} e^{iky}g(y)\, dy \int_{-\infty}^{\infty} e^{ikx'}f(x')\, dx', \tag{C.18a}$$

$$= \bar{f}(k)\bar{g}(k). \tag{C.18b}$$

This is the convolution theorem for Fourier transforms: The transform of a convolution is the product of the transforms.

Fourier transforms can also be applied in more than one dimension. For three dimensions, for example, the transform is defined by

$$\bar{f}(\mathbf{k}) = \int d^3r\, e^{i\mathbf{k}\cdot\mathbf{r}}f(\mathbf{r}), \tag{C.19a}$$

and the inversion formula is

$$\hat{f}(\mathbf{r}) = \left(\frac{1}{2\pi}\right)^3 \int d^3k\, e^{-i\mathbf{k}\cdot\mathbf{r}}\bar{f}(\mathbf{k}). \tag{C.19b}$$

This actually can be thought of as the product of three Fourier transforms with transform variables k_x, k_y, and k_z. However, sometimes it is convenient to work in k polar coordinates. Thus, the equation

$$-\nabla^2\phi + \kappa^2\phi = \frac{Q}{D}\delta(\mathbf{r}), \tag{C.20}$$

i.e., the diffusion equation in three dimensions, can be solved by multiplying by $e^{i\mathbf{k}\cdot\mathbf{r}}$. and integrating. Note from partial integration that[1]

$$\int e^{i\mathbf{k}\cdot\mathbf{r}}\nabla^2\phi\, d^3r = -k^2 \int e^{i\mathbf{k}\cdot\mathbf{r}}\phi(\mathbf{r})\, d^3r \tag{C.21a}$$

$$= -k^2\bar{\phi}(\mathbf{k}), \tag{C.21b}$$

[1] To prove this result most simply, use cartesian coordinates.

[since $\phi(\pm\infty)=0$], and so the transformed equation becomes

$$(\kappa^2 + k^2)\bar{\phi}(\mathbf{k}) = \frac{Q}{D}, \tag{C.22a}$$

or

$$\bar{\phi}(\mathbf{k}) = \frac{Q}{D}\frac{1}{\kappa^2 + k^2}. \tag{C.22b}$$

Comparing with Eq. (C.7*b*), we see that the Fourier transforms of the one- and three-dimensional fluxes are identical! However, the inversion formula gives

$$\phi(\mathbf{r}) = \frac{Q}{(2\pi)^3}\int d^3k \frac{e^{-i\mathbf{k}\cdot\mathbf{r}}}{\kappa^2 + k^2}. \tag{C.23}$$

The integration is easily carried out in polar coordinates. Writing

$$d^3k = 2\pi k^2\, dk\, d(\cos\theta) \tag{C.24a}$$

and

$$\mathbf{k}\cdot\mathbf{r} = kr\cos\theta, \tag{C.24b}$$

we find

$$\phi(\mathbf{r}) = \frac{Q}{4\pi^2 D}\int_{-\infty}^{\infty}\frac{k^2\, dk}{\kappa^2 + k^2}\int_{-1}^{1} d(\cos\theta)\, e^{-ikr\cos\theta} \tag{C.25a}$$

$$= \frac{Q}{2\pi^2 Dr}\int_{-0}^{\infty}\frac{k\sin kr}{\kappa^2 + k^2}dk \tag{C.25b}$$

$$= \frac{Q}{\pi^2 Dr}\int_{-\infty}^{\infty}\frac{k\sin kr}{\kappa^2 + k^2}. \tag{C.25c}$$

This integral can also be evaluated by residue techniques after expressing $\sin kr = (1/2i)(e^{ikr} - e^{-ikr})$. The details will be left as an exercise.

Related to the Fourier transforms are the Laplace transforms. We define the Laplace transform $\hat{f}(s)$ of a function $f(t)$ by

$$\hat{f}(s) = \int_0^{\infty} e^{-st} f(t)\, dt. \tag{C.26}$$

Because the range of integration is from 0 to ∞ rather than from $-\infty$ to $+\infty$ as in Fourier transforms, Laplace transforms are generally applied to initial-value problems. Consider then the simple problem

$$\frac{df(t)}{dt} = \lambda f(t), \tag{C.27a}$$

with the initial condition

$$f(0) = 1. \tag{C.27b}$$

We obtain the Laplace transform of (C.27) by multiplying by e^{-st} and integrating from 0 to ∞. Observe that

$$\int_0^\infty e^{-st} \frac{df}{dt}\, dt = f(t)e^{-st}\Big|_0^\infty + s\int_0^\infty e^{-st} f(t)\, dt, \tag{C.28}$$

and so the transformed equation becomes

$$\hat{f}(s) - 1 = \lambda \hat{f}(s), \tag{C.29}$$

where we have used Eq. (C.27*b*). Thus

$$\hat{f}(s) = \frac{1}{s - \lambda}. \tag{C.30}$$

The inversion formula, i.e., the rule for obtaining $f(t)$ from $\hat{f}(s)$, is

$$f(t) = \frac{1}{2\pi i}\int_{\gamma - i\infty}^{\gamma + i\infty} e^{st}\hat{f}(s)\, ds, \tag{C.31}$$

where the line of integration is a line in the complex s plane parallel to the imaginary axis but to the right of all singularities of the integrand. In most cases (and all the cases dealt with in this book) the contour can be closed with an infinite semicircle to the left of all singularities, so that by applying the residue theorem,

$$f(t) = \sum_{\text{residues}} e^{st}\hat{f}(s). \tag{C.32}$$

In our particular problem, $\hat{f}(s) = 1/(s - \lambda)$; there is a single pole at $s = \lambda$, and the residue is $e^{\lambda t}$. Thus

$$f(t) = e^{\lambda t} \tag{C.33}$$

is the solution of Eq. (C.20).

There is also a convolution theorem for Laplace transforms. It is left as an exercise to the reader to verify that if the convolution is defined by

$$f * g = \int_0^t f(t')g(t - t')\, dt', \tag{C.34a}$$

then

$$\widehat{f * g} = \hat{f}(s)\hat{g}(s). \tag{C.34b}$$

appendix D

Special Functions of Reactor Physics

In solving the diffusion equation in various systems of coordinates, certain special functions are encountered. Some of the more important of these are described here.

D.1 SPHERICAL COORDINATES

The Helmholtz equation $\nabla^2\psi + B^2\psi$ becomes, in spherical coordinates,

$$\left[\frac{1}{r}\frac{\partial^2}{\partial r^2} + \frac{2}{r}\frac{\partial}{\partial r} + \frac{1}{r^2 \sin\theta}\frac{\partial}{\partial\theta}\left(\sin\theta\frac{\partial}{\partial\theta}\right) + \frac{1}{r^2\sin^2\theta}\frac{\partial^2}{\partial\phi^2}\right]\psi + B^2\psi = 0. \tag{D.1}$$

Separation of variables leads to an angular equation and a radial equation. The angular equation is

$$\frac{1}{\sin\theta}\frac{\partial}{\partial\theta}\left(\sin\frac{\partial Y}{\partial\theta}\right) + \frac{1}{\sin^2\theta}\frac{\partial^2 Y}{\partial\phi^2} + \lambda^2 Y = 0, \tag{D.2}$$

where λ^2 is the separation constant.

Eigensolutions of (D.2) exist only for

$$\lambda^2 = l(l+1), \qquad l = 0, 1, 2, \ldots . \tag{D.3}$$

For a given value of l, there are $2l + 1$ linearly independent solutions of (D.2), which are called *spherical harmonics of order l* and are denoted by Y_{lm}, $m = -l, -l+1, \ldots, +l$. The normalized Y_{lm} are given by

$$Y_{lm}(\theta,\phi) = (-1)^m \left[\frac{(2l+1)}{4\pi}\frac{(l-m)!}{(l+m)!}\right]^{1/2} P_{lm}(\cos\theta)\, e^{im\phi}, \quad m > 0, \tag{D.4a}$$

$$Y_{l,-m} = (-1)^m Y_{lm}^*. \tag{D.4b}$$

Here, the P_{lm} are called *associated Legendre functions*,

$$P_{lm}(x) = (1-x^2)^{1/2m} \frac{d^m}{dx^m} P_l(x), \tag{D.5}$$

where, finally, $P_l(x)$ is a Legendre polynomial of order l,

$$P_l(x) = \frac{1}{2^l l!} \frac{dl}{dx^l} (x^2 - 1)^l. \tag{D.6}$$

The Legendre polynomials are themselves of considerable interest in reactor physics. They are solutions of the Legendre differential equation

$$\left[(1-x^2)\frac{d^2}{dx^2} - 2x\frac{d}{dx} + l(l+1)\right] P_l(x) = 0. \tag{D.7}$$

$P_l(x)$ is a polynomial of degree l. Furthermore it contains only odd or even powers of x according as l is even or odd. Also P_l obeys the recursion relation

$$(2l+1)xP_l(x) = (l+1)P_{l+1}(x) + lP_{l-1}(x). \tag{D.8}$$

From this, all other P_l may be found from

$$P_0(x) = 1, \tag{D.9a}$$

$$P_1(x) = x. \tag{D.9b}$$

For example,

$$P_2(x) = \tfrac{1}{2}(3x^2 - 1), \tag{D.10}$$

and so forth. The $P_l(x)$ are orthogonal,

$$\int_{-1}^{1} P_l(x) P_{l'}(x)\, dx = \frac{2}{2l+1} \delta_{ll'}. \tag{D.11}$$

Also,

$$P_l(1) = 1, \tag{D.12a}$$

$$P_l(-1) = (-1)^l. \tag{D.12b}$$

Returning to the spherical harmonics, we have the orthogonality relation

$$\int Y^*_{lm}(\theta,\phi)\, Y'_{l'm'}(\theta,\phi)\, d\mathbf{\Omega} = \delta_{ll'}\, \delta_{mm'}, \tag{D.13a}$$

where

$$d\mathbf{\Omega} = \sin\theta\, d\theta\, d\phi. \tag{D.13b}$$

A few spherical harmonics are given below:

$$Y_{00} = \frac{1}{4\pi}, \qquad Y_{10} = \frac{3}{4\pi}\cos\theta, \qquad Y_{1,\pm1} = \mp\sqrt{\frac{3}{8\pi}}\sin\theta e^{\pm i\phi}, \tag{D.14}$$

and so forth. Finally we note

$$Y_{l0}(\theta,\phi) = \sqrt{\frac{2l+1}{4\pi}}\, P_l\,(\cos\theta) \tag{D.15}$$

and the addition theorem

$$P(\hat{a}\cdot\hat{b}) = \frac{4\pi}{2l+1}\sum_{m=-l}^{l} Y^*_{lm}(\hat{a})Y_{lm}(\hat{b}), \tag{D.16}$$

where $\hat{a}$ and $\hat{b}$ are unit vectors, and $Y_{lm}(\hat{a})$ means $Y_{lm}(\theta_a,\phi_a)$, the terms θ_a and ϕ_a being the spherical polar angles of the vector $\mathbf{\hat{a}}$.

Returning to (D.1), the radial equation is

$$\left[\frac{1}{r}\frac{d^2}{dr^2} + \frac{2}{r}\frac{d}{dr} + B^2 - \frac{l(l+1)}{r^2}\right]R(r) = 0. \tag{D.17}$$

This equation, called the *spherical Bessel equation*, has two linearly independent solutions for every l. These are denoted by $j_l(\rho)$ and $n_l(\rho)$, were $\rho \equiv Br$. Near $\rho = 0$, we have

$$\underset{\rho\to 0}{L}\; j_l(\rho) = \frac{\rho^l}{(2l+1)!!} \tag{D.18a}$$

and

$$\underset{\rho\to 0}{L}\; n_l(\rho) = (2l+1)!!\rho^{l-1}, \tag{D.18b}$$

where the notation $(2l+1)!!$ means $1\cdot3\cdot5\cdot7\cdots(2l+1)$, that is, the product of the odd numbers up to $2l+1$. Thus, we see that n_l misbehaves at $\rho = 0$. At infinity, both n_l and j_l approach zero.

The n_l and j_l can be expressed as the ordinary Bessel functions, which we discuss in the next section,

$$j_l(\rho) = \left(\frac{\pi}{2\rho}\right)^{1/2} J_{l+1/2}(\rho),$$

$$n_l(\rho) = (-1)^l \left(\frac{\pi}{2\rho}\right)^{1/2} J_{-l-1/2}(\rho). \tag{D.19}$$

A few of the j_l and n_l are

$$j_0 = \frac{\sin\rho}{\rho}, \qquad n_0 = \frac{\cos\rho}{\rho},$$

$$j_1 = \frac{\sin\rho}{\rho^2} - \frac{\cos\rho}{\rho}, \qquad n_1 = \frac{\cos\rho}{\rho^2} + \frac{\sin\rho}{\rho}. \tag{D.20}$$

D.2 CYLINDRICAL COORDINATES

In cylindrical coordinates, the Helmholtz equation is

$$\left(\frac{\partial^2}{\partial\rho^2} + \frac{1}{\rho}\frac{\partial}{\partial\rho} + \frac{\partial^2}{\partial z^2} + \frac{1}{\rho^2}\frac{\partial^2}{\partial\phi^2}\right)\psi + B^2\psi = 0. \tag{D.21}$$

Separation of variables leads to a radial equation known as the *cylindrical Bessel equation* or, more informally, *Bessel's equation*:

$$\left(\frac{d^2}{d\rho'^2} + \frac{1}{\rho'}\frac{d}{d\rho'} + 1 - \frac{n^2}{\rho'^2}\right)R(\rho') = 0, \tag{D.22}$$

where $\rho' = B_\rho \rho$ (B_ρ is the radial buckling).

The two linearly independent solutions of this equation are called $J_n(\rho')$ and $Y_n(\rho')$. There are extensive tables of the J_n and Y_n functions. The following expansions for J_n and Y_n are useful in determining the behavior of the J_n's and Y_n's at the origin:

$$J_n(\rho) = \left(\frac{\rho}{2}\right)^n \sum_{m=0}^{\infty} \frac{(-1)^m}{m!(n+m)!}\left(\frac{\rho}{2}\right)^{2m}, \tag{D.23a}$$

$$\underset{\rho\to 0}{L}\, Y_n(\rho) = \begin{cases} \dfrac{2}{\pi} J_0(\rho) \log\dfrac{\rho}{2}, & n = 0 \\[2ex] \dfrac{(n-1)!}{\pi}\left(\dfrac{\rho}{2}\right)^{-n}, & n > 0. \end{cases} \tag{D.23b}$$

We see that J_n is well behaved at the origin and Y_n is singular. The derivatives of the J_n's can be determined from these expansions as

$$\frac{dJ_n(\rho)}{d\rho} = \begin{cases} \frac{1}{2} J_{n-1}(\rho) - J_{n+1}(\rho), & n > 0, \quad \text{(D.24}a\text{)} \\ -J_1(\rho), & n = 0. \quad \text{(D.24}b\text{)} \end{cases}$$

In several applications we need the solution of the following equation:

$$\left(\frac{d^2}{d\rho^2} + \frac{1}{\rho}\frac{d}{d\rho} - 1 + \frac{n^2}{\rho^2}\right)R(\rho) = 0. \quad \text{(D.25)}$$

The solution of this equation is an arbitrary linear combination of the J_n's and Y_n's of imaginary argument, namely,

$$I_n(\rho) = i^{-n} J_n(i\rho) \quad \text{(D.26}a\text{)}$$

and

$$K_n(\rho) = \underset{v \to n}{L} \frac{\pi}{2} \frac{I_{-v}(\rho) - I_v(\rho)}{\sin v\pi}. \quad \text{(D.26}b\text{)}$$

The expansions for these functions at $\rho \approx 0$ are

$$\underset{\rho \to 0}{L} K_0(\rho) = -\log(\rho/2) \quad \text{(D.27}a\text{)}$$

$$\underset{\rho \to 0}{L} I_0(\rho) = 1. \quad \text{(D.27}b\text{)}$$

The functions at ∞ are

$$\underset{\rho \to \infty}{L} K_0(\rho) = \frac{\pi}{2\rho} e^{-\rho}, \quad \text{(D.28}a\text{)}$$

$$\underset{\rho \to \infty}{L} I_0(\rho) = \frac{e^\rho}{2\pi\rho}. \quad \text{(D.28}b\text{)}$$

Now from (D.20) and the connections between the J's and the K's we see that

$$\frac{d}{d\rho} K_0(\rho) = -K_1(\rho). \quad \text{(D.29)}$$

appendix E
Dynamics of Elastic Collisions

It is helpful to describe the collision of a neutron with a nucleus of mass number A in both the center-of-momentum (CMCS) and laboratory coordinate (LCS) systems. In particular, the relations among the scattering angle in the laboratory system, $\cos^{-1}\mu_0$, the scattering angle in the center-of-momentum system, $\cos^{-1}\mu_c$, and the energy transfer in the collision are to be found. The laboratory system LCS is, as the name implies, that in which measurement can be made, i.e., the system in which the nucleus may be assumed at rest while the neutron approaches it with a velocity $v_0 = (2E)^{1/2}$, if the neutron mass is taken to be unity (see Fig. E.1). After the scattering event, the neutron goes off at an angle $\cos^{-1}\mu_0 = \theta_0$ (the scattering angle in LCS) with speed u_0, after having imparted a speed U_0 to the nucleus, which goes off at angle ϕ.

In the CMCS, on the other hand, the center of mass of nucleus-neutron system is assumed to be at rest (see Fig. E.2). Before the collision, the neutron is moving to the right with a speed $v_0 - V_M$, where V_M is the speed of the

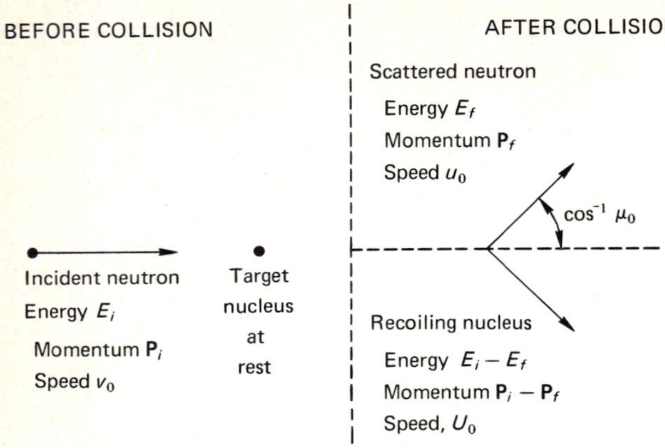

Fig. E.1 Elastic scattering event in the laboratory coordinate system.

center-of-momentum of the neutron-nuclear system relative to the laboratory system,

$$V_M = \frac{v_0}{A+1}. \tag{E.1}$$

The neutron speed in the CMCS is thus

$$v_0 - V_M = \frac{Av_0}{A+1} = v_c. \tag{E.2}$$

The nucleus moves to the left with a speed V_M. In the CMCS, the neutron and nucleus appear to scatter each other since they are both moving before the collision. After the collision the neutron has speed u_c, and the nucleus speed U_c. The scattering angle in this coordinate system is $\cos^{-1}\mu_c = \theta_c$; the neutron and nucleus come off in opposite directions since the total momentum before the collision in CMCS is

$$(v_0 - V_M) - AV_M = v_0\left[\left(1 - \frac{1}{A+1}\right) - \frac{A}{A+1}\right] = 0, \tag{E.3}$$

and must, because of momentum conservation, remain zero after the collision. Clearly, this is possible only if the neutron and nucleus are moving in opposite directions after the collision, and furthermore only if

$$u_c = AU_c. \tag{E.4}$$

Energy conservation in CMCS provides another relation between u_c and U_c:

$$\frac{1}{2}A\left(\frac{v_0}{A+1}\right)^2 + \frac{1}{2}\left(\frac{Av_0}{A+1}\right)^2 = \frac{1}{2}u_c^2 + \frac{1}{2}AU_c^2. \tag{E.5}$$

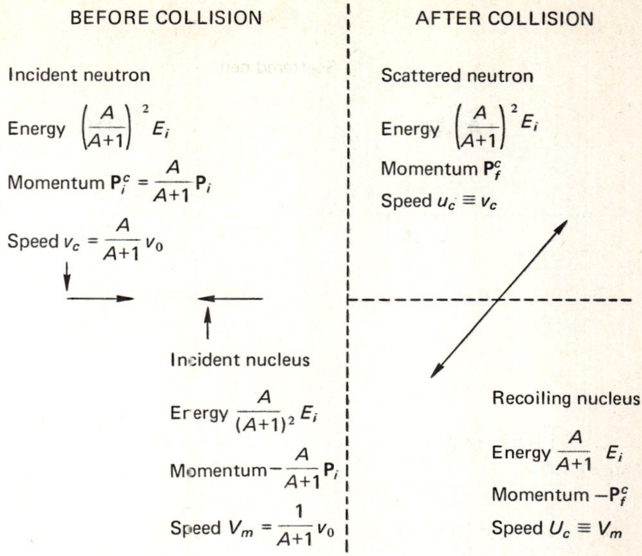

Fig. E.2 Elastic scattering event in the center-of-momentum coordinate system.

Upon solving the last two equations for u_c and U_c in terms of v_0, one finds that in the CMCS the speeds of neutron and nucleus are unchanged by an elastic collision; that is,

$$v_0 - V_M = u_c = \frac{Av_0}{A+1} \tag{E.6}$$

and

$$V_M = U_c = \frac{v_0}{A+1}. \tag{E.7}$$

We are now in a position to derive a relation between μ_0 and μ_c. Noting that the velocity of the neutron in LCS is the vector sum of its velocity in C and the velocity of the center-of-momentum in LCS, one can write (see Fig. E.3)

$$u_0 \mu_0 = V_M + u_c \mu_c. \tag{E.8}$$

Substituting from Eqs. (E.6) and (E.7) for u_c and V_M,

$$\left(\frac{u_0}{v_0}\right)\mu_0 = \frac{1}{A+1} + \frac{A}{A+1}\mu_c. \tag{E.9}$$

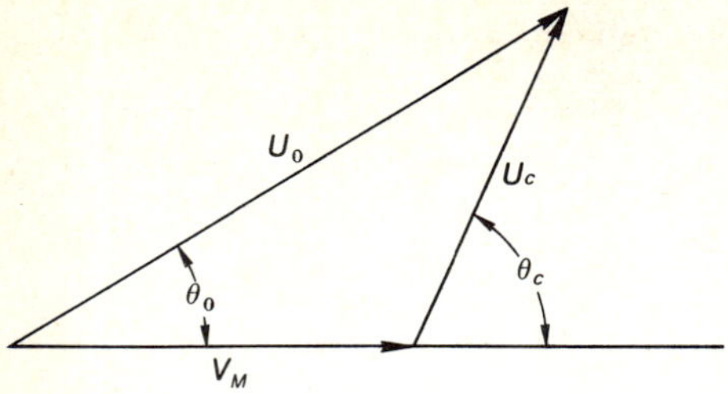

Fig. E.3 Relation between laboratory and center-of-momentum velocities.

The ratio u_0/v_0 may also be obtained by applying the law of cosines to Fig. E.3,

$$u_0{}^2 = u_c{}^2 + V_M{}^2 + 2u_c V_M \mu_c, \tag{E.10}$$

or by using Eqs. (E.6) and (E.7) again,

$$\frac{u_0{}^2}{v_0{}^2} = \frac{A^2 + 2A\mu_c + 1}{(A + 1)^2}. \tag{E.11}$$

Equations (E.9) and (E.11) yield the desired relation between μ_0 and μ_c:

$$\mu_0 = \frac{A\mu_c + 1}{(A^2 + 2A\mu_c + 1)^{1/2}}, \tag{E.12a}$$

or in terms of $\gamma = A^{-1}$,

$$\mu_0 = \frac{\mu_c + \gamma}{(\gamma^2 + 2\gamma\mu_c + 1)^{1/2}}. \tag{E.12b}$$

Next, consider the energy change in scattering. In LCS, the ratio of the energy after collision, E_f, to that before the collision, E_i, is just $u_0{}^2/v_0{}^2$, already given in Eq. (E.11):

$$\frac{E_f}{E_i} = \frac{\gamma^2 + 2v\mu_c + 1}{(\gamma + 1)^2}, \tag{E.13}$$

or in terms of the parameter α,

$$\alpha = \left(\frac{1 - \gamma}{1 + \gamma}\right)^2 = \left(\frac{A - 1}{A + 1}\right)^2, \tag{E.14}$$

the ratio of the energies is

$$\frac{E_f}{E_i} = \tfrac{1}{2}[(1 + \alpha) + (1 - \alpha)\mu_c]. \tag{E.15}$$

Note that Eq. (E.15) implies $\alpha E' \leq E \leq E'$, since $-1 \leq \mu_c \leq 1$. Thus, a neutron cannot be speeded up and cannot lose more than a fraction $(1 - \alpha)$ of its energy in a single elastic collision.

In terms of the lethargy u, which is, we recall, defined as

$$u = \ln (E_0/E), \qquad E = E_0 e^{-u}, \tag{E.16}$$

(E_0 is some convenient reference energy, generally chosen to be 10 MeV), the above results become somewhat simpler. For example, Eq. (E.15) becomes

$$e^{-\Delta u} = \tfrac{1}{2}[(1 + \alpha) + (1 - \alpha)\mu_c], \tag{E.17a}$$

where

$$\Delta u = u_f - u_i \tag{E.17b}$$

is the lethargy gained by the neutron in the collision. Inverting and writing the result in terms of γ, we have

$$\mu_c(\Delta u) = 1 - \frac{(1 + \gamma)^2}{2\gamma}(1 - e^{-\Delta u}). \tag{E.18}$$

Combining this with Eq. (E.16), one finds

$$\mu_0(\Delta u) = \frac{1}{2\gamma}[(1 + \gamma)e^{-\Delta u/2} - (1 - \gamma)e^{\Delta u/2}]. \tag{E.19}$$

Here, μ_0 and μ_c have been written explicitly as functions of the lethargy change Δu to call attention to the fact that the angle of scattering is uniquely related to the lethargy change (gain) in a collision. For some purposes, later in the development it will be convenient to consider μ_0 as the independent variable; at other times, the choice of Δu will be more convenient. The maximum and minimum possible neutron-energy transfers place restrictions on minimum and maximum lethargy change,

$$0 \le \Delta u \le \ln\frac{1}{\alpha} \tag{E.20a}$$

or

$$u_i \le u_f \le u_0 + \ln\frac{1}{\alpha}. \tag{E.20b}$$

appendix F
The Telegrapher's Equation

Here we shall eliminate **J** between Eqs. (2.47) and (2.32) *without* using the approximation that $(1/u)(\partial \mathbf{J}/\partial t)$ is zero in (2.47). For simplicity we shall assume that the cross sections are not dependent upon position or time.

If we take the divergence of (2.47), we obtain

$$\frac{3D}{u}\frac{\partial}{\partial t}\nabla \cdot \mathbf{J} + \nabla \cdot \mathbf{J} = -D\,\nabla^2 \phi, \tag{F.1}$$

where we have used the definition of the diffusion constant (2.49*b*). Now we can take the time derivative of (2.32) to obtain

$$\frac{1}{u}\frac{\partial^2 \phi}{\partial t^2} + \frac{\partial}{\partial t}\nabla \cdot \mathbf{J} + \Sigma_a \frac{\partial \phi}{\partial t} = \frac{\partial S}{\partial t}. \tag{F.2}$$

We see that (F.2) gives us a relation between $(\partial/\partial t)\nabla \cdot \mathbf{J}$ and ϕ and S. Equation (2.32) gives a relation between $\nabla \cdot \mathbf{J}$ and ϕ and S,

$$\nabla \cdot \mathbf{J} = S - \frac{1}{u}\frac{\partial \phi}{\partial t} - \Sigma_a \phi. \tag{F.3}$$

We can use (F.2) and (F.3) in (F.1) to obtain the telegrapher's equation

$$\frac{3D}{u^2}\frac{\partial^2\phi}{\partial t^2} + \frac{1 + 3D\Sigma_a}{u}\frac{\partial\phi}{\partial t} + \Sigma_a\phi - D\,\nabla^2\phi = \frac{3D}{u}\frac{\partial S}{\partial t} + S. \tag{F.4}$$

We see, that to obtain the diffusion equation, we must have

$$\frac{1}{u}\frac{\partial^2\phi}{\partial t^2} + \Sigma_a\frac{\partial\phi}{\partial t} - \frac{\partial S}{\partial t} = 0. \tag{F.5}$$

This condition must be met as well as the conditions implied by the assumption that the flux depends linearly on $\mathbf{\Omega}$, which was used to obtain (2.47). The conditions implied by the latter assumption may be consistent with the assumption that one of the terms in (F.5) is small. For example, an almost isotropic flux distribution implies that Σ_a is small, which would seem to imply that the second term on the left-hand side of (F.5) is small. However, we see that we must also assume that the flux and the source are not strongly time dependent.

Bibliography

GENERAL

1. Samuel Glasstone and Milton C. Edlund: "The Elements of Nuclear Reactor Theory," Van Nostrand Reinhold Company, New York, 1952. This was one of the first textbooks in reactor physics, and in many ways is still the best. It suffers, of course, from being twenty years out of date. However its clarity of exposition and the physical insight that it provides still make it a useful addition to any student's library.
2. Alvin M. Weinberg and Eugene P. Wigner: "The Physical Theory of Neutron Chain Reactors," The University of Chicago Press, Chicago, 1958. This is, of course, the *magnum opus* of reactor physics, and the only textbook in the field written, or likely to be written, by a Nobel laureate. The first chapter, all 139 pages worth, is a beautiful introduction to nuclear physics, and every student of reactor physics would do well to study it. The rest of the book is of more historical interest than current practical interest, since it contains very little information on modern calculational techniques. Still, as a basic reference for the advanced student, it stands alone.
3. Robert V. Meghreblian and David K. Holmes: "Reactor Analyses," McGraw-Hill Book Company, New York, 1960. This book is mainly an exposition of the methods taught at ORSORT (Oak Ridge School of Reactor Technology) during the 1950s. While it contains a wealth of useful information, it is not oriented toward the computer and, in fact, stresses a great many long-since forgotten analytical techniques.

Like items 1 and 2 above, it is essentially oriented towards thermal reactors, particularly graphite piles. Strictly for the advanced student.

4. George I. Bell and Samuel Glasstone: "Nuclear Reactor Theory," Van Nostrand Reinhold Company, New York, 1970. This recent, and excellent, book details all that the advanced student is likely to know, or want to know, about reactor theory. It includes many modern topics not found in most other texts, such as reactor noise, xenon oscillations, methods for dealing with gas-cooled reactors, and many others.
5. Raymond L. Murray: "Nuclear Reactor Physics," Prentice-Hall, Inc., Englewood Cliffs, N.J., 1957. Students have always loved this little book for its clarity and wealth of examples. Quite elementary, quite out of date, and quite nice.
6. John R. Lamarsh: "Introduction to Nuclear Reactor Theory," Addison-Wesley Publishing Company, Inc., Reading, Mass., 1966. By far the most popular text for introductory reactor physics courses today. In fact, this book is our principal competitor, so what are we supposed to say about it?

For a review of mathematics at a level most appropriate to this text, it is hard to beat:

7. John W. Dettman: "Mathematical Methods in Physics and Engineering," 2d ed., McGraw-Hill Book Company, New York, 1969.

Nuclear and modern physics topics are well covered by:

8. Irving Kaplan: "Nuclear Physics," Addison-Wesley Press, Inc., Reading, Mass., 1955.

In addition, chap. 1 of Ref. 2 has already been cited.

CHAPTER 1

The nuclear data cited in this chapter were taken from Ref. 2 and from:

9. Argonne National Laboratory, Reactor Physics Constants, *USAEC Rep. ANL* 5800, 2d ed., 1963. This compendium of nuclear data, reactor data, calculational methods, etc., represented a pretty good exposition of the state of the art as of 1963. Like most multiply-authored handbooks, however, its quality is far from even, and it is a little difficult to dig out just what one wants. It is worth owning, though, for anyone working in the field.
10. Brookhaven National Laboratory, *USAEC Rep. BNL* 325 *and Suppls.* The place to go to fill all your cross-section needs.

For a more extensive discussion of breeding, see pages 124–130 of Ref. 2. A beautiful discussion of the physical description of the chain reactor (Sec. 1.3) is found in chap. 2 of Ref. 2. (We seem to keep coming back to Ref. 2!)

CHAPTER 2

11. Joel H. Ferziger and P. F. Zweifel: "The Theory of Neutron Slowing Down in Nuclear Reactors," Pergamon Press, New York, and The M.I.T. Press, Cambridge, Mass., 1966. In this text the neutron transport equation is derived, and the diffusion approximation is obtained. The discussion parallels that given in the present text.

The transport equation is also discussed extensively in:

12. Kenneth M. Case and P. F. Zweifel: "Linear Transport Theory," Addison-Wesley Publishing Company, Inc.
and
13. B. Davison: "Neutron Transport Theory," Oxford University Press, London, 1951.

Analytical solutions of the diffusion equation are obtained, *ad nauseam*, in all reactor physics books. For especially nice treatments, see Refs. 1 and 5. Actually, the problem of solving the diffusion equation is just a special case of the more general problem of *separation of variables* in the solution of boundary value problems. See chap. 4 of Ref. 7.

A novel application of transport theory to the control of reactors has been given by:

14. C. Lingus: "Neutron Transport in Magnetic Materials," *Rep. ORO* 3858-1, p. 669, 1969.

Diffusion length measurements are discussed in more detail in chap. 4 of:

15. M. M. R. Williams: "The Slowing Down and Thermalization of Neutrons," North-Holland Publishing Co., Amsterdam, 1966.

Also see Ref. 4 and, particularly, chap. 7 of Ref. 3.

CHAPTER 3

The material in this chapter is standard. Similar treatments can be found in many texts on reactor physics. An exception is Sec. 3.8 on integral equations. This approach is discussed in chap. 12 of Ref. 1 and chap. 1 of Ref. 11.

CHAPTER 4

The standard reference in the field of reactor kinetics is:

16. G. R. Keepin: "Physics of Nuclear Kinetics," Addison-Wesley Publishing Company, Inc., Reading, Mass., 1965. An advanced text.

A somewhat more modern book, stressing stability and control, is:

17. Lynn E. Weaver: "Reactor Dynamics and Control," American Elsevier Publishing Company, Inc., New York, 1960. This book is particularly nice because it uses the notation of linear operators introduced in Chap. 5 of this book. An advanced text.

A fairly general description of the pulsed-neutron experiment may be found in Ref. 15, chap. 5.

A book which is still in press as this bibliography is written, has such eminent authors that its appearance (June, 1971) is eagerly awaited. It is:

18. A. Ziya Akcasu, Gerald S. Lellouche, and Lewis M. Shotkin: "Mathematical Methods in Nuclear Reactor Dynamics," Academic Press, Inc., New York.

Another excellent reference is:

19. Milton Ash: "Nuclear Reactor Kinetics," McGraw-Hill Book Company, New York, 1965. The treatment of specific reactor types (chap. 9) is particularly nice.

CHAPTER 5

The theory of linear operators is well described by Dettman, Ref. 7. The sophisticated student might enjoy reading:

20. Paul R. Halmos: "A Hilbert Space Problem Book," Van Nostrand Reinhold Company, New York, 1967. This book proves that mathematics can be fun, but it is definitely only for the advanced student.

Linear operator theory is called *functional analyses* by many mathematicians. An example of the use of functional analyses in reactor theory has been given by:

21. Basil Nicolaenko: *J. Math. Phys.*, **11**, 174 (1970).

In recent years, variational methods have become very popular in reactor physics. The following references (22–26) have been prepared by Dr. A. G. Bullard:

22. M. Becker: "The Principles and Applications of Variational Methods," The M.I.T. Press, Cambridge, Mass., 1964. A generalized view of variational methods. An interesting application to fuel depletion in nuclear reactors is made.

23. W. M. Stacey: "Modal Approximations." The M.I.T. Press, Cambridge, Mass., 1964. The relation of variational methods to orthogonal expansions. Applications to fast-reactor physical calculations.

24. G. C. Pomraning: *J. Nucl. Energy*, **20**, 617 (1966). The Roussopoulos variational formalism is applied to equations with a first-order time derivative.

25. E. L. Wachspress: "Variational Methods and Neutron Flux Synthesis." *Proc. Conf. Effective Use Computers Nucl. Ind.*, April, 1969. Review of recent developments in variational synthesis methods.

26. D. S. Selengut: "Variational Analysis of a Multidimensional System", *Handford Lab., HW* 59126, 1959. Classic study of the application of the Roussopoulos formalism to reactor physics and transport theory problems.

CHAPTER 6

The most modern textbook on the subject of numerical solution of the equations of reactor physics is:

27. E. Wachspress: "Iterative Solution of Elliptic Systems," Prentice-Hall, Inc., Englewood Cliffs, N.J., 1966.

An older and less advanced text which is highly readable and perhaps more appropriate to the beginning graduate student is:

28. M. Clark and K. F. Hansen: "Numerical Methods of Reactor Analysis," Academic Press, Inc., New York, 1969.

An excellent advanced text, more general in its scope than either of the above references, is:

29. Richard S. Varga: "Matrix Iterative Analysis," Prentice-Hall, Inc., Englewood Cliffs, N.J., 1962.

In addition, an excellent treatment is found in chap. 3 of Ref. 4.

A compendium of test cases with which numerical comparisons have been made has recently been published by:

30. C. Lingus: "Analytical Test Cases for Neutron and Radiation Transport Codes," *USAEC Rep. Conf.* 710107, 1971.

CHAPTER 7

The general physics of reactor control can be found in a number of the earlier bibliography items (e.g., Refs. 1–6, 9, and 17).

In addition, a goodly number of reactor *engineering* texts have nice discussions of control problems. An elementary treatment is:

31. Arthur R. Foster and Robert L. Wright, Jr.: "Basic Nuclear Engineering," Allyn and Bacon, Inc., 1968.

Also:

32. Samuel Glasstone: "Principles of Nuclear Reactor Engineering," Van Nostrand Reinhold Company, New York, 1955.

As is mentioned in the text, first collision probabilities are tabulated extensively by:

33. K. M. Case, F. de Hoffman, and G. Placzek: "Introduction to the Theory of Neutron Diffusion," U.S. Government Printing Office, Washington, D.C., 1953.

CHAPTER 8

The basic references in reactor physics, Refs. 1–6, especially Ref. 4, have good discussions of multi-group theory. See particularly Ref. 11, chap. 4. The reader may be interested in referring to the classic original paper on the subject. It is:

34. R. Ehrlich and H. Hurwitz, Jr.: *Nucleonics*, **23**, 12 (1952). This paper, however, refers mostly to age-diffusion group theory (i.e., for heavy moderators).

For hydrogenous systems an early review can be found in:

35. H. Goldstein, G. Foster, and P. F. Zweifel: "The Slowing Down of Neutrons in Hydrogenous Media: Status of Theory of Experiment", *Geneva Conf. Peaceful Uses Atomic Energy, Paper* 2375, 1958.

CHAPTER 9

For a detailed discussion of slowing-down theory, Refs. 11 and 15 (chaps. 8–10) should be consulted. A description of the work carried out during the war years is given in:

36. R. E. Marshak: *Rev. Mod. Phys.*, **19**, 185 (1947). It is a "must" for the serious student.

An early description of the slowing-down of neutrons by hydrogenous moderator was given by:

37. H. Hurwitz, Jr., and P. F. Zweifel: *J. Appl. Phys.*, **26**, 8 (1955).

In addition, Refs. 3 and 5 have nice treatments of slowing down in hydrogen.

Resonance capture is treated very well in a number of the previous references, particularly Refs. 4 and 11. However, the classic book is:

38. L. Dresner: "Resonance Absorption of Neutrons in Nuclear Reactors," Pergamon Press, New York, 1961.

CHAPTER 10

Dr. A. G. Bullard has provided three references (39–41) to which the student can turn for application of physical calculational methods to specific reactor systems. They are:

39. M. M. El-Wakil: "Nuclear Power Engineering," McGraw-Hill Book Company, New York, 1962.

40. J. George Wills: "Nuclear Power Plant Technology," John Wiley & Sons, Inc., New York, 1967.

41. Peaceful Uses of Atomic Energy, 1958, "Nuclear Power Plants," pt. 1, vol. 8, and pt. 2, vol. 9.

APPENDICES

For further study on the mathematical topics discussed in Appendices A to D, Ref. 7 is recommended. In addition, advanced discussions of matrix algebra can be found in Ref. 25 and in:

42. H. Margenau and G. M. Murphy: "The Mathematics of Physics and Chemistry", Van Nostrand Reinhold Company, New York, 1945.

The delta distribution, discussed in Appendix B, is only one example of a *Schwartz distribution*, using the term *distribution* in its technical mathematical sense. It is hard to conceive of a better treatment of the subject than that of:

43. I. M. Gel'fand and G. E. Shilov, "Generalized Functions," vol. 1, Academic Press, Inc., New York.

All of the special functions treated in Appendix D are discussed at length in chap. 4 of:

44. Arnold Sommerfeld: "Partial Differential Equations in Physics," Academic Press, Inc., New York, 1964.

The really serious student will, of course, refer to "scripture:"

45. G. M. Watson: "A Treatise on the Theory of Bessel Functions," Cambridge University Press, London, 1922.

Appendix E was taken almost verbatum from Ref. 11. See also any text in reactor physics. Finally, the discussion of the telegrapher's equation, Appendix F, is also mentioned in Ref. 2. p. 235.

Name Index

Subject Index